Advances in Experimental Medicine and Biology

Volume 1437

Advances in Experimental Medicine and Biology provides a platform for scientific contributions in the main disciplines of the biomedicine and the life sciences. This series publishes thematic volumes on contemporary research in the areas of microbiology, immunology, neurosciences, biochemistry, biomedical engineering, genetics, physiology, and cancer research. Covering emerging topics and techniques in basic and clinical science, it brings together clinicians and researchers from various fields.

Advances in Experimental Medicine and Biology has been publishing exceptional works in the field for over 40 years, and is indexed in SCOPUS, Medline (PubMed), EMBASE, BIOSIS, Reaxys, EMBiology, the Chemical Abstracts Service (CAS), and Pathway Studio.

2022 CiteScore: 6.2

Yong Gu • Adam Zaidel
Editors

Advances of Multisensory Integration in the Brain

Springer

Editors
Yong Gu
Systems Neuroscience
Institute of Neuroscience
Chinese Academy of Sciences
Shanghai, China

Adam Zaidel
Gonda Multidisciplinary Brain Research Center
Bar-Ilan University
Ramat Gan, Israel

ISSN 0065-2598 ISSN 2214-8019 (electronic)
Advances in Experimental Medicine and Biology
ISBN 978-981-99-7610-2 ISBN 978-981-99-7611-9 (eBook)
https://doi.org/10.1007/978-981-99-7611-9

This Springer imprint is published by the registered company Springer Nature Singapore Pte Ltd.
The registered company address is: 152 Beach Road, #21-01/04 Gateway East, Singapore 189721, Singapore

Paper in this product is recyclable

Preface

Everyday, our brains receive massive, yet often noisy and redundant information from different sensory channels. In the last two decades, numerous psychophysical experiments have demonstrated that humans integrate different sources of sensory information in a statistically optimal or near-optimal way to improve perceptual decisions. These studies have covered many modalities including vision, audition, somatosensory, vestibular, proprioceptive cues, etc., suggesting it is a general principle of brain function. In contrast to the dramatic progress achieved in psychophysics, how multisensory integration is mediated by the brain remains far less understood. Fortunately, in recent years, owing to the state of the art in non-invasive and invasive techniques developed and applied in humans and animals, respectively, we have begun to shed light on the black box of multisensory integration in the brain across spatial and temporal domains. This book aims to highlight recent findings in neurophysiology that may underlie multisensory integration functions in the brain.

In Chaps. 1–3, by combining animal behavioural performance, neurophysiological recordings, and computational modelling, researchers discuss multisensory integration effects and their potential neural correlates. Multisensory integration, by comparison to the underlying single cues, often lies close to predictions from Bayesian optimal integration theory, yet sometimes these can also deviate.

In Chaps. 4 and 5, researchers examine more complex environments, when there are conflicting signals, how the brain decides to combine them (assuming that they arise from the same source) or segregate them (if concluding that they arise from different sources). Such "causal inference" happens frequently in our daily lives for example, during a cocktail party we face multiple sources of visual and auditory cues at the same time and need to correctly pair the auditory and visual information from the person we are talking to.

Chapters 6–8 discuss how different cues are associated through brain oscillations, to form a unified percept of an object.

Chapters 9 and 10 discuss neural plasticity of multisensory processing, in different temporal scales, from seconds and minutes, to as long as across the life span.

Thus, this book covers broad topics including cue integration, perceptual decision-making, causal inference, spatial reference frames, synthesis, and plasticity. Importantly, it emphasizes recent findings from neurophysiology and their link with behavioural perception. This is achieved by combining psychophysics, neurophysiological recordings, neural activity manipulations, and computational modelling, which has been applied by many researchers in the field. With these efforts, research of multisensory brain function has provided an indispensable building block towards our future understanding of the brain. Critically, this basic research can also benefit future clinical applications and the design of brain-inspired artificial intelligence (AI), particularly, in complex, and dynamic environments.

Shanghai, China | Yong Gu
Ramat Gan, Israel | Adam Zaidel

Contents

1 Decentralized Neural Circuits of Multisensory Information Integration in the Brain

Wen-Hao Zhang

Abstract

The brain combines multisensory inputs together to obtain a complete and reliable description of the world. Recent experiments suggest that several interconnected multisensory brain areas are simultaneously involved to integrate multisensory information. It was unknown how these mutually connected multisensory areas achieve multisensory integration. To answer this question, using biologically plausible neural circuit models we developed a decentralized system for information integration that comprises multiple interconnected multisensory brain areas. Through studying an example of integrating visual and vestibular cues to infer heading direction, we show that such a decentralized system is well consistent with experimental observations. In particular, we demonstrate that this decentralized system can optimally integrate information by implementing sampling-based Bayesian inference. The Poisson variability of spike generation provides appropriate variability to drive sampling, and the interconnections between multisensory areas store the correlation prior between multisensory stimuli. The decentralized system predicts that optimally integrated information emerges locally from the dynamics of the communication between brain areas and sheds new light on the interpretation of the connectivity between multisensory brain areas.

Keywords

Multisensory integration · Decentralized architecture · Sampling-based Bayesian inference · Coupled network dynamics

1.1 Introduction

Our brain is bombarded with inputs from different sensory modalities, including visual, auditory, touch, vestibular, etc. These different forms of inputs usually contain the information of the same object in the world, and thus they can be combined in the brain to form a coherent and complete picture of the world (Knill and Pouget 2004; Ernst et al. 2004). For example, when walking, the visual inputs x_1 (optic flow) and vestibular inputs x_2 (body movement) both contain the information of self-motion direction (Bertin and Berthoz 2004), and the brain can integrate them to increase the reliability of the estimate of self-motion direction. The information integration seems ubiquitous across different sensory

W.-H. Zhang (✉)
Lyda Hill Department of Bioinformatics and O'Donnell Brain Institute, UT Southwestern Medical Center, Dallas, TX, USA
e-mail: wenhao.zhang@utsouthwestern.edu

Y. Gu, A. Zaidel (eds.), *Advances of Multisensory Integration in the Brain*, Advances in Experimental Medicine and Biology 1437, https://doi.org/10.1007/978-981-99-7611-9_1

modalities and/or different features. For example, the brain is able to optimally integrate visual and auditory inputs to estimate the object location (Alais and Burr 2004), integrating the motion and texture information for depth estimation (Jacobs 1999), the integration of visual and proprioceptive inputs for hand position (van Beers et al. 1999), and the visual and haptic information integration for estimating the object height (Ernst and Banks 2002), and so on.

1.1.1 The Bayesian Model of Multisensory Integration

A vital problem the brain faces in multisensory integration is the ubiquitous noises in the external world and the ones in the brain. Here the noise refers to the disturbances corrupting the signal, rather than sound. The noises eliminate a fixed relation between the stimulus and observed inputs, making the inputs stochastic and can be naturally described by a probabilistic model. Using the visual and vestibular integration for heading direction as an example, denote the visual and vestibular stimulus direction as s_1 and s_2, respectively. Each stimulus s_m $(m = 1, 2)$ independently generates a cue input x_m observed by multisensory neurons in the brain, which can be regarded as stochastic sensory transmission from external stimulus to the responses of neurons in unisensory cortex (comparing Fig. 1.1a, b). The stochastic process from s_m to x_m can be described by a likelihood distribution, $p(x_m|s_m)$, which describes how much likely that each observed x_m is generated from s_m.

The ubiquitous noises pose a computational challenge that the inputs are unreliable, and the brain needs to adjust the importance of multisensory inputs based on their reliability (noiseness level). Supposing you are in a dark environment

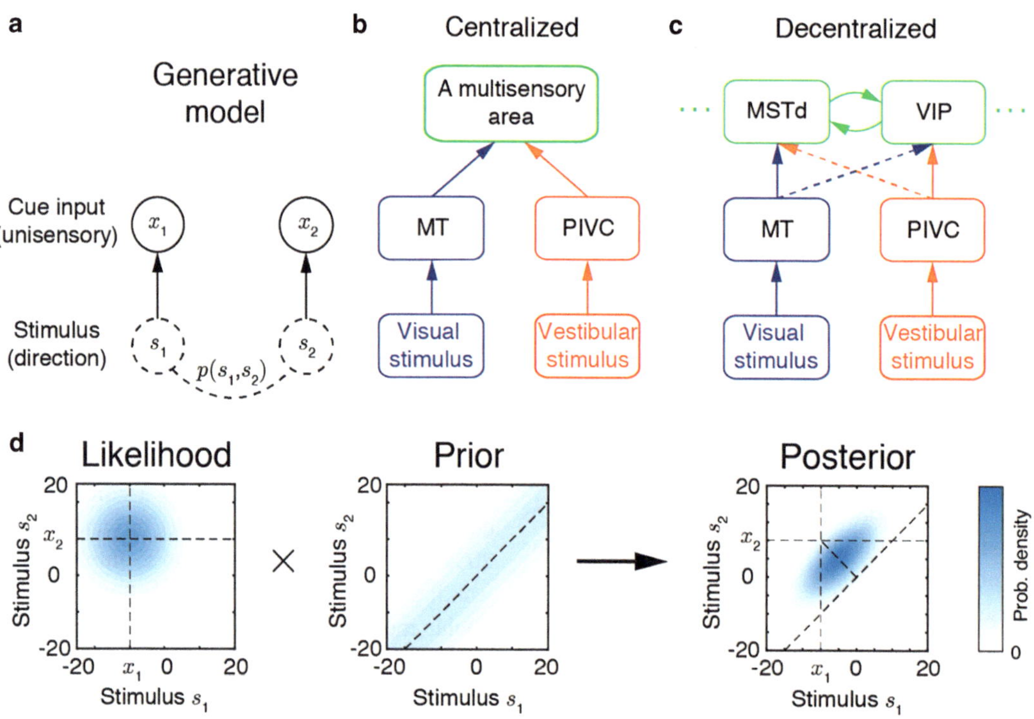

Fig. 1.1 (**a**) The probabilistic generative model describing the generation of unisensory inputs. (**b**) A conventional centralized framework to integrate multisensory inputs, where a single dedicated multisensory brain area combines the feedforward inputs from unisensory areas. (**c**) A decentralized architecture of multisensory integration where several interconnected multisensory areas achieve the multisensory integration concurrently. (**d**) An example of computing the posterior of stimuli s_1 and s_2 by multiplying the likelihood function and prior distribution

and then the visual inputs are less reliable than the vestibular inputs, the estimate of your self-motion direction should more rely on the vestibular inputs, and vice versa. This implies that the multisensory integration in the brain is not a fixed process but should be dynamically adjusted based on input statistics. In theory, Bayesian inference is the optimal way to deal with the uncertainty from noises, and numerous studies suggest that the brain performs Bayesian inference to combine multisensory inputs in estimating latent stimulus (e.g., Ernst and Banks 2002).

Performing Bayesian inference to estimate the latent stimuli (s_1 and s_2) requires the brain stores an internal model of the world (Fig. 1.1a), which includes two parts: One is the likelihood function, $p(x_m | s_m)$, as described above (Fig. 1.1d, left); and another is the prior distribution, $p(s_1, s_2)$, describing how the two latent stimuli are distributed and correlated in the world (Fig. 1.1d, middle). When the multisensory brain area receives two cue inputs, i.e., x_1 and x_2, it utilizes the internal model to invert the generative process (Fig. 1.1a) and synthesize the posterior distribution of the latent stimuli (Fig. 1.1d, right),

$$p(s_1, s_2 | x_1, x_2) \propto p(x_1 | s_1) p(x_2 | s_2) p(s_1, s_2). \tag{1.1}$$

The posterior distribution is regarded as the perception of the stimulus, as reported by extensive studies (e.g., Knill and Pouget 2004; Ernst et al. 2004; Yong et al. 2008). It is worth noting that Bayesian inference outputs a whole posterior distribution depicting the probability of every possible value of s, rather than merely a single estimate of stimulus. In a concrete example, suppose the likelihood and prior are both modeled as Gaussian distributions (Fig. 1.1d),

$$p(x_m | s_m) = \mathcal{N}\left(x_m | s_m, \Lambda_m^{-1}\right),$$

$$p(s_1, s_2) \propto \mathcal{N}\left(s_1 - s_2 | 0, \Lambda_s^{-1}\right). \tag{1.2}$$

where $\mathcal{N}\left(x_m | s_m, \Lambda_m^{-1}\right)$ denotes a Gaussian distribution of x_m with mean s_m, and the precision (inverse of variance) Λ_m. Λ_m controls the reliability of the sensory transmission, with a larger value denoting smaller noise associated with x_m. Note that the prior only captures the correlation between two stimuli which is determined by Λ_s, while the marginal prior of each stimulus, $p(s_m)$, can be calculated as a uniform distribution for simplicity. A larger Λ_s indicates the two stimuli are more correlated, which is exhibited by a narrower band along the diagonal line (Eq. 1.1, middle). Then the posterior of stimulus s_1, $p(s_1 | x_1, x_2)$ is also a Gaussian distribution (Fig. 1.1d, right), whose mean, s_1 and precision, Ω_1, are (see details in Eq. 1.16),

$$\Omega_1 = \Lambda_1 + \left(\Lambda_1^{-1} + \Lambda_2^{-2}\right)^{-1},$$

$$\hat{s}_1 = \Omega^{-1}\left[\Lambda_1 x_1 + \left(\Lambda_1^{-1} + \Lambda_2^{-2}\right)^{-1} x_2\right]. \tag{1.3}$$

The posterior of s_2 can be similarly obtained by changing indices in the above equation. We see the posterior precision is larger than the precision of likelihood, indicating the estimation accuracy is increased after integrating multisensory inputs (the spread of posterior is smaller than likelihood, Fig. 1.1d), which is a benefit of multisensory integration. Moreover, the posterior mean is weighted average of two inputs, with the weight of each input is determined by its precision (reliability). Hence, the Bayesian inference naturally considers the importance of each cue input by its reliability. Apart from multisensory integration, the Bayesian inference has been framed as a computational basis underlying a wide spectrum of perceptual processes, including decision-making (Gold and Shadlen 2001), sensorimotor learning (Körding and Wolpert 2004), object recognition (Kersten et al. 2004), visual processing (Yuille and Kersten 2006), and so on.

1.1.2 Towards the Neural Architecture of Multisensory Integration

How exactly the neural circuits implement the Bayesian inference in multisensory integration remains not very clear. There are two main challenges in answering this question. First, how the abstract probability distribution involved in the Bayesian inference of multisensory integration (Eq. 1.1) is *represented* in concrete neuronal population activities which could be recorded in an experiment? Second, how the neural circuit computes the posterior distribution by combining the information in likelihood and prior as shown in Eq. (1.1)? The above two challenges are related to representation and algorithm in the famous framework of three levels of analysis proposed by David Marr (2010). Different forms of representation and algorithm will directly lead to different neural architecture to implement the computation. Below let's summarize two possible neural architecture which can implement multisensory integration.

A conventional perspective (implicitly) assumes a *centralized* architecture (Fig. 1.1b), where a single multisensory brain area receives the feedforward inputs from unisensory brain areas to combine multisensory inputs to infer the posterior of latent stimulus (Ma et al. 2006; Alvarado et al. 2008; Magosso et al. 2008; Ursino et al. 2009; Ohshiro et al. 2011; Makin et al. 2013). A neural coding theory (Ma et al. 2006; Jazayeri et al. 2006) suggested that the multisensory integration (Bayesian inference) can be achieved by linearly summing feedforward inputs in a multisensory area, as long as the inputs from unisensory brain areas are independent Poisson spike trains. Although the centralized architecture is simple and has been widely used in engineering (e.g., Srivastava et al. 2012), it suffers from the burden of the computational center (multisensory area) and the susceptibility of being paralyzed once the multisensory area fails.

In contrast to the centralized architecture, the computation of multisensory integration (posterior) can be distributed to several multisensory brain areas, which is called *decentralized* architecture (Fig. 1.1c) (Zhang et al. 2016; Zhang and Wu 2013; Durrant-Whyte and Henderson 2016). The decentralized architecture is composed of several interconnected multisensory areas, and no area is located at the center of the system topology, and hence the whole system is robust to local damage. Each multisensory area firstly computes a local result and then communicates with each other to update their results. Eventually, the final result of the computation emerges from the interaction loop between multisensory areas. This decentralized architecture is consistent with recent findings that many interconnected multisensory areas concurrently contribute to the visual and vestibular integration to infer heading direction (Yong et al. 2008; Chen et al. 2011a, b, 2013; Yong et al. 2016). It was found that the neurons in dorsal medial superior temporal (MSTd) area (Yong et al. 2006, 2008), and ventral intraparietal (VIP) area (Chen et al. 2011a, 2013) optimally integrate the information (Fig. 1.1c). And there are abundant reciprocal connections between the two multisensory areas (Boussaoud et al. 1990; Baizer et al. 1991). Apart from MSTd and VIP, the area of the posterior sylvian fissure (VPS) (Chen et al. 2011b), frontal eye field (FEF) (Yong et al. 2016), also show similar multisensory responses (not shown in Fig. 1.1c). Furthermore, inactivating either MSTd or VIP will not block the integrative behavior of the animal, but only decrease the accuracy of the estimate of heading direction (Yong et al. 2012; Chen et al. 2016). This is strong evidence in supporting the decentralized architecture (Fig. 1.1c) but rejecting the centralized architecture (Fig. 1.1b).

1.2 A Decentralized Network of Multisensory Integration

To study how the decentralized architecture with interconnected multisensory areas implements the Bayesian inference in multisensory integration, we built a neural circuit model with sufficient biological details. For simplicity, we first consider a minimal decentralized system which is only composed of two reciprocally connected networks (Fig. 1.2a, b), mimicking the interactions between MSTd and VIP. The decentralized architecture has the capability of extending to more networks as will be shown later. Both networks receive the feedforward inputs from unisensory (visual and vestibular) areas (Fig. 1.2c). Moreover, we ignored the cross-feedforward connections in anatomy (dashed arrows in Fig. 1.1c) which allows us focus on how in a decentralized system the reciprocal connections between two areas achieve the integration. This simplification doesn't influence the result substantially, in that it

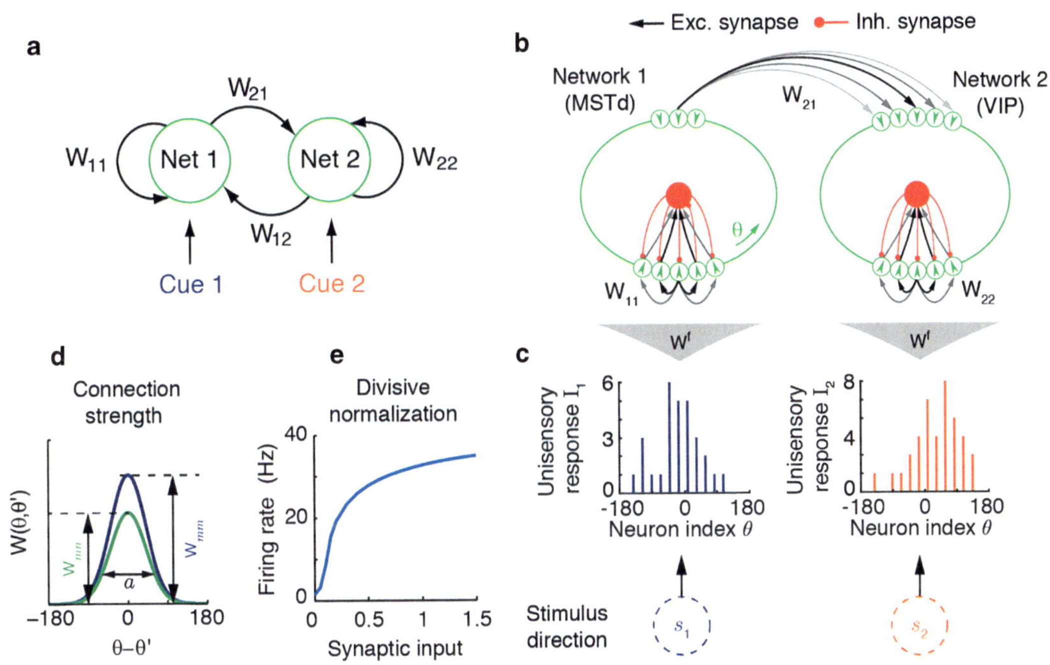

Fig. 1.2 (**a**) A minimal decentralized network for multisensory integration consists of two interconnected networks where each network receives an independent input from a sensory cue. (**b**) The detailed network structure of the decentralized system in (**a**). Each network is modeled as a continuous attractor network. Circles on the green ring represent excitatory neurons whose preferred stimulus direction θ is indicated by the direction of the arrow inside. Another pool of inhibitory neurons (red disk) is driven by all excitatory neurons and generates negative feedback with a math form of divisive normalization. Black arrows indicate excitatory synaptic connections whose weight is represented by the grey scale; red lines are inhibitory connections. (**c**) The evoked responses in unisensory areas provide feedforward inputs to multisensory neurons. The unisensory responses are modeled as independent Poisson spikes whose firing rate has Gaussian profile tuning over the stimulus direction. (**d**) Recurrent and reciprocal connections in the network model are translation-invariant in the feature space, that is, the connection strength between two neurons depends on their distance in feature space only. (**e**) The divisive normalization provided by inhibitory neurons transfers the synaptic input into firing rate

was found MSTd (VIP) is dominated by visual (vestibular) inputs (Yong et al. 2016; Chen et al. 2011c). With this simplification, the recurrent connections between two networks (multisensory areas) are the only media to convey the inputs from another sensory modalities.

1.2.1 The Network Model

The multisensory neurons in both MSTd and VIP have bell-shaped tuning functions (mean firing rate) over the heading (stimulus) direction s (Yong et al. 2008; Chen et al. 2013). To reproduce this neuronal tuning in the network, each network (multisensory area) was modeled as a continuous attractor network (CAN), which is a canonical neural circuit model to explain the circuit mechanism of encoding and processing of continuous stimulus features such as moving direction (Zhang 1996), orientation (Ben-Yishai et al. 1995; Rubin et al. 2015), spatial location (Samsonovich and McNaughton 1997; Burak and Fiete 2009), and so on. For simplicity, I assumed each network in the model has the same structure, and the same number of neurons. Each network contains a population of excitatory (E) neurons (neurons arranged on a ring, Fig. 1.2b), and a pool of inhibitory (I) neurons (red circle, Fig. 1.2b). Every neuron in the network has *Poisson* variability mimicking the variability associated with spike generation (Eq. 1.17), which implies that the neuronal response is stochastic even if the feedforward input is fixed.

Later, I will show the Poisson spiking variability is important for the network to implement the Bayesian inference in multisensory integration. Below I only describe the main characteristics and mechanisms of the network model, and the mathematical equations of the network are shown in Appendix for readers who have interests.

Each E neuron in the network is uniquely identified by its preferred value θ over the stimulus direction s. And it was assumed that the preferred values of all E neurons are uniformly distributed on the range of stimulus direction (the ring in Fig. 1.2b). This ensures the information of stimulus direction s can be always encoded by a portion of E neurons no matter what the value of s is. Since stimulus direction is a periodic variable, i.e., $s \in (-\pi, \pi]$, the E neurons are effectively arranged on a ring space (Fig. 1.2b), thus the network is also called as a ring network.

In contrast, the inhibitory (I) neurons in the network are not tuned to the heading direction. That is the firing rate of I neurons will not change with stimulus direction s but will change with the input intensity. In the network, the I neurons are only driven by E neurons, and then provide negative feedback to E neurons to keep the stability of network activities (Fig. 1.2b, red arrows). For simplicity, in the model the effect of inhibitory neurons is implicitly modeled as divisive normalization, which is a widely observed phenomenon across cortex (Eq. 1.19, Carandini and Heeger 2012). The divisive normalization is resulted from PV inhibitory neurons (Niell 2015) and acts as an activation function which turns the instantaneous synaptic inputs of E neurons into their instantaneous firing rate (Fig. 1.2e).

1.2.2 Recurrent Connections in the Network

The tuning of E neurons in the network is an emergent property from the recurrent connections between E neurons. The connection strength between two E neurons within the same network only depends on their distance on the ring (the disparity between their preferred stimulus direction θ), specifically, which is modeled as a Gaussian function decaying with preferred stimulus difference between two E neurons (the blue line in Fig. 1.2d, Eq. 1.18).

The E neurons across networks are also interconnected with each other, whose connection profile is also assumed to have the same Gaussian profile with recurrent connections within the same network for simplicity (Fig. 1.2d, green line; Eq. 1.18). That is, the connections between E neurons within the same network or across different networks have the same width a but could have different peak connection strength (Fig. 1.2d). I will use w_{mn} $(m, n = 1, 2)$ to denote the peak connection strength in the network model, where w_{mm} denotes the peak recurrent strength within the same network, and w_{mn} $(m \neq n)$ denotes the one from network n to network m. Note that there are no inhibitory connections across networks, and also no connections from E neurons in a network to the I neurons in another network (Fig. 1.2b). For simplicity, the connection strength in two networks, and the reciprocal connections between two networks are set to be symmetric with each other, i.e., $w_{11}^r = w_{22}^r \equiv w_{rc}, w_{21}^r = w_{12}^r \equiv w_{rp}$. However, the feedforward inputs applied to each network can be different as shown below.

1.2.3 Feedforward Inputs from Unisensory Brain Areas

Each network receives the feedforward inputs from a corresponding unisensory area, for example, MSTd neurons receive the feedforward inputs from MT, and VIP neurons receive the feedforward inputs from PIVC (Figs. 1.1c and 1.2a). Only E neurons in each network directly receive the feedforward inputs, while the I neurons in the network do not. For simplicity, I assumed the number of neurons in a unisensory area is the same as the number of E neurons in a multisensory area, N_E. The feedforward connections from unisensory neurons to the multisensory neurons in the network is also modeled to have the same Gaussian profile with the recurrent connections in the network (Fig. 1.2d, Eq. 1.18), with peak connection strength denoted as w^f.

The stimulus s_m evokes the responses of unisensory neurons $\mathbf{I}_m$, for example, the visual (vestibular) stimulus evokes the responses of MT (PIVC) neurons. Given the stimulus s_m, the responses of unisensory neurons $\mathbf{I}_m$ are modeled as independent Poisson spike counts and have Gaussian tuning (mean firing rate) over s_m (Ma et al. 2006; Jazayeri et al. 2006). Therefore, the probability of observing a unisensory neurons' response given stimulus can be described by the following probabilistic model,

$$p(\mathbf{I}_m|s_m) = \prod_{j=1}^{N_E} \text{Poisson}\left[\langle \mathbf{I}_{mj}(s_m)\rangle\right],$$

$$\langle \mathbf{I}_{mj}(s_m)\rangle = R_m^f e^{-(\theta_j - s_m)^2/2a^2}, \ (m=1,2). \quad (1.4)$$

R_m^f denotes the intensity (peaking firing rate) of unisensory neurons' responses, which is determined by the stimulus strength, for example, the intensity of inputs from visual area (MT) increases with the motion coherence of random moving dots (optic flow) of the visual stimulus. θ_j is the preferred stimulus (direction) s_m of jth unisensory neuron providing inputs to network m, which is also assumed to be uniformly distributed on the range of s_m for simplicity.

1.3 Neuronal Responses in the Decentralized Network

It is instructive to show the neuronal responses in the network model to demonstrate its biological plausibility. Mimicking experimental protocols (Yong et al. 2008; Chen et al. 2013), I firstly compared the network activities in response to a stimulus from either sensory modality, and the stimuli from both modalities simultaneously (Fig. 1.3a). Given a pair of stimuli (s_1 and s_2), the unisensory

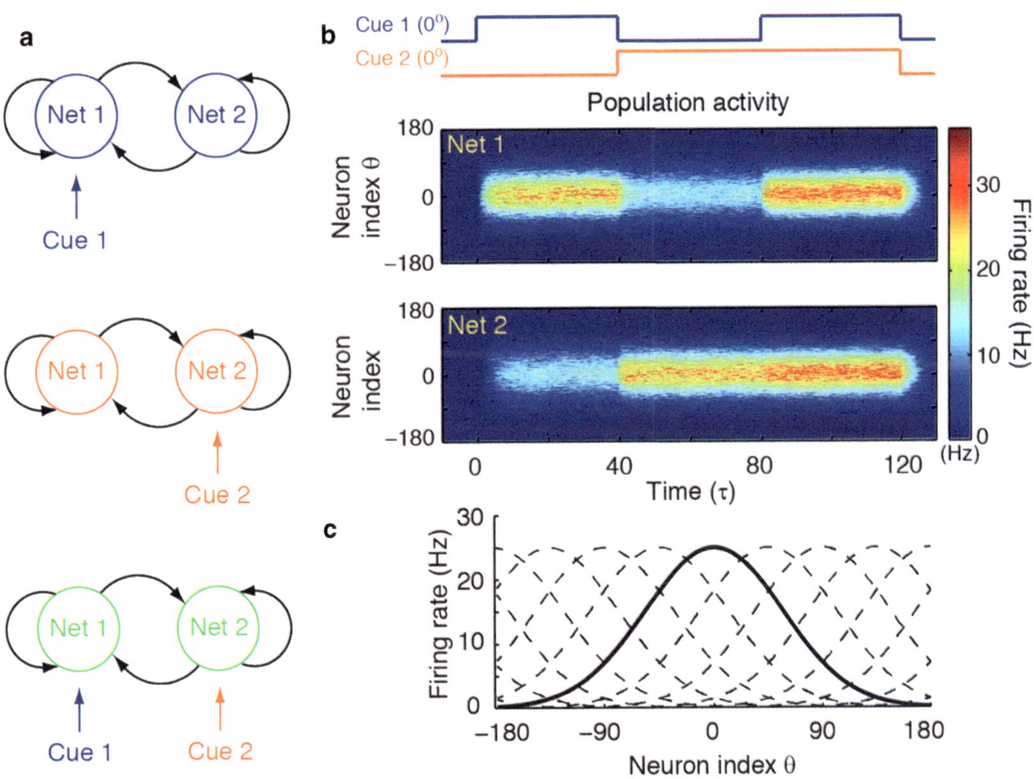

Fig. 1.3 Neuronal population responses in the decentralized network. (**a**) Illustration of three stimulus conditions applied to the network. (**b**) An example of population activities in two networks in response to the stimulus conditions shown in (**a**) in temporal order. Both cues are static and located at 0°. The color encodes the firing rate of the population activity. (**c**) The population activity is a family of bell-shaped bumps, with the position on the stimulus subspace (x-axis) determined by the direction of cue inputs. The figure is adapted from Zhang et al. (2016)

neurons' responses were randomly generated (Eq. 1.4), and then applied to the decentralized network. The population response in each network m, $\boldsymbol{r}_m$, has a Gaussian profile (Fig. 1.3b, c), as a consequence of Gaussian feedforward inputs, and the Gaussian-profile recurrent connections in the network, which is consistent with experimental data. When only the stimulus s_1 is presented, the network 1 has stronger responses than network 2 since it directly receives the feedforward inputs from corresponding unisensory neurons. When both stimuli are simultaneously presented to the network, the responses in both networks are enhanced compared with the one in either single stimulus conditions.

Although the multisensory integration is achieved collectively by all neurons in the network, it was found that single neurons in multisensory areas optimally integrate information, which is exhibited by the fact that the neurometric function built from single neurons' activities is consistent with the prediction of Bayesian inference (Eq. 1.3) (Yong et al. 2008; Chen et al. 2013). The neurometric measures the proportion of discriminating the stimulus direction based solely on a single neurons' response (Fig. 1.4c). In particular, the slope of the neurometric function characterizes the accuracy of the estimate, whose inverse, called the discrimination threshold, is related to the posterior variance (Ernst and Banks 2002). Therefore, we could compare the discrimination threshold in combined cue conditions (when two cue inputs are presented simultaneously) with the Bayesian prediction of the variance (Eq. 1.3). Figure 1.4c shows the neurometric functions under three stimulus conditions of the same example neuron which prefers −40°

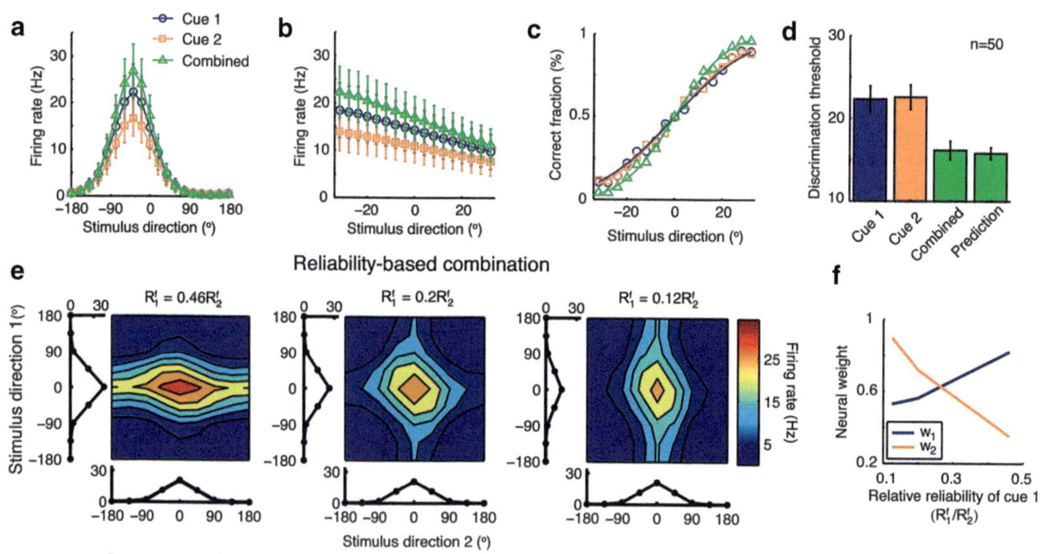

Fig. 1.4 Single neurons' responses in the decentralized network. (**a**) The tuning curve (mean firing rate over the stimulus direction) of an example in network 1 under three cueing conditions. The example neurons prefer stimulus direction of −40°. Error bar denotes one standard deviation of firing rate fluctuation across trials. (**b**) Responses of the example neuron in a narrow range of stimulus direction values. (**c**) The example neuron's neurometric functions which denote the correct fraction of judging the stimulus direction to be larger than 0°. Lines: cumulative Gaussian fit of the neurometric function. (**d**) The neuronal discrimination thresholds of the example neuron in three stimulus conditions compared with Bayesian prediction. The actual neuronal discrimination thresholds in the case of combined cues are not significantly different with the Bayesian prediction ($p = 0.044$, $n = 50$, unpaired t-test). (**e**) The bimodal tuning curve of neurons changes with the reliability of inputs. The curves at the left and the bottom of each contour are unimodal tuning curves in response to either cue. The cue 1's intensity, R_1^f, decreases from left to right plots, whereas cue 2's intensity, R_2^f, is fixed. (**f**) The two-dimensional tuning curve fitted as a linear model of the two unimodal tuning curves is shown on the left and at the bottom. The figure shows the combination weight of two cues with respect to the relative intensity of cue 1. The figure is adapted from Zhang et al. (2016)

direction in network 1. The neurometric function was computed based on the firing activities of the single neurons (Fig. 1.4b) via the analysis called receiver operating characteristics (ROC) (Yong et al. 2008; Chen et al. 2013; Britten et al. 1992). The slope of the neurometric function in combined cue conditions (Fig. 1.4c, green line) is steeper than those under single cue conditions. Moreover, the discrimination threshold of the neurometric function under combined cues is consistent with the Bayesian prediction, suggesting the single neurons in the decentralized network optimally integrate the multisensory information, which is consistent with experimental findings (Yong et al. 2008; Chen et al. 2013).

Since optimal multisensory integration weighs each input based on its reliability (Eq. 1.3), it is expected that a neuron's activity will be dominated by the input which is more reliable, called reliability-based combination (Morgan et al. 2008). Note that the dynamical weight of cue input is not a result of synaptic plasticity (learning), in that the cue input weight is changed trial by trial which is in a time scale of milliseconds, and thus is much shorter than the time scale of synaptic plasticity which has a time scale of hours. To reproduce this in the decentralized network, we fix the reliability of cue input 2 while increasing one of cue input 1. In the model, the stimulus reliability is determined by the firing rate of unisensory responses ($\mathrm{R}_m^{\mathrm{f}}$ in Eq. 1.4), which is supported by the fact that the firing rate of MT neurons (Fig. 1.1c) increases with motion coherence of the visual optic flow. With low reliability of cue input 1 (low firing rate of cue input 1), the shape of the bimodal tuning is dominated by cue input 2 (Fig. 1.4e, left). As the reliability of cue input 1 increases, the bimodal tuning gradually dominated by stimulus 1 (Fig. 1.4e). When fitting the bimodal tuning as a linear model of two unimodal tuning curves (Fig. 1.4e, marginal plot), the weight of cue input 1 increases with the reliability of cue 1, while that of cue input 2 decreases with cue 1's reliability (Fig. 1.4f). All of above results suggest the decentralized network is a biologically plausible circuit model in reproducing the phenomena observed in neurophysiology experiments (Zhang et al. 2016).

1.4 Bayesian Inference in the Decentralized Network

Next I study the representational and algorithmic mechanism that how the decentralized network implements the Bayesian inference in multisensory integration. That is, how the network computes the posterior distribution and represents it distributively in neuronal population activities? I will perform theoretical analysis to unveil the underlying mechanism of the network and use numerical simulation to confirm the analysis.

1.4.1 Feedforward Inputs Convey the Likelihood Distribution

The feedforward inputs from the unisensory neurons convey the stimulus information. Inverting the probabilistic model of the unisensory neurons' responses (Eq. 1.4) can tell us how much likely an observed $\mathbf{I}_m$ is evoked by a stimulus, i.e., the likelihood (probability) of a value of s_m which induces the $\mathbf{I}_m$. Combining the two lines in Eq. (1.4), we have (see detailed derivation in Appendix),

$$p\left(\mathbf{I}_m|s_m\right) \propto \mathcal{N}\left(s_m|x_m, \Lambda_m^{-1}\right), \ (m=1,2) \qquad (1.5)$$

where

$$x_m = \frac{\sum_j \mathbf{I}_{mj}\theta_j}{\sum_j \mathbf{I}_{mj}}, \ \Lambda_m = a^{-2}\sum_j \mathbf{I}_{mj}. \qquad (1.6)$$

Comparing Eqs. (1.1) and (1.5), we see a single snapshot of unisensory response $\mathbf{I}_m$ with Poisson variability across trials (Eq. 1.4) represents the whole likelihood function of stimulus direction s_m, and thus it encodes the Gaussian likelihood used in the abstract probabilistic model (Eqs. 1.1–1.3). The Gaussian likelihood encoded by $\mathbf{I}_m$ is determined by the Gaussian tuning of unisensory neurons (Eq. 1.4, Ma et al. 2006; Jazayeri et al. 2006).

The mean, x_m, and the precision (inverse of variance), Λ_m, of the likelihood can be read out linearly from $\mathbf{I}_m^{\mathrm{f}}$ by using a linear decoder called population vector (Georgopoulos et al. 1986)

(Eq. 1.6). Geometrically, the mean x_m represents the position (center of mass) of $\mathbf{I}_m$ on the ring space of the moving direction; and the precision is proportional to the magnitude (or the sum of spikes) of $\mathbf{I}_m$. Note that the explicit decoding of likelihood parameters from unisensory neurons' responses (Eq. 1.6) is only served as a way to evaluate the stimulus information conveyed by unisensory neurons, but which doesn't require the neural network has such explicit mechanism to read out the likelihood.

The observed cue input, x_m, and the whole likelihood distribution, $p(x_m|s_m)$ is represented by a single snapshot of $\mathbf{I}_m$. Meanwhile, the posterior, $p(s_1,s_2|x_1,x_2)$ (Eq. 1.1), is conditioned on given observed cue inputs (x_1 and x_2). Therefore, next we study how the decentralized network in responses to temporally *fixed* unisensory responses $\mathbf{I}_m$ and computes the posterior.

1.4.2 The Network Dynamics on the Stimulus Subspace

The low-dimensional stimulus evokes the activities of a population of neurons, which indicates the stimulus information is distributed across the whole network. Furthermore, this implies that there must be a low-dimensional subspace (manifold) in the high-dimensional neuronal response which corresponds to the change of stimulus. Finding the network dynamics on this stimulus subspace is critical to understand how the network computes and represents the posterior distribution. Therefore, we performed perturbative analysis to analytically derive the direction corresponding to stimulus subspace (the ring as shown in Fig. 1.2b), and then project the high-dimensional neural dynamics onto the stimulus subspace (Zhang et al. 2016, 2020) (Fig. 1.5).

Given fixed feedforward inputs, the network dynamics on the stimulus subspace is governed by a Langevin dynamics (see detailed derivation in Appendix),

$$\frac{d\mathbf{s}_t^r}{dt} = -\frac{\rho}{\sqrt{2}}\left(\tau\mathbf{D}_U\right)^{-1}\left[\mathbf{L}\mathbf{s}_t^r + w^f\mathbf{D}_f\left(\mathbf{s}_t^r - \mathbf{x}\right)\right] + \sigma_s\left(\tau\mathbf{D}_U\right)^{-1/2}\xi_t \tag{1.7}$$

$\mathbf{s}_t^r = \left(s_{1t}^r,\ s_{2t}^r\right)^{\mathrm{T}}$ denotes the instantaneous position of population activity of both networks on the stimulus subspace at time t. Each s_{mt}^r can be read out from instantaneous neuronal activity in network m, $\boldsymbol{r}_{mt}$, by using a linear decoder called population vector (Georgopoulos et al. 1986), which is resulted from the Gaussian profile population activity in each network,

$$s_{mt}^r = \sum_j \boldsymbol{r}_{mt}\theta j \Big/ \sum_j \boldsymbol{r}_{mtj}. \tag{1.8}$$

In Eq. (1.7), $\mathbf{L}$ specifies the peak reciprocal input strength between networks and determines the interaction strength between s_{mt}^r in each network,

$$\mathbf{L}_{mn} = -w_{mn}^r R_n,\ \mathbf{L}_{mm} = -\sum\nolimits_{n\neq m}\mathbf{L}_{mn} \tag{1.9}$$

where R_n is the peak firing rate of network n (Eq. 1.23), and w_{mn}^r is the peak strength of the connections from network n to network m (Eq. 1.18).

The second term inside the bracket in Eq. (1.7) characterizes the influences of the feedforward inputs from unisensory neurons. $\boldsymbol{x} = (x_1,x_2)^{\mathrm{T}}$ is the observed stimulus (direction) conveyed by feedforward inputs (Eq. 1.6). $\mathbf{D}_{\mathrm{f}} = diag\left(R_1^f, R_2^f\right)$ is a diagonal matrix denoting the peak value of input neurons' responses (Eq. 1.4) and represents the reliability of cue input (comparing Eqs. 1.4 and 1.6). w^f is a scalar variable denoting the feedforward connection strength (Eq. 1.18). In addition, $\mathbf{D}_U = diag\ (U_1, U_2)$ is also a diagonal matrix denoting the peak value of synaptic input bump in every network (Eq. 1.23).

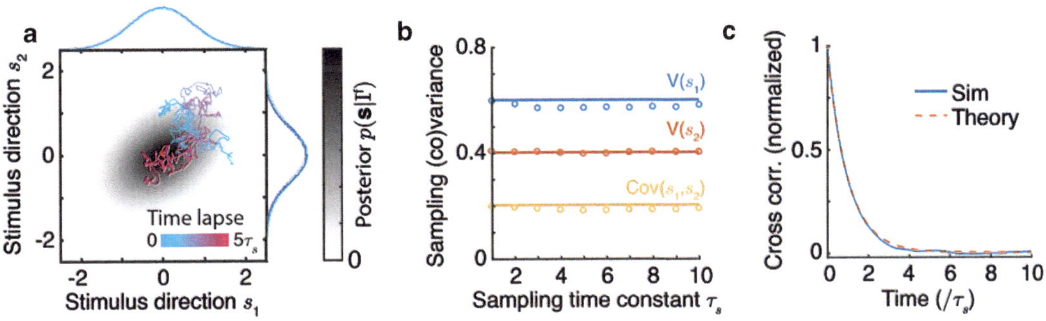

Fig. 1.5 (**a**) The Langevin sampling which can approximately infer the posterior of two-dimensional stimulus features. Black contour: the posterior. The trajectory colors indicate the elapsed time. The curves on the top and right denote marginal posteriors (solid line: empirical distribution; shaded line: posterior from theory). (**b**) The equilibrium variance of samples does not change with sampling time constant τ_s (circle: empirical result; solid line: theoretical derivation). (**c**) The auto correlation function of samples over time. The figure is adapted from Zhang et al. (2020)

The Gaussian-white variability (Eq. 1.17, the last term) is exclusively from the internal Poisson variability of spike generation since the feedforward inputs are fixed over time. It is worth noting that the Poisson variability on single neurons becomes the Gaussian-white variability on the stimulus subspace, which is resulted from the Gaussian profile of recurrent connections in the network (Eq. 1.23). $\sigma_s^2 = 8aF/\left(3\sqrt{3\pi}\right)$ is a constant unchanged with respect to the feedforward inputs and network responses, with F characterizes the Fano factor of neurons. It determines how effective the Poisson variability on single neurons is transferred to the Gaussian-white variability in the stimulus feature subspace.

1.4.3 Approximate Inference of the Posterior by Sampling

The decentralized network has stochastic neuronal responses from internal Poisson variability, even if the feedforward inputs are fixed over time. The stochastic neuronal response suggests that the network may implement sampling-based inference to approximate the posterior. The sampling is a type of Markov chain Monte Carlo (MCMC) algorithms widely used to numerically approximate the posterior (Bishop 2006). Previous theoretical studies also suggest the stochastic neural dynamics may implement sampling-based inference, for example, (Hoyer and Hyvärinen 2003; Fiser et al. 2010; Hennequin et al. 2014; Savin and Deneve 2014; Aitchison and Lengyel 2016; Orbán et al. 2016), with the neuronal response variability as a hallmark of sampling.

The sampling-based inference can be implemented by using a Langevin dynamics (Welling and Teh 2011; Neal et al. 2011), which has a similar form with Eq. (1.7). Mathematically, the Langevin sampling performs stochastic gradient ascent on the manifold of the log-posterior of stimulus, which is written as,

$$\frac{d\mathbf{s}_t}{dt} = \tau_s^{-1}\frac{d}{d\mathbf{s}}\ln p\left(\mathbf{s}|\mathbf{I}^f\right) + \left(\frac{\tau_s}{2}\right)^{-\frac{1}{2}}\xi_t,$$

$$= \tau_s^{-1}\left[\Lambda_s\mathbf{s} + \Lambda\left(\mathbf{s}-\mathbf{x}\right)\right] + \left(\tau_s/2\right)^{-1/2}\xi_t. \tag{1.10}$$

where ξ_t is a multivariate independent Gaussian-white noise which induces fluctuations to drive the sampling process and satisfies that $\langle \boldsymbol{\xi}_t(i)\boldsymbol{\xi}_{t'}(j)\rangle = \delta_{ij}\delta(t-t')$, with δ_{ij} the Kronecker delta function and $\delta(t-t')$ the Dirac delta function. Figure 1.5a plots an example trajectory of stimulus samples walking randomly on the surface of the posterior distribution, which is characterized by the temporal correlation of stimulus feature samples decays exponentially over time (Fig. 1.5c). τ_s is a positive number or a positive definite matrix which doesn't influence the equilibrium distribution of samples s_t (Fig. 1.5b), but only determines the temporal speed of sampling (Fig. 1.5c).

The sampling-based inference is a stochastic algorithm to numerically approximate the posterior through time. The algorithm randomly draws *stimulus samples*, $\mathbf{s}_t$ over time. After collecting sufficient number of samples, their empirical distribution is able to approximate the posterior (Eq. 1.1) with certain accuracy, i.e., $p(\mathbf{s}|\mathbf{x}) \approx T^{-1}\sum_t \delta(s - s_t)$. Indeed, it can be computed that the mean and covariance matrix of stimulus samples $\mathbf{s}_t$ generated by Eq. (1.10) is the same as the posterior mean and covariance, respectively (Eq. 1.16).

There are two crucial features of Langevin sampling so as to correctly sample the posterior, i.e., the empirical equilibrium distribution of samples is the same as the posterior. First, the injected variability to drive sampling should be independent (white) Gaussian (Eq. 1.10, last term). Comparing Eqs. (1.7) and (1.10), we see the Gaussian white variability in the stimulus subspace is guaranteed in the network model, which is resulted from the Gaussian profile of population activities (Fig. 1.3b). Second, the coefficients of the first and the second term on the right-hand side (RHS) of Eq. (1.10) should be simultaneously determined by the same τ_s as the way described above. We see in Eq. (1.7) the first and second term share the same factor $\tau \mathbf{D}_{\mathrm{U}}$, which is similar with the τ_s in Eq. (1.10). We found the decentralized network can optimally sample the posterior once the connection weight is appropriately adjusted,

$$w^f = \left(\frac{2}{\sqrt{3}}\right)^3 \mathrm{F},\; w_{mn}^r = \frac{aw^f}{\sqrt{2\pi}\rho}\frac{\Lambda_s}{R_n} \tag{1.11}$$

1.4.4 Sampling-Based Inference in the Stimulus Subspace in Decentralized Network

To confirm the theoretical analysis as shown above, we simulated the decentralized network model whose connections were set according to Eq. (1.11). In the decentralized network, each network samples a corresponding stimulus, for example, the network 1 samples stimulus s_1 and the network 2 samples stimulus s_2. Therefore, given the population activities of the network, we individually read out each stimulus sample s_{mt}^{r} from the instantaneous population responses from network m, $\mathbf{r}_{mt}$ by using population vector (Eq. 1.8, Fig. 1.6a, b). And then the concatenation of stimulus samples individually read out from each network, $\mathbf{s}_t^r = \left(s_{1t}^r, s_{2t}^r\right)^{\mathrm{T}}$, approximate the bivariate posterior (Eq. 1.15).

As an illustration, we plotted the sampling distributions generated by the decentralized network under three stimulus conditions (Fig. 1.6b). When only cue 1 was presented, the samples generated from both networks are centered at x_1 (the observed direction), and the samples of stimulus s_1 (generated by network 1) has smaller variance the sample of stimulus s_2 (generated from network 2). This is because the network 2 receives the inputs indirectly via network 1. This behavior is in accordance with the posterior computed from the Bayesian inference (Eq. 1.3): under cue 1 condition, the sampling variance of stimulus 1 is $V\left(s_1|x_1\right) = \Lambda_1^{-1}$, which is smaller than the variance of stimulus 2, $V\left(s_2|x_1\right) = \Lambda_1^{-1} + \Lambda_2^{-1}$. The mean and variance of samples generated from both networks are reversed when only cue 2 is presented in comparison with cue 1 condition. In the combined cue condition, the mean of samples shifts towards a position in between x_1 and x_2 and have the smallest variance. Note that the sampling mean of both stimuli are different in the combined cue condition when the two cues are disparate, which is consistent with the Bayesian observer.

We further systematically tested whether the decentralized network is able to correctly sample the posterior. For simplicity, the connection strength in two networks, and the reciprocal connections between two networks are symmetric with each other, i.e., $w_{11}^r = w_{22}^r \equiv w_{rc}$, $w_{21}^r = w_{12}^r \equiv w_{rp}$. But the two networks can receive different feedforward cue inputs. We changed the intensity of each cue which is controlled by the peak firing rate of feedforward input (Eq. 1.4), as well as the reciprocal connection strength between networks w_{rp}.

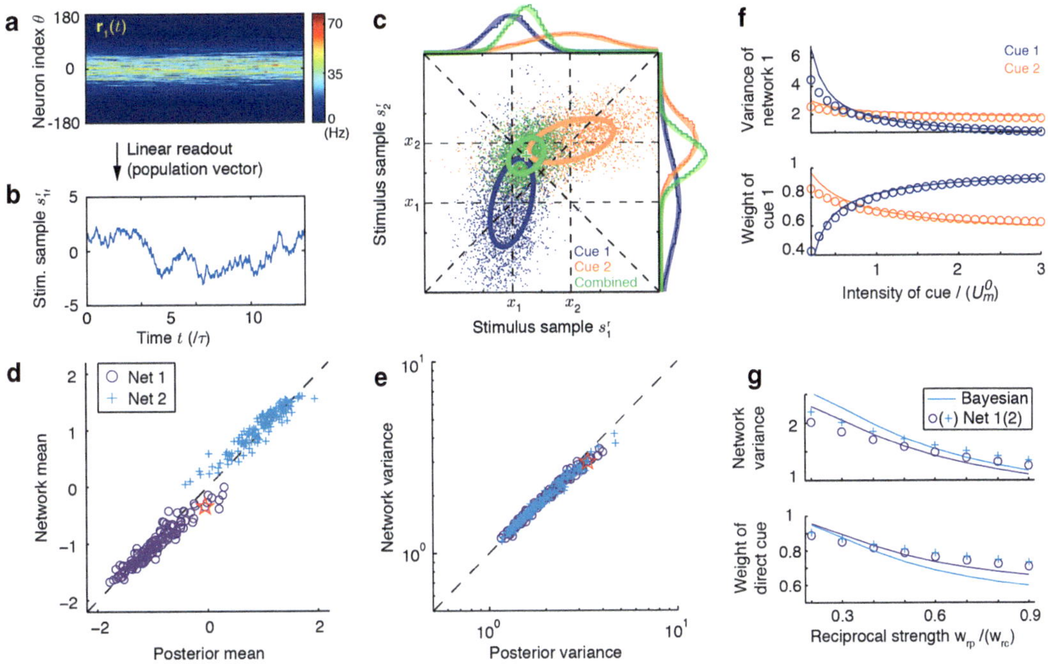

Fig. 1.6 The sampling-based Bayesian inference in decentralized network. (**a**) The spatiotemporal responses of network 1. (**b**) Each sample of stimulus 1 can be read out from the instantaneous population responses in network 1 by a linear decoder of population vector. The samples of stimulus 2 can be similarly read out from network 2. (**c**) The sampling distributions generated by the decentralized network under three stimulus condition, with the marginal distributions plotted on the margin. (**d**, **e**) The comparison of the sampling mean (D) and variance (**e**) under the combined cue condition with the Bayesian prediction. (**f**) The mean and variance stimulus s_1 (read out from network 1) when changing the intensity of either cue or fixing the intensity of another. Symbols: network results; lines: Bayesian model. (**g**) The sampling mean and variance of both stimuli with reciprocal connection strength between two networks. The figure is adapted from Zhang et al. (2016)

The cue intensity spans a large interval which ranges from superadditive to near-saturation region of network responses. Figure 1.6d, e compares the mean and variance of samples individually read out from each network with the marginal posterior of corresponding stimulus (Eq. 1.15). Indeed, the simulation results show that each network individually samples a corresponding stimulus, and the bivariate posterior is approximated by samples from both networks.

In order to confirm the input intensity encodes the cue reliability (Eq. 1.6), we varied the reliability of one cue (controlled by the firing rate of unisensory neurons, Eq. 1.4) while fixing another cue as well as network connections. With increasing cue 1's reliability, the sampling variance of stimulus 1 decreases in a way consistent with Bayesian model (Fig. 1.6d, blue lines), and the weight of cue 1 becomes larger. Analogous results are obtained when changing the reliability of cue 2.

1.4.5 The Stimulus Prior Is Stored in Reciprocal Connections Between Networks

The theoretical analysis suggests that the reciprocal connection strength between two networks store the prior precision (Eq. 1.11), $\Lambda_s \propto w^r_{mn} R_n$, which suggests that a decentralized network with stronger reciprocal connections between networks stores a more correlated prior. This result was also confirmed by network simulation (Fig. 1.6g). Increasing reciprocal connection strength, w_{rp}, between networks decreases the sampling variance of both stimuli, as well as the weight of direct cue in both networks (the cue input directly received by feedforward inputs). These behaviors are consistent with increasing the prior precision Λ_s in the Bayesian model.

The storage of prior precision in the reciprocal connections between networks can be understood intuitively (Fig. 1.7). Suppose the prior $p(\mathbf{s})$ is uniform ($\Lambda_s = 0$), which means there is no correlation between two stimuli, and then the posterior is the same as the likelihood distribution (Fig. 1.7a). In this case, the cue input generated by one stimulus, for example, s_1, doesn't have useful information for the estimation of another, for example, s_2, and thus there is no integration of two cue inputs. As a consequence, the two networks in the decentralized system don't need to cross-talk with each other, and only need to reply to the likelihood information conveyed by feedforward inputs.

In another extreme case, suppose the two stimuli are always exactly the same (fully correlated, $\Lambda_s = \infty$), and the two cue inputs will be fully integrated into a single estimate for both stimuli. Geometrically, the bivariate posterior of two stimuli will collapse onto the diagonal line in the 2d plane (Fig. 1.7c, right). Since the estimate of two stimuli should be exactly the same, the two networks should be fully synchronous with each other, which requires the effective reciprocal connection strength between networks to be sufficiently strong. And then the two networks effectively merge into a single big network, similar with the centralized architecture shown in Fig. 1.1b.

In the intermediate case ($0 < \Lambda_s < \infty$, Fig. 1.7b), the two stimuli are correlated which implies the cue input 1 has information useful for estimating stimulus s_2. Therefore, the estimate of two stimuli is correlated which is exhibited by a tilted posterior (Fig. 1.7b, right), but they don't need to be necessarily the same. In this case, the two networks are coupled to communicate the information from another cue, but the responses and stimulus samples in two networks don't need to be exactly the same. This corresponds to the situ-

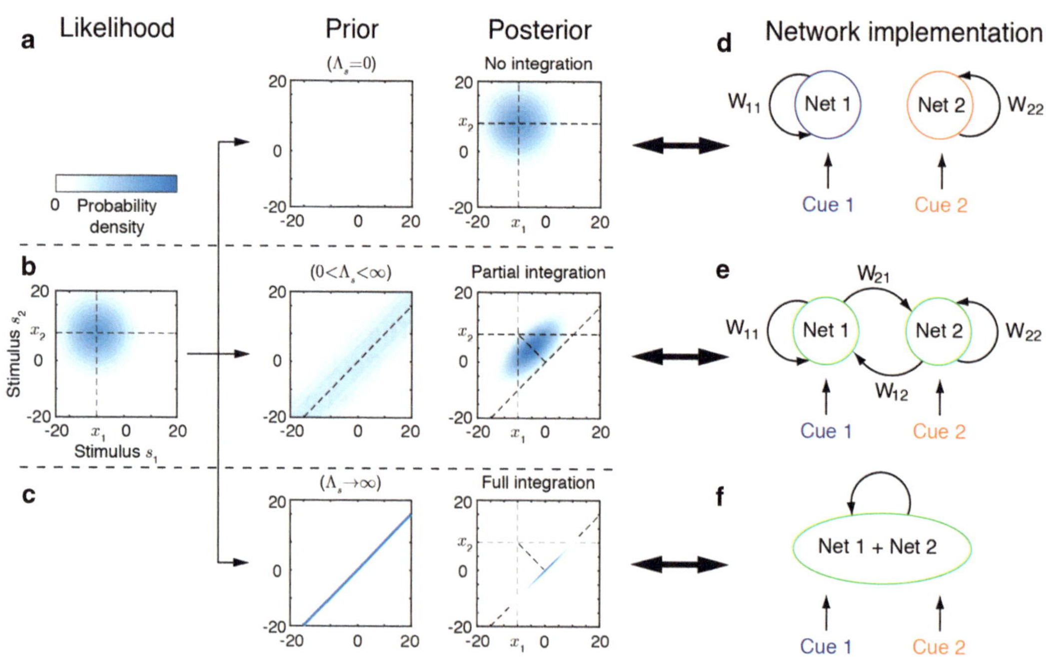

Fig. 1.7 The reciprocal connections between networks store the prior precision characterizing the stimulus correlation. (**a**–**c**) The posterior computed based on the same likelihood but with different prior distribution. The priors have different values of precision Λ_s (Eq. 1.13), which leads to a uniform prior (**a**), a correlated prior (**b**), and a prior the two stimuli are always exactly the same (**c**). (**d**–**f**) The network implementation corresponds to different priors. Increasing prior precision increases the reciprocal connections between networks

ation that the MSTd and VIP are interconnected (Fig. 1.1c), where MSTd infers the visual direction and VIP infers the vestibular direction.

1.5 A Large-Scale Decentralized Network Is Robust to Local Failure

In real life, the brain often needs to integrate information from more than two sensory cues (Wozny et al. 2008). A decentralized architecture is flexible to the number of sensory cues due to its modular structure, and thus can be easily upgraded to arbitrary number of coupled networks with each of them receiving and processing an individual cue (Zhang et al. 2016, 2020). For example, the integration of three cues can be implemented by inserting a third network, which receives cue 3, to the aforementioned system (Fig. 1.8a). Furthermore, in the general case of integration of M cues, a decentralized system can be further e to comprise M coupled networks and each network receives feedforward inputs from its corresponding sensory cue (Zhang et al. 2020).

In the decentralized system with two coupled networks, the reciprocal connections between two coupled networks store the prior between two stimuli. Similarly, we found in a larger decentralized system with more coupled networks, the reciprocal connections between each pair of networks store a prior distribution between the stimuli estimate by the two networks. And the whole high-dimensional prior stored in the large decentralized system is the product of each part of prior,

$$p(s_1, s_2, \ldots, s_M) \propto \prod_{i \neq j} p(s_i, s_j) \propto \exp\left[-\frac{1}{2}\sum_{i \neq j} \Lambda_{ij} (s_i - s_j)^2\right], \tag{1.12}$$

where Λ_{ij} measures the correlation between stimulus s_i and s_j and is stored by the reciprocal connection strength between network i and network j.

Since there is no network located at the center of the whole decentralized system, the whole network is robust to local damage. Suppose we have a decentralized system which consists of three coupled networks. The damage of one of few networks doesn't impair the rest networks optimally integrate the information, i.e., the sampling distribution of the network under combined cue condition is consistent with Bayesian prediction. For example, the network 1 is still able to optimally integrate cue 1 and cue 2 no matter whether the network 3 exists or not although the sampling variance of the stimulus s_1 in network 1 increases (Fig. 1.8b, green bars). The increase of sampling variance can be understood from two perspectives. From the perspective of network dynamics, the existing connections between networks help to average out noise (Zhang and Wu 2012; Kilpatrick et al. 2013). The more excitatory connections a network receives, the smaller sampling variance would be, and therefore losing a connection due to damage subsequently increases the sampling variance. From the perspective of Bayesian inference, the loss of network 3 changes the effective prior of s_1 and s_2 stored in the decentralized system, from $p_3(s_1, s_2) = \int p(s_1, s_2, s_3) ds_3$ in three coupled networks to $p(s_1, s_2)$ in two coupled networks. From Eq. (1.12), it can be calculated that the prior precision is decreased from $\Lambda_{12} + \left(\Lambda_{23}^{-1} + \Lambda_{13}^{-1}\right)^{-1}$ in three coupled networks to Λ_{12} in two coupled networks, suggesting the prior correlation between s_1 and s_2 is decreased. A decrease of the prior precision will cause the two cues are integrated in a less degree, and hence the sampling variance of s_1 increases. However, even if the prior is disrupted after blocking a network in the decentralized system, the whole system is able to optimally combine the likelihood information from input and the prior stored in the system.

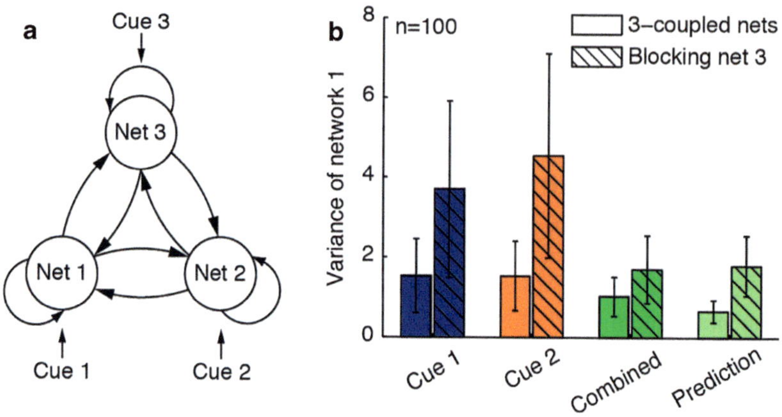

Fig. 1.8 Robust multisensory integration in a large, decentralized network. (**a**) Architecture of a decentralized network consisting of three reciprocally connected networks. The insertion of a third network is done simply by reciprocally connecting network 3 with other networks. (**b**) After blocking network 3 (shaded bars), the sample of stimulus 1 (read out from network 1) in the combined cue condition is nevertheless still similar to the Bayesian prediction although its variance increases. Error bars plot one standard deviation of the sampling variance obtained from 100 trials. The figure is adapted from Zhang et al. (2016)

Appendix

The Probabilistic Model of Multisensory Integration

To ease of computation, I write the likelihood and prior (Eq. 1.13) in matrix form,

$$p(\mathbf{x}|\mathbf{s}) = \mathcal{N}\left(\mathbf{x}|\mathbf{s}, \Lambda^{-1}\right), \; p(\mathbf{s}) \propto \mathcal{N}\left(\mathbf{s}|\mu_s, \Lambda_s^{-1}\right), \tag{1.13}$$

where $\mathbf{s} = (s_1, s_2)^{\mathrm{T}}$ and $\mathbf{x} = (x_1, x_2)^{\mathrm{T}}$. $\mathbf{\Lambda} = diag\,(\Lambda_1, \Lambda_2)$ is a diagonal precision matrix (the inverse of the covariance matrix) of the likelihood function. The diagonal precision matrix suggests the generation of each x_m is conditionally independent given s_m, i.e., $p(\mathbf{x}|\mathbf{s}) = \prod_{m=1}^{2} p(x_m|s_m)$. The precision matrix $\mathbf{\Lambda}_s$ of the prior $p(\mathbf{s})$ is,

$$\mathbf{\Lambda}_s = \Lambda_s \begin{pmatrix} 1 & -1 \\ -1 & 1 \end{pmatrix}, \tag{1.14}$$

with the scalar Λ_s denotes the correlation between s_1 and s_2. It is worth noting that the prior $p(\mathbf{s})$ is not centered on μ_s (Eq. 1.13) since it can be checked that $p(\mathbf{s}) \propto \exp\,[-\Lambda_s(s_1 - s_2)^2/2]$ which doesn't contain μ_s. Notably, according to Eq. (1.13), the marginal prior of each stimulus feature is uniform, i.e., $p(s_m) = 1/w_s$, which is resulted from that the determinant of the precision matrix is zero, i.e., $|\mathbf{\Lambda}_s| = 0$. This form of prior was also used in previous studies of multisensory integration (Bresciani et al. 2006; Roach et al. 2006; Sato et al. 2007). According to Bayes' theorem, the posterior distribution of the latent stimulus **s** given the observation **x** is,

$$p(\mathbf{s}|\mathbf{x}) \propto p(\mathbf{x}|\mathbf{s})\, p(\mathbf{s}) = \mathcal{N}\left(\mathbf{s}|\mu_s, \mathbf{\Omega}^{-1}\right), \tag{1.15}$$

which is also a bivariate Gaussian distribution, with the precision matrix $\mathbf{\Omega}$ and mean $\hat{s}$ given by,

$$\mathbf{\Omega} = \mathbf{\Lambda} + \mathbf{\Lambda}_s, \; \hat{s} = \mathbf{\Omega}^{-1} \rangle\; \mathbf{x}. \tag{1.16}$$

By expanding the matrix form in the above equation, we could arrive Eq. (1.3) in the main text.

Neural Dynamics of the Decentralized Network

The decentralized network is composed of two or several interconnected networks (Zhang et al. 2016), with each network modeled as a continuous attractor network (CAN) (Wu et al. 2016). For simplicity, the two networks have the same number of N excitatory neurons with preferred stimulus values $\{\theta_j\}_{j=1}^N$, where θ_j is the preference of the j-th E neuron. We take $\{\theta_j\}_{j=1}^N$ to uniformly cover the range of stimulus s_m, where $\theta \in (-\pi, \pi]$ satisfying the periodic boundary condition.

CAN is a canonical neural circuit model widely used to elucidate the network mechanism in coding and processing continuous stimulus features (see, e.g., Ben-Yishai et al. 1995; Knierim and Zhang 2012; Si et al. 2008). In the continuum limit ($\theta_j \to \theta$), the dynamics of coupled CANs is written as,

$$\tau \frac{\partial \mathbf{u}_{mt}(\theta)}{\partial t} = -\mathbf{u}_{mt}(\theta) + \rho \sum_n \mathbf{W}_{mn}^r * \mathbf{r}_{nt}(\theta) + \rho \mathbf{W}_m^f * \mathbf{I}_m(\theta) + \sqrt{\tau \mathrm{F} \mathbf{u}_{mt}(\theta)}\xi_{mt}(\theta), \tag{1.17}$$

where $\mathbf{u}_{mt}(\theta)$ and $\mathbf{r}_{mt}(\theta)$ denote, respectively, the total synaptic inputs and firing rates of the neurons at time t with preferred stimulus $s_m = \theta$ in network m $(m = 1,2)$. τ characterizes the time constant of synaptic input, and $\rho = N/(2\pi)$ is the density of neurons covering the stimulus (direction) space. F denotes the Fano factor of the internal Poisson variability capturing the variability of spike generation. $\mathbf{W}_{mm}^r$ is the kernel of recurrent connections between neurons within the same network, $\mathbf{W}_{mn}^r$ for $m \neq n$ is the kernel of reciprocal connections between neurons from network n to network m, and $\mathbf{W}_m^f$ is the kernel of feedforward connection. As an example, these kernels all have the Gaussian profile and are written as,

$$\mathbf{W}_{mn}^r(\theta) = w_{mn}^r \mathbf{g}(\theta), \mathbf{W}_m^f = w_m^f \mathbf{g}(\theta), \ \mathbf{g}(\theta) = \left(\sqrt{2\pi}a\right)^{-1} \exp\left(-\theta^2 / 2a^2\right). \tag{1.18}$$

The symbol $*$ represents the convolution, i.e., $\mathbf{W}(\theta) * \mathbf{r}(\theta) = \int \mathbf{W}(\theta - \theta')\mathbf{r}(\theta')d\theta'$, which implies the connection pattern between neurons is translation-invariant in the stimulus space, a key property of CANs. For simplicity, the feedforward connection weight w^f is assumed to be the same across networks, and this doesn't affect the results substantially.

In the model, the neurons convert its synaptic input into the firing rate instantaneous, whose relation is modeled as divisive normalization (Carandini and Heeger 2012; Deneve et al. 1999), which is given by

$$\mathbf{r}_{mt}(\theta) = \frac{\left[\mathbf{u}_{mt}(\theta)\right]_+^2}{1 + k\rho \int \left[\mathbf{u}_{mt}(\theta')\right]_+^2 d\theta'}, \tag{1.19}$$

Here k controls the strength of global inhibition and $[\cdot]_+$ denotes negative rectification. Divisive normalization is extensively observed in the cerebral cortex and is considered as a canonical operation of the cortex. It could be implemented via parvalbumin (PV) inhibitory neurons (Niell 2015).

$\mathbf{I}_m(\theta)$ in Eq. (1.17) denotes the response of unisensory neurons (neurons in MT or PIVC, Fig. 1.1c) which provides feedforward inputs to network m. $\mathbf{I}_m(\theta)$ is modeled as independent spike train with Gaussian tunning over stimulus s_m (Eq. 1.4). Since the neural dynamics is based on rate-based neurons (Eq. 1.17), in network simulation $\mathbf{I}_m(\theta)$ is approximated by Gaussian noise with the variance proportional to its mean rate, i.e., $\mathbf{I}_m(\theta) = \langle \mathbf{I}_m(\theta)\rangle + \langle \mathbf{I}_m(\theta)\rangle \boldsymbol{\xi}_t$, where $\boldsymbol{\xi}_t$ is an N-dim independent Gaussian white noise.

Unisensory Neurons' Response Convey the Likelihood Function

I present the math details in deriving that a single snapshot of the population responses of unisensory neurons (MT or PIVC, Fig. 1.1c) convey the whole likelihood function over stimulus direction. Substituting the firing rate $\langle \mathbf{I}_m(s_m)\rangle$ (Eq. 1.4, last line) into the Poisson spike generation (Eq. 1.4, first line),

$$p(\mathbf{I}_m|s_m) = \prod_{j=1}^{N} \text{Poisson}\left(\mathbf{I}_{mj}|\mathbf{I}_{mj}(s_m)\right)$$

$$\propto \left(R^f\right)^{\sum_j \mathbf{I}_{mj}} \exp\left[-\sum_j \mathbf{I}_{mj}\frac{\left(s_m-\theta_j\right)^2}{2a^2}\right]\exp\left[-R^f\sum_j e^{-\left(s_m-\theta_j\right)^2/2a^2}\right]. \quad (1.20)$$

Since the preferred stimulus value of E neurons, $\{\theta_j\}_{j=1}^N$ of unisensory neurons is uniformly distributed in the range of stimulus s_m, the summation inside the last bracket in the above equation can be treated as a constant irrelevant with s_m. Moreover, omitting the first term in the above equation since it doesn't contain the term of s_m, and normalizing the distribution over s_m, it arrives

$$p(\mathbf{I}_m|s_m) \propto \exp\left[-\sum_j \mathbf{I}_{mj}\frac{\left(s_m-\theta_j\right)^2}{2a^2}\right] \propto \mathcal{N}\left(\tilde{s}_m|x_m,\Lambda_m\right), \quad (1.21)$$

where

$$x_m = \frac{\sum_j \mathbf{I}_{mj}\theta_j}{\sum_j \mathbf{I}_{mj}}, \Lambda_m = a^{-2}\sum_j \mathbf{I}_{mj} \quad (1.22)$$

This is presented in Eqs. (1.5) and (1.6) in the main text.

Theoretical Analysis of the Network Dynamics on the Stimulus Subspace

Let's first compute the mean responses of each network in equilibrium state that can be analytically calculated. Since the recurrent connections (Eq. 1.18) and the feedforward inputs from unisensory neurons (Eq. 1.4) both have Gaussian profile, the mean network response also has a Gaussian profile (Fig. 1.3c) (Zhang and Wu 2013; Si et al. 2008; Alan Fung et al. 2010),

$$\langle \mathbf{u}_{mt}(\theta)\rangle = \mathrm{U}_m \exp\left[-\left(\bar{s}_m-\theta_j\right)^2/4a^2\right],$$

$$\langle \mathbf{r}_{mt}(\theta)\rangle = \mathrm{R}_m \exp\left[-\left(\bar{s}_m-\theta_j\right)^2/2a^2\right]. \quad (1.23)$$

where $\langle \mathbf{u}_{mt}(\theta)\rangle$ and $\langle \mathbf{r}_{mt}(\theta)\rangle$ denotes the mean synaptic inputs and firing rate of E neurons preferring stimulus (direction) θ in network m. $\bar{s}_m$ denotes the mean position of the population response on the stimulus subspace.

In order to study how the decentralized network performs sampling-based inference to approximate the posterior, we performed perturbative analysis to derive the network dynamics on the stimulus subspace. For each network m, an instantaneous neuronal response is considered to be a perturbed response from its equilibrium mean, $\mathbf{u}_{mt}(\theta) = \langle \mathbf{u}_{mt}(\theta) \rangle + \delta\mathbf{u}_{mt}(\theta)$, with $\delta\mathbf{u}_{mt}(\theta)$ denoting the perturbation. By performing eigen-analysis of the perturbative dynamics (the dynamics of $\delta\mathbf{u}_{mt}(\theta)$), the direction corresponding to stimulus s_m embedded in neuronal response $\mathbf{u}_{mt}(\theta)$ can be analytically derived as the derivative of the mean neuronal response with stimulus (Alan Fung et al. 2010),

$$\phi_m\left(\theta|s_m^{\mathrm{r}}\right) = \frac{d\mathbf{u}_{mt}(\theta)}{ds_m^{\mathrm{r}}} \propto \left(\theta - s_m^{\mathrm{r}}\right)\exp\left[-\left(\theta - s_m^{\mathrm{r}}\right)^2 / 4a^2\right], \tag{1.24}$$

It was found that $\phi_m\left(\theta|s_m^{\mathrm{r}}\right)$ corresponds to the eigenvector of the perturbative dynamics with the largest eigenvalue, which indicates that the network response is dominated by the change along the stimulus subspace.

Next we project the high-dimensional dynamics of each network onto the stimulus direction (Eq. 1.24), where the projection is computing the inner product between ϕ_m and $\mathbf{u}_{mt}$,

$$\langle \phi_m, \mathbf{u}_{mt} \rangle = \int \phi_m(\theta)\mathbf{u}_{mt}(\theta)d\theta, \tag{1.25}$$

After projection, the network dynamics on the stimulus subspace can be derived which is shown in Eq. (1.7),

$$\frac{d\mathbf{s}_t^r}{dt} = -\frac{\rho}{\sqrt{2}}\left(\tau\mathbf{D}_U\right)^{-1}\left[\mathbf{L}\mathbf{s}_t^r + w^f\mathbf{D}_f\left(\mathbf{s}_t^r - \mathbf{x}\right)\right] + \sigma_s\left(\tau\mathbf{D}_U\right)^{-1/2}\xi_t, \tag{1.26}$$

It can be calculated that the distribution of stimulus features $\mathbf{s}^r$ in equilibrium in Eq. (1.26) is a multivariate Gaussian distribution, denoted as $\mathcal{N}\left(\mathbf{s}^{\mathrm{r}}|\overline{\mathbf{s}}^{\mathrm{r}}, \Sigma_{\mathbf{s}}^{\mathrm{r}}\right)$, whose mean $\overline{\mathbf{s}}^{\mathrm{r}}$ and covariance of $\Sigma_{\mathbf{s}}^{\mathrm{r}}$ satisfy the two equations given below:

$$\left(\mathbf{L} + w^f\mathbf{D}_f\right)\overline{\mathbf{s}}^r = w^f\mathbf{D}_f\mathbf{x}, \tag{1.27}$$

$$\left(\mathbf{L} + \mathrm{w}^f\mathbf{D}_f\right)\Sigma_{\mathbf{s}}^{\mathrm{r}} + \Sigma_{\mathbf{s}}^{\mathrm{r}}\left(\mathbf{L} + \mathrm{w}^f\mathbf{D}_f\right)^{\mathrm{T}} = \sqrt{2}\sigma_s^2\rho^{-1} \tag{1.28}$$

where $\overline{\mathbf{s}}^r = \left\langle \mathbf{s}_t^r \right\rangle$, and $\Sigma_{\mathbf{s}}^{\mathrm{r}} = \left\langle \left(\mathbf{s}_t^r - \overline{\mathbf{s}}^r\right)\left(\mathbf{s}_t^r - \overline{\mathbf{s}}^r\right)^{\mathrm{T}} \right\rangle$. By equating the equilibrium mean and covariance of stimulus features samples generated by the network to that of the posterior of latent stimulus features, i.e., $\overline{\mathbf{s}}^r = \hat{s}$ and $\Sigma_{\mathbf{s}}^{\mathrm{r}} = \boldsymbol{\Omega}^{-1}$ (see Eq. 1.3), we can derive the network connections which are necessary to correctly sample the posterior. It can be verified that $\mathbf{L} + \mathrm{w}^f\mathbf{D}_f = \sigma_s^2\left(\sqrt{2}\rho\right)^{-1}\boldsymbol{\Omega}$ can satisfy the requirement $\Sigma_{\mathbf{s}}^{\mathrm{r}} = \boldsymbol{\Omega}^{-1}$. Substituting this condition into Eqs. (1.27) and (1.28), we get the required network connections which can enable the network dynamics utilizing Langevin dynamics to sample posterior of stimulus features, which are,

$$\mathrm{w}_{mn}^r = \frac{aw^f}{\sqrt{2\pi}\rho}\frac{L_{mn}}{R_n}, \; w^f = \frac{\sqrt{\pi}}{a}\sigma_s^2 = \left(\frac{2}{\sqrt{3}}\right)^3 \mathrm{F}. \tag{1.29}$$

References

Aitchison L, Lengyel M (2016) The Hamiltonian brain: efficient probabilistic inference with excitatory-inhibitory neural circuit dynamics. PLoS Comput Biol 12(12):e1005186

Alais D, Burr D (2004) The ventriloquist effect results from near-optimal bimodal integration. Curr Biol 14(3):257–262

Alan Fung CC, Michael Wong KY, Wu S (2010) A moving bump in a continuous manifold: a comprehensive study of the tracking dynamics of continuous attractor neural networks. Neural Comput 22(3):752–792

Alvarado JC, Rowland BA, Stanford TR, Stein BE (2008) A neural network model of multisensory integration also accounts for unisensory integration in superior colliculus. Brain Res 1242:13–23

Baizer JS, Ungerleider LG, Desimone R (1991) Organization of visual inputs to the inferior temporal and posterior parietal cortex in macaques. J Neurosci 11(1):168–190

Ben-Yishai R, Lev Bar-Or L, Sompolinsky H (1995) Theory of orientation tuning in visual cortex. Proc Natl Acad Sci 92(9):3844–3848

Bertin RJV, Berthoz A (2004) Visuo-vestibular interaction in the reconstruction of travelled trajectories. Exp Brain Res 154(1):11–21

Bishop CM (2006) Pattern recognition and machine learning. Springer, New York

Boussaoud D, Ungerleider LG, Desimone R (1990) Pathways for motion analysis: cortical connections of the medial superior temporal and fundus of the superior temporal visual areas in the macaque. J Comp Neurol 296(3):462–495

Bresciani J-P, Dammeier F, Ernst MO (2006) Vision and touch are automatically integrated for the perception of sequences of events. J Vis 6(5):2

Britten KH, Shadlen MN, Newsome WT, Movshon JA (1992) The analysis of visual motion: a comparison of neuronal and psychophysical performance. J Neurosci 12(12):4745–4765

Burak Y, Fiete IR (2009) Accurate path integration in continuous attractor network models of grid cells. PLoS Comput Biol 5(2):e1000291

Carandini M, Heeger DJ (2012) Normalization as a canonical neural computation. Nat Rev Neurosci 13(1):51–62

Chen A, DeAngelis GC, Angelaki DE (2011a) Representation of vestibular and visual cues to self-motion in ventral intraparietal cortex. J Neurosci 31(33):12036–12052

Chen A, DeAngelis GC, Angelaki DE (2011b) Convergence of vestibular and visual self-motion signals in an area of the posterior sylvian fissure. J Neurosci 31(32):11617–11627

Chen A, DeAngelis GC, Angelaki DE (2011c) A comparison of vestibular spatiotemporal tuning in macaque parietoinsular vestibular cortex, ventral intraparietal area, and medial superior temporal area. J Neurosci 31(8):3082–3094

Chen A, DeAngelis GC, Angelaki DE (2013) Functional specializations of the ventral intraparietal area for multisensory heading discrimination. J Neurosci 33(8):3567–3581

Chen A, Yong G, Liu S, DeAngelis GC, Angelaki DE (2016) Evidence for a causal contribution of macaque vestibular, but not intraparietal, cortex to heading perception. J Neurosci 36(13):3789–3798

Deneve S, Latham PE, Pouget A (1999) Reading population codes: a neural implementation of ideal observers. Nature Neurosci 2(8):740–745

Durrant-Whyte H, Henderson TC (2016) Multisensor data fusion. In: Springer handbook of robotics. Springer, pp 867–896

Ernst MO, Banks MS (2002) Humans integrate visual and haptic information in a statistically optimal fashion. Nature 415(6870):429–433

Ernst MO, Heinrich H, Bu¨lthoff. (2004) Merging the senses into a robust percept. Trends Cogn Sci 8(4):162–169

Fiser J, Berkes P, Orbán G, Lengyel M (2010) Statistically optimal perception and learning: from behavior to neural representations. Trends Cogn Sci 14(3):119–130

Georgopoulos AP, Schwartz AB, Kettner RE (1986) Neuronal population coding of movement direction. Science 233(4771):1416–1419

Gold JI, Shadlen MN (2001) Neural computations that underlie decisions about sensory stimuli. Trends Cogn Sci 5(1):10–16

Hennequin G, Aitchison L, Lengyel M (2014) Fast sampling-based inference in balanced neuronal networks. In: NIPS, vol 27. Citeseer, pp 2240–2248

Hoyer PO, Hyvärinen A (2003) Interpreting neural response variability as Monte Carlo sampling of the posterior. In: Advances in neural information processing systems, pp 293–300

Jacobs RA (1999) Optimal integration of texture and motion cues to depth. Vis Res 39(21):3621–3629

Jazayeri M, Anthony J, Movshon. (2006) Optimal representation of sensory information by neural populations. Nat Neurosci 9(5):690–696

Kersten D, Mamassian P, Yuille A (2004) Object perception as Bayesian inference. Annu Rev Psychol 55:271–304

Kilpatrick ZP, Ermentrout B, Doiron B (2013) Optimizing working memory with heterogeneity of recurrent cortical excitation. J Neurosci 33(48):18999–19011

Knierim JJ, Zhang K (2012) Attractor dynamics of spatially correlated neural activity in the limbic system. Annu Rev Neurosci 35:267–285

Knill DC, Pouget A (2004) The Bayesian brain: the role of uncertainty in neural coding and computation. Trends Neurosci 27(12):712–719

Körding KP, Wolpert DM (2004) Bayesian integration in sensorimotor learning. Nature 427(6971):244–247

Ma WJ, Beck JM, Latham PE, Pouget A (2006) Bayesian inference with probabilistic population codes. Nat Neurosci 9(11):1432–1438

Magosso E, Cuppini C, Serino A, Di Pellegrino G, Ursino M (2008) A theoretical study of multisensory integration in the superior colliculus by a neural network model. Neural Netw 21(6):817–829
Makin JG, Fellows MR, Sabes PN (2013) Learning multisensory integration and coordinate transformation via density estimation. PLoS Comput Biol 9(4):e1003035
Marr D (2010) Vision: A computational investigation into the human representation and processing of visual information, 1982. MIT Press
Morgan ML, DeAngelis GC, Angelaki DE (2008) Multisensory integration in macaque visual cortex depends on cue reliability. Neuron 59(4):662–673
Neal RM et al (2011) MCMC using Hamiltonian dynamics. Handbook Markov Chain Monte Carlo 2(11):2
Niell CM (2015) Cell types, circuits, and receptive fields in the mouse visual cortex. Annu Rev Neurosci 38:413–431
Ohshiro T, Angelaki DE, DeAngelis GC (2011) A normalization model of multisensory integration. Nat Neurosci 14(6):775–782
Orbán G, Berkes P, Fiser J, Lengyel M (2016) Neural variability and sampling based probabilistic representations in the visual cortex. Neuron 92(2):530–543
Roach NW, Heron J, McGraw PV (2006) Resolving multisensory conflict: a strategy for balancing the costs and benefits of audio-visual integration. Proc Royal Soc London B: Biol Sci 273(1598):2159–2168
Rubin DB, Van Hooser SD, Miller KD (2015) The stabilized supralinear network: a unifying circuit motif underlying multi-input integration in sensory cortex. Neuron 85(2):402–417
Samsonovich A, McNaughton BL (1997) Path integration and cognitive mapping in a continuous attractor neural network model. J Neurosci 17(15):5900–5920
Sato Y, Toyoizumi T, Aihara K (2007) Bayesian inference explains perception of unity and ventriloquism aftereffect: identification of common sources of audiovisual stimuli. Neural Comput 19(12):3335–3355
Savin C, Deneve S (2014) Spatio-temporal representations of uncertainty in spiking neural networks. Adv Neural Inform Process Syst 27:2024–2032
Srivastava N, Salakhutdinov R et al (2012) Multimodal learning with deep boltzmann machines. In: NIPS, vol 1. Citeseer, p 2
Ursino M, Cuppini C, Magosso E, Serino A, Di Pellegrino G (2009) Multisensory integration in the superior colliculus: a neural network model. J Comput Neurosci 26(1):55–73
Van Beers RJ, Sittig AC, Gon JJDVD (1999) Integration of proprioceptive and visual position-information: an experimentally supported model. J Neurophysiol 81(3):1355–1364
Welling M, Teh YW (2011) Bayesian learning via stochastic gradient Langevin dynamics. In: Proceedings of the 28th international conference on machine learning (ICML-11). Citeseer, pp 681–688
Wozny DR, Beierholm UR, Shams L (2008) Human trimodal perception follows optimal statistical inference. J Vis 8(3):24
Wu S, Hamaguchi K, Amari S-i (2008) Dynamics and computation of continuous attractors. Neural Comput 20(4):994–1025
Wu S, Wong KM, Fung CA, Mi Y, Zhang W (2016) Continuous attractor neural networks: candidate of a canonical model for neural information representation. F1000 Res 5. https://doi.org/10.12688/f1000research.7387.1
Yong G, Watkins PV, Angelaki DE, DeAngelis GC (2006) Visual and nonvisual contributions to three-dimensional heading selectivity in the medial superior temporal area. J Neurosci 26(1):73–85
Yong G, Angelaki DE, DeAngelis GC (2008) Neural correlates of multisensory cue integration in macaque MSTD. Nat Neurosci 11(10):1201–1210
Yong G, DeAngelis GC, Angelaki DE (2012) Causal links between dorsal medial superior temporal area neurons and multisensory heading perception. J Neurosci 32(7):2299–2313
Yong G, Cheng Z, Yang L, DeAngelis GC, Angelaki DE (2016) Multisensory convergence of visual and vestibular heading cues in the pursuit area of the frontal eye field. Cereb Cortex 26(9):3785–3801
Yuille A, Kersten D (2006) Vision as Bayesian inference: analysis by synthesis? Trends Cogn Sci 10(7):301–308
Zhang K (1996) Representation of spatial orientation by the intrinsic dynamics of the head direction cell ensemble: a theory. J Neurosci 16(6):2112–2126
Zhang W, Wu S (2012) Neural information processing with feedback modulations. Neural Comput 24(7):1695–1721
Zhang W-H, Wu S (2013) Reciprocally coupled local estimators implement Bayesian information integration distributively. In: Advances in neural information processing systems. Curran Associates, Inc. pp 19–27
Zhang W-H, Chen A, Rasch MJ, Wu S (2016) Decentralized multisensory information integration in neural systems. J Neurosci 36(2):532–547
Wenhao Zhang, Tai Sing Lee, Brent Doiron, Si Wu Distributed sampling-based Bayesian inference in coupled neural circuits. bioRxiv, 2020

From Multisensory Integration to Multisensory Decision-Making

2

Qihao Zheng and Yong Gu

Abstract

Organisms live in a dynamic environment in which sensory information from multiple sources is ever changing. A conceptually complex task for the organisms is to accumulate evidence across sensory modalities and over time, a process known as multisensory decision-making. This is a new concept, in terms of that previous researches have been largely conducted in parallel disciplines. That is, much efforts have been put either in sensory integration across modalities using activity summed over a duration of time, or in decision-making with only one sensory modality that evolves over time. Recently, a few studies with neurophysiological measurements emerge to study how different sensory modality information is processed, accumulated, and integrated over time in decision-related areas such as the parietal or frontal lobes in mammals. In this review, we summarize and comment on these studies that combine the long-existed two parallel fields of multisensory integration and decision-making. We show how the new findings provide insight into our understanding about neural mechanisms mediating multisensory information processing in a more complete way.

Keywords

Multisensory integration · Decision-making · Vestibular · Optic flow

Animals live in a sensory world where they are exposed to different types of information from the surrounding environment, as well as from oneself. However, noise is always accompanied with information, more or less, causing uncertainties when animals detect or discriminate signals. The brain has evolved two strategies to overcome the uncertainty embedded in the sensory channels. One is to reduce the uncertainty of individual modalities by integrating sensory inputs from different modalities, and the other is to reduce the interference of transient noise by integrating signals over some time. The process of information integration across modalities is the so-called multisensory integration (Stein et al. 2020; Hou and Gu 2020; Chandrasekaran 2017; Ursino et al. 2014; Seilheimer et al. 2014; Fetsch et al. 2013; Angelaki et al. 2009; Stein and Stanford 2008), whereas the process of information integration over time is the so-called decision-making, or more specifically, perceptual

Q. Zheng
Center for Excellence in Brain Science and Intelligence Technology, Institute of Neuroscience, Chinese Academy of Sciences, Shanghai, China

Y. Gu (✉)
Systems Neuroscience, SInstitute of Neuroscience, Chinese Academy of Sciences, Shanghai, China
e-mail: guyong@ion.ac.cn

Y. Gu, A. Zaidel (eds.), *Advances of Multisensory Integration in the Brain*, Advances in Experimental Medicine and Biology 1437, https://doi.org/10.1007/978-981-99-7611-9_2

decision-making (Najafi and Churchland 2018; Hanks and Summerfield 2017; Raposo 2016; Summerfield and de Lange 2014; Shadlen and Kiani 2013; Gold and Shadlen 2007). The combination of these two processes is thus defined as multisensory decision-making (Bizley et al. 2016; Raposo 2016).

The two fields of multisensory integration and decision-making have developed in parallel for a long time in the past. In particular, the field of multisensory integration mainly focuses on comparison of neuronal activity across different stimulus conditions in polysensory cortices (Smith et al. 2017; Gu et al. 2016; Yau et al. 2015; Fetsch et al. 2013; Chen et al. 2008, 2011a, b, c, 2013; Angelaki et al. 2011; Gu et al. 2006, 2008). By contrast, the field of perceptual decision-making commonly uses a single sensory modality (e.g., visual for most of the time) paradigm to study accumulation of sensory evidence over a temporal domain in sensory-motor transformation cortices such as the posterior parietal cortex or the prefrontal cortex (Li et al. 2016; Churchland et al. 2011; Kravitz et al. 2011; Bisley and Goldberg 2010; Churchland et al. 2008; Gold and Shadlen 2007; Roitman and Shadlen 2002; Shadlen and Newsome 1996, 2001). Recently, the two fields start to fuse, developing the cross-cutting field of "multisensory decision-making" (Coen et al. 2021; Hou et al. 2019; Nikbakht et al. 2018; Raposo et al. 2012, 2014; Sheppard et al. 2013).

This chapter reviews recent research advances in the field of multisensory decision-making at the behavioral, neurophysiological, and computational modeling levels and provides an outlook on the future of this field.

2.1 Computational Modeling

The computational essence of multisensory decision-making can be understood with the help of normative models (Fetsch et al. 2013) that has been used in multisensory integration and perceptual decision-making separately.

Firstly, multisensory integration is considered as a Bayesian inference process (Ma 2019). Assuming that the probability of occurrence of stimulus s is $P(s)$, an observer makes two estimates of the same stimulus s using both senses, denoted as r_1 and r_2, respectively. Due to sensory uncertainty, the observation of r_1 and r_2 in the case of stimulus s is also a probabilistic event, denoted as $P(r_1, r_2|s)$. Now, how should the observer infer the value of stimulus s based on senses r_1 and r_2? Mathematically, this can be solved by maximizing the posterior probability: $P(s|r_1, r_2)$. The posterior could be expanded by Bayesian formula as $P(r_1, r_2|s)P(s)/P(r_1, r_2)$, where $P(r_1, r_2|s)$ is likelihood function and $P(s)$ is the prior. Since our ultimate goal is to solve for s, and $P(r_1, r_2)$ does not contain s, this term can be ignored. Thus, the formula can be further written as $P(s|r_1, r_2) \propto P(r_1, r_2|s)P(s)$. If it is assumed that r_1 and r_2 are conditionally independent with respect to s and s is uniformly distributed, then the above equation is further simplified as: $P(s|r_1, r_2) \propto P(r_1|s)P(r_2|s)$. Under the uniform prior, the maximum posterior probability method is also known as the maximum likelihood estimation because the posterior probability has been reduced to a likelihood function. As long as we take the s that maximizes this probability, then we have completed the solution and can prove that this approach is optimal. Such an observer is also known as an "ideal observer" (Landy et al. 2011; Doya et al. 2006; Knill and Pouget 2004; Knill and Richards 1996).

If we further assume that the likelihood functions $P(r_1|s)$ and $P(r_2|s)$ follow a Gaussian distribution, then using the maximum likelihood method, we can obtain that s inferred from a single cue r_i also follows a Gaussian distribution, which can be noted as a Gaussian distribution with mean μ_i and variance σ_i^2 $i = 1, 2$, respectively. And the s inferred from the two modality cues also follow a Gaussian distribution, denoted as a Gaussian distribution with mean μ_{comb} and variance σ_{comb}^2, with the following relations (Ursino et al. 2014; Fetsch et al. 2013; Angelaki et al. 2009; Ernst and Banks 2002):

$$\mu_{\text{comb}} = \frac{\sigma_2^2}{\sigma_1^2 + \sigma_2^2}\mu_1 + \frac{\sigma_1^2}{\sigma_1^2 + \sigma_2^2}\mu_2$$

$$\frac{1}{\sigma_{\text{comb}}^2} = \frac{1}{\sigma_1^2} + \frac{1}{\sigma_2^2}$$

To date a large number of behavioral experiments have demonstrated that integration between different species and modalities approximately satisfies Bayesian optimal integration theory. For example:

1. Visual-tactile integration in humans (Knill and Saunders 2003; Ernst and Banks 2002).
2. Visual-auditory integration in humans (Sheppard et al. 2013; Raposo et al. 2012; Alais and Burr 2004; Battaglia et al. 2003).
3. Visual-proprioceptive integration in humans (Sober and Sabes 2005; van Beers et al. 1999).
4. Visual-vestibular integration in macaques (Zheng et al. 2021; Hou et al. 2019; Chen et al. 2013; Fetsch et al. 2011; Gu et al. 2008).
5. Integration of visual and electrical stimulation-evoked proprioception in macaques (Dadarlat et al. 2015).
6. Visual-auditory integration in rats (Raposo et al. 2012, 2014; Sheppard et al. 2013).

These studies suggest that multisensory integration mechanisms may be conserved across species and modality combinations.

Bayesian integration model explains the behavior of multisensory integration very well in many cases; however, this model ignores an important factor—time. Bayesian models do not consider the time course of multisensory integration (Drugowitsch and Pouget 2012) and only predict two time-independent parameters, psychophysical thresholds and biases. In fact, however, the stimuli and the corresponding reliability may vary over time in the experiment, and parameters like reaction-time is an important performance indicator in many situations.

In perceptual decision models, time is an essential and critical parameter. The process of making a decision based on sensory input is considered as the temporal integration of the decision variable with noisy sensory input and finally reaching a decision bound to make a decision. This decision model is called the drift-diffusion model (DDM) (Ratcliff and McKoon 2008; Ratcliff and Smith 2004; Ratcliff and Rouder 1998; Ratcliff 1978). For example, in a typical random-dots direction discrimination task, subjects observe stimulus whose intensity is modulated by a parameter called coherence, which is the proportion of dots that represent consistent motion direction. Higher the coherence, easier for subjects to discriminate motion directions. In addition, while observing the stimuli, subjects were disturbed by Gaussian noise with mean 0 and variance σ^2. According to the DDM, the slope of the accumulation curve is equal to the coherence. This accumulation process is called "drift." Due to the noise, the accumulated signal is disturbed by the noise, and the disturbance causes the fluctuation of the decision variable, which is called "diffusion." The DDM describes the decision process and explains decision-related phenomena. DDM has been used successfully to explain the ramping activities of neurons in LIP (Roitman and Shadlen 2002; Shadlen and Newsome 1996, 2001), FEF (Kim and Shadlen 1999), SC (Horwitz and Newsome 1999, 2001), and so forth, suggesting that these brain areas may encode decision variables.

Since Bayesian integration model describes integration across modalities and drift-diffusion model describes integration across time, can these two models be combined and used to describe multisensory decision-making? The answer is yes. In fact, Drugowitsch and colleagues used the combination of these two models (i.e., extended DDM) to explain subjects' speed-accuracy tradeoff behavior in a reaction-time version of heading discrimination task based on visual and vestibular cues (Drugowitsch et al. 2014). In particular, velocity profile of the motion was set as a Gaussian profile. The transient evidence of vestibular input was assumed to be generated by acceleration $a(t)$, whereas the transient evidence of visual input was generated by velocity cues $v(t)$. Visual and vestibular cues contain spatially consistent heading information, and both are presented synchronously in time. Therefore, both visual and vestibular cues are time-varying signals, and thus their reliability also vary over time. Let $X_{\text{vis}}(t)$ be the cumulative evidence of visual cues (optic flow) and $X_{\text{ves}}(t)$ be the cumulative evidence of vestibular cues, then $X_{vis}(t)$ and $X_{ves}(t)$ are the time integrals of the noisy velocity and acceleration signals at $[0,t]$, respectively. Let k_{vis} be a constant representing the intensity of the visual signal and k_{ves} be a constant representing the intensity of the vestibular signal, the joint signal $X_{comb}(t)$ is the sum of these two signals weighted by the corresponding reliability.

$$X_{comb}(t)=\frac{k_{vis}}{\sqrt{k_{vis}^2+k_{ves}^2}}X_{vis}(t)+\frac{k_{ves}}{\sqrt{k_{vis}^2+k_{ves}^2}}X_{ves}(t)$$

In short, the extended DDM does the time integration of the sensory signals weighted by the time reliability, and then calculate the weighted sum of the two signals according to the modal reliability. This extended DDM is so far, to our knowledge, the only normative model for optimal multisensory decision-making (Drugowitsch et al. 2014).

2.2 Physiological Studies

2.2.1 Brain Regions Associated with Multisensory Decision-Making

Physiological studies of multisensory integration began with Barry Stein and colleagues in anesthetized cats in the 1980s. The researchers identified neurons in the superior colliculus (SC), a structure located in the midbrain that processes multiple types of information such as visual and auditory. Three empirical principles were proposed (Stein and Stanford 2008; Stein and Meredith 1993), including the temporal principle (Meredith et al. 1987), the spatial principle (Meredith and Stein 1986a), and the principle of inverse effectiveness (Meredith and Stein 1986b).

In addition to SC, studies on awake macaques using visual (e.g., optic flow) and vestibular stimuli to study how multisensory heading perception is mediated by neurons in the cerebral cortex. A series of multisensory brain regions have thus been identified (Fig. 2.1), including the dorsal medial superior temporal area (MSTd) (Maciokas and Britten 2010; Morgan et al. 2008; Takahashi et al. 2007; Gu et al. 2006; Page and Duffy 2003; Bremmer et al. 1999; Duffy 1998), the ventral intraparietal area (VIP) (Chen et al. 2011a, b, c, 2013; Zhang and Britten 2004; Zhang et al. 2004; Bremmer et al. 2002a, b; Schlack et al. 2002), the visual posterior sylvian area (VPS) (Chen et al. 2011b), the smooth eye movement area of the frontal eye field (FEFsem) (Yang and Gu 2017; Gu et al. 2016), and area 7a (Avila et al. 2019). Among all these, MSTd has received the highest attention. Numerous studies investigated basic integration properties in this area, as well as how they are linked to perception when the animals are trained to perform behavioral discrimination task at the same time (see following sections for more detail).

In addition to visuo-vestibular, other multisensory signals have been found including tactile and visual signals in mice' dorsal striatum (Reig and Silberberg 2014), or in macaques' VIP (Avillac et al. 2007). These regions are essentially part of the association cortex, but there is evidence that the neural basis of multisensory integration extends into early sensory processing (Ghazanfar and Schroeder 2006). For example, primary auditory cortex (A1) in macaques encodes not only auditory but also visual (Ghazanfar et al. 2005) and tactile signals (Lemus et al. 2010). Primary somatosensory cortex (S1) in macaques encodes not only tactile but also proprioceptive (Kim et al. 2015) and auditory (Lemus et al. 2010) sense.

In parallel, many brain regions are related to perceptual decision-making, particularly under the oculomotor paradigm such as the lateral intraparietal area (Roitman and Shadlen 2002; Shadlen and Newsome 1996, 2001), the frontal eye field (FEF) (Kim and Shadlen 1999), the superior colliculus (SC) (Horwitz and Newsome 1999, 2001), and the caudate (CD) in the basal ganglia (Ding and Gold 2010). In these sensory-motor transformation areas, ramping activities have been found during delay period and its rising slope depends on task difficulty and therefore

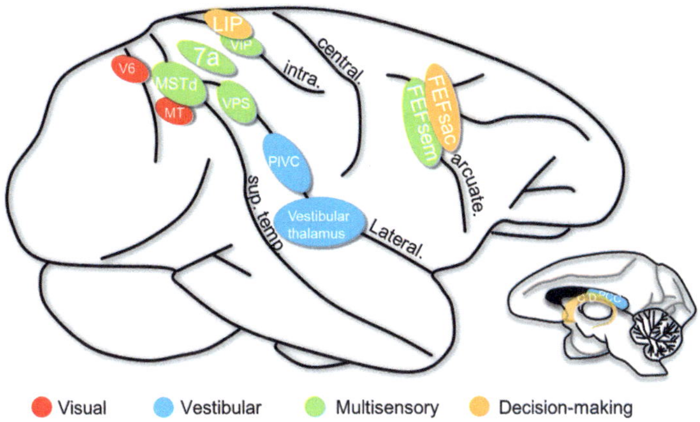

Fig. 2.1 Self-motion perception-related brain areas in macaque. Red: sensory area encoding visual information only, blue: sensory area encoding vestibular information only, green: sensory area encoding both visual and vestibular information, yellow: decision-related areas. *MT* middle temporal cortex, *V6* sixth visual area, *PIVC* parieto-insular vestibular cortex, *PCC* posterior cingulate cortex, *MSTd* dorsal medial superior temporal area, *VIP* ventral intraparietal area, *VPS* visual posterior sylvian area, *FEFsem* smooth eye movement/pursuit area of the frontal eye field, *7a* Brodmann area 7a, *LIP* lateral intraparietal area, *FEFsac* saccade area of the frontal eye field, *CD* caudate

cannot be interpreted as a pure sensory signal or a pure motor signal (Shadlen and Newsome 1996). Instead, evidence from manipulation experiments such as electrical microstimulation (Hanks et al. 2006) and pulse perturbation (Huk and Shadlen 2005) suggest that this signal may reflect a process of accumulation of sensory information that favors one of the decisions over the other choices, i.e., decision variables, and thus these brain areas are considered to be decision-related.

Recently, a few studies combining the two fields, that is, multisensory integration in decision-making-related areas, begin to emerge. These studies have been conducted in posterior parietal cortex (PPC) in rats (Nikbakht et al. 2018; Raposo 2016; Raposo et al. 2014), LIP (Hou et al. 2019) and FEF (Zheng et al. 2021) in macaques, and the secondary motor area (MO) in mice (Coen et al. 2021). Although there are not many of them, these studies already make interesting and valuable insights into how cross-modal information is accumulated, and integrated in sensory-motor transformation areas, reflecting how multisensory signals are propagated and decoded along the sensory channels.

2.2.2 Modality and Category-Free Coding

In sensory cortex, it is frequently seen that single neurons in the same brain region tend to be homogeneous by exhibiting clear tuning that can be described by some descriptive models like basis functions. For example, in MSTd, many neurons have visual or vestibular heading tuning (Fetsch et al. 2007; Gu et al. 2006; Duffy 1998) that can be modeled by a cosine or wrapped Gaussian function. Interestingly, many neurons coded a labeled-line of left and rightward heading preference (Gu et al. 2006), leading to the strongest discriminability of headings varied in fine steps around straightforward (Gu et al. 2010). Furthermore, temporal dynamics of MSTd neurons typically show single-peaked peristimulus time histogram (PSTH), indicating a velocity quantity (Gu 2018; Laurens et al. 2017; Chen et al. 2011a; Gu et al. 2008).

The spatial relation between the two sensory modality is also clear in most of cases on individual neurons. Difference in heading preference between visual and vestibular for multisensory MSTd neurons shows a bimodal pattern (Gu et al. 2006, 2008; Morgan et al. 2008; Takahashi et al. 2007), such that the majority of cells are either "congruent" or "conflict." The congruent and conflict patterns are so clear that it is straightforward to speculate that two categories may execute different functions. Specifically, congruent cells may facilitate multisensory integration, for example, during natural navigation when optic flow arisen from self-motion registers well with vestibular signals. By contrast, conflict cells may be associated with multisensory segregation, for example, when estimate of self-motion is confounded by independently moving objects in the environment (Zhang et al. 2019; Gu et al. 2008).

In sum, single neurons in sensory cortices tend to show clear modulations, and these properties across neurons tend to show a high degree of homogeneity and aggregation. For example, similar tuning preference for both visual and vestibular tend to be spatially clustered in a local range in MSTd, or VIP (Shao et al. 2018; Yu and Gu 2018; Gu et al. 2012; Zhang and Britten 2004).

By contrast to the above-mentioned sensory domains, single neurons in decision-related regions are neither homogeneous nor clustered very much, a phenomenon of which is known as "category-free." Even worse, studying multisensory integration in decision-related regions involves a higher dimensional space in which both decision-related and modality signals need to be represented in certain ways. One of the earliest studies of this issue is from Raposo and colleagues (Raposo et al. 2014), who trained rats to perform an audio-visual frequency recognition task while recording the responses of PPC neurons. They found that PPC neurons are heterogeneous, with different neurons exhibiting different dynamics, and carrying different degree of decision and modality signals. With exception of a minor population of cells presenting only decision or only modality signals, most cells encode a mixture of the two signals, which is referred as mixed selectivity (Rigotti et al. 2013). Overall, the preference for choice (i.e., decision) and modality is randomly distributed across individual neurons and thus no correlation between the two exists per se. In conclusion, it is very difficult to immediately understand the function of a single neuron when just eyeballing them individually.

Despite that task-related variables are mixed at the level of individual neurons, encoding of task parameters at the PPC population level remains linearly separable, and the PPC network is able to decode the desired variables according to task demands (Raposo et al. 2014). Category-free phenomenon is recently observed in the macaque LIP in a task when the animals are trained to discriminate fine headings-based optic flow, vestibular, or the combination of both cues (Hou et al. 2019). In a motion direction-color mixed discrimination task, this phenomenon has also been observed in several frontal and parietal regions, allowing for flexible decision for certain type of information according to task needs (Siegel et al. 2015; Ibos and Freedman 2014; Mante et al. 2013).

2.2.3 Modality-Dependent Dynamics of Decision Signals

The "decision" part of multisensory decision-making emphasizes the importance of time, over which pieces of instantaneous information is integrated. Looking into temporal dynamics of different modality signals may provide useful insight into how sensory signals are decoded and are ultimately integrated across time and modality. A good example comes from the visuo-vestibular studies. First of all, it is notable that vestibular channel is unique among all sensory systems in terms of its dramatic temporal dynamics. In particular, peripheral vestibular organs, especially the otolith apparatus, only encodes self-motion with the physical quantity of acceleration (Fernandez and Goldberg 1976; Goldberg and Fernandez 1971). The acceleration information is then temporally integrated to different extent when being propagated to the central nervous system, resulting in a broad distribution from acceleration to velocity in the brain (Laurens et al. 2017). In contrast to plentiful vestibular dynamic signals, visual signals are typically represented in velocity rather than acceleration (Gu et al. 2006; Liu and Newsome 2005; Lisberger and Movshon 1999; Rodman and Albright 1987; Maunsell and Van Essen 1983). Thus, a question arises: does the brain integrate visual and vestibular with temporally identical signal of velocity, or with temporally incongruent signal (velocity and acceleration) during self-motion? Notably, it would be difficult to address this question by looking only into the psychophysical data. Instead, neurophysiological data would give us useful hints.

Firstly, evidence from the polysensory area of MSTd appears to support the temporal-identical (i.e., velocity for both visual and vestibular) integration hypothesis. In particular, vestibular signals in MSTd are predominantly velocity, making it seemingly ideal for integration with visual (Gu 2018; Laurens et al. 2017; Gu et al. 2006, 2008). Secondly, however, recent studies in macaques have found that the temporal dynamics of visuo-vestibular signals in decision-related region like LIP and FEF are not consistent across modalities (Zheng et al. 2021; Hou et al. 2019). Specifically, vestibular signals ramp up early, roughly corresponding to the peak of the acceleration profile of the stimulus, and the visual signals ramp up relatively late, roughly corresponding to the peak of the velocity profile. This divergent temporal dynamics between the two types of signals suggests that the brain accumulates different physical quantities for different modalities. To further verify this, researchers broaden the motion profiles such that the peak moment of velocity remains the same, which is still the middle point of the stimulus duration, whereas the peak moment of the acceleration was advanced by some time compared to that before being adjusted. If decision neurons do accumulate velocity signals in the visual condition and acceleration signals in the vestibular condition, then the timing for steepest rising slope should remain unchanged for visual, yet should move earlier for the vestibular. Indeed, LIP responses are consistent with these expectations, supporting that LIP accumulates momentary evidence from vestibular acceleration and visual velocity (Hou et al. 2019).

To distinguish how exactly the brain employs visual and vestibular signals for estimate of heading, researchers introduced temporal offset at a step of 250 ms between the two sensory inputs (Zheng et al. 2021). The rationale is that if the brain has used a temporal-identical model to combine heading cues, introducing cross-modal temporal offset would misalign the two signals and presumably reduce the efficiency of integration. Otherwise, temporal-offset manipulation could artificially better align the two dynamics and may instead boost cue integration. Surprisingly, in conditions when optic flow was adjusted to lead vestibular by 250–500 ms, benefit of multisensory integration in macaques' heading performance was indeed further improved compared to that under zero-offset condition. Such an effect was not observed under other offset conditions, for example, when visual stimuli led vestibular by 750 ms, or lagged vestibular by 250 ms, suggesting that the 250–500 ms offset window was specific with respect to further-enhanced cue combination effect. Moreover, simultaneous recordings revealed that under this specific offset window (250–500 ms visual leading), difference in temporal dynamics between the two modality signals in FEF and LIP was reduced. Thus with the more synchronous signals across modality, neural activity was enhanced more than that under zero-offset condition, which exactly explains the behavior. In sum, these results support that under natural conditions, the brain combines optic flow and vestibular with inconsistent physical quantities for multisensory heading perception (Fig. 2.2).

2.2.4 Causality Issue

Causal manipulation studies are few, especially when compared to the large number of studies based on a correlation measurement as described above. The causal experiments, however, are critically important for pinning down circuits that mediate multisensory processing. For example, a few possible models have been envisioned for multisensory decision-making (Bizley et al. 2016):

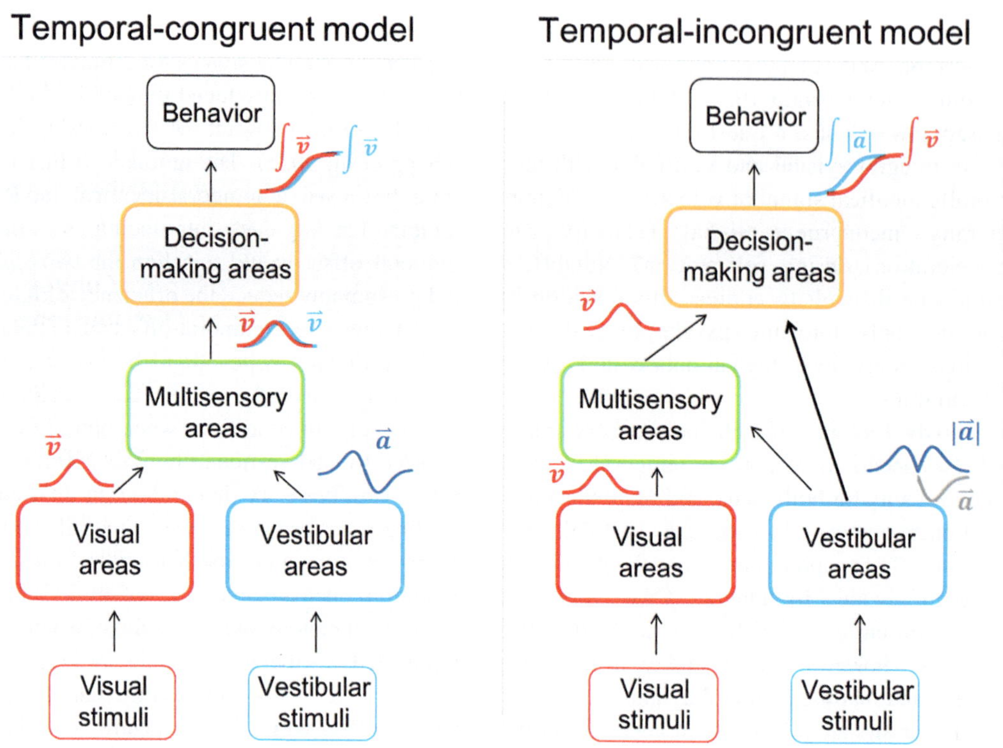

Fig. 2.2 Two models of visual-vestibular integration for self-motion perception. Velocity (v) follows Gaussian-shape profile; acceleration (a) is derivative of velocity and follows a peak-and-trough profile. In temporal-congruent model, multisensory areas convert vestibular acceleration signal to velocity signal and then decision-making areas receive and accumulate this vestibular velocity signal. In temporal-incongruent model, decision-making areas receive and accumulate absolute vestibular acceleration signal from vestibular areas

1. Late integration model: signals from different modalities are first transmitted to decision-making areas along each individual pathway, after which unimodal decision signals are integrated.
2. Early integration model: signals from different modalities are first converged in the same sensory area for integration, and then the integrated signals are transmitted to the decision-making area.

So in self-motion perception, how exactly are visual and vestibular signals converged and integrated in the brain along hierarchical level? A number of casual experiments have been conducted to examine whether and how optic flow and vestibular signals in cortices contribute to macaques' heading judgments. For example, inactivating MSTd with reversible drugs dramatically impairs vision-based heading perception but has a much weaker effect on vestibular discriminability. Consistent with this result, microcurrent stimulation applied in MSTd biases the animals' perceptual judgment on heading based on optic flow, but not on vestibular (Gu et al. 2012). By contrast, inactivating PIVC causes large deficits in vestibular heading perception (Chen et al. 2016). Surprisingly, inactivation of VIP does not generate any significant influence on either visual or vestibular-based heading judgments although this area contains robust vestibular and visual motion signals (Chen et al. 2016). Thus, it is likely that the coexistence of visuo-vestibular signals in MSTd and VIP may be for some other functions rather than heading estimate. Such techniques and methods should be applied in other areas to examine their roles in self-motion perception.

To our knowledge, there is so far only one study reporting that causal manipulations affect both modality signals. Specifically, optogenetic inactivation of the secondary motor cortex (MO), a multisensory region in rodents that encodes both visual and auditory information, impairs mice's performance in spatial localization tasks based on either modality (Coen et al. 2021). This result suggests that MO in mice is critical for processing cross modality information, and consequently, for its integration and evidence accumulation.

Finally, it should be noted that results from causality manipulation experiments should be explained with cautiousness. For example, due to the trait of highly redundant information coding in the brain, inactivating one of the nodes could be quickly compensated by other nodes within the network, and thus may not cause observable changes in the behavior. For activation experiments on the other hand, it is likely to activate passing fibers, leading to results that could be misleadingly explained by links with the manipulated areas. Furthermore, region of interest (ROI) may be involved in different task context. For example, while inactivation and electrical microstimulation in VIP fail to evoke significant effects on the animals' heading judgments during central fixation (Yu and Gu 2018; Chen et al. 2016), microstimulation does produce more salient effects when smooth pursuit eye movements are accompanied during heading perception (Zhang and Britten 2011). In decision-related areas, inactivation of LIP (Katz et al. 2016) or PPC (Erlich et al. 2015; Raposo et al. 2014) typically does not influence the animals' ability much in accumulation of sensory evidence, yet this effect is much stronger when novel stimuli are involved, suggesting that LIP may be involved more in the early phase of task-learning (Zhong et al. 2019). In sum, cross-discipline methods, as well as more task contexts need to be used to reveal a clearer picture about neural circuits mediating multisensory decision-making in the future.

References

Alais D, Burr D (2004) The ventriloquist effect results from near-optimal bimodal integration. Curr Biol 14(3):257–262. https://doi.org/10.1016/j.cub.2004.01.029

Angelaki DE, Gu Y, DeAngelis GC (2009) Multisensory integration: psychophysics, neurophysiology, and computation. Curr Opin Neurobiol 19(4):452–458. https://doi.org/10.1016/j.conb.2009.06.008

Angelaki DE, Gu Y, Deangelis GC (2011) Visual and vestibular cue integration for heading perception in extrastriate visual cortex. J Physiol 589(Pt 4):825–833. https://doi.org/10.1113/jphysiol.2010.194720

Avila E, Lakshminarasimhan KJ, DeAngelis GC, Angelaki DE (2019) Visual and vestibular selectivity for self-motion in macaque posterior parietal area 7a. Cereb Cortex 29(9):3932–3947. https://doi.org/10.1093/cercor/bhy272

Avillac M, Hamed BS, Duhamel JR (2007) Multisensory integration in the ventral intraparietal area of the macaque monkey. J Neurosci 27(8):1922–1932. https://doi.org/10.1523/JNEUROSCI.2646-06.2007

Battaglia PW, Jacobs RA, Aslin RN (2003) Bayesian integration of visual and auditory signals for spatial localization. J Opt Soc Am A Opt Image Sci Vis 20(7):1391–1397. https://doi.org/10.1364/josaa.20.001391

Bisley JW, Goldberg ME (2010) Attention, intention, and priority in the parietal lobe. Annu Rev Neurosci 33(1):1–21. https://doi.org/10.1146/annurev-neuro-060909-152823

Bizley JK, Jones GP, Town SM (2016) Where are multisensory signals combined for perceptual decision-making? Curr Opin Neurobiol 40:31–37. https://doi.org/10.1016/j.conb.2016.06.003

Bremmer F, Kubischik M, Pekel M, Lappe M, Hoffmann KP (1999) Linear vestibular self-motion signals in monkey medial superior temporal area. Ann N Y Acad Sci 871(1 OTOLITH FUNCT):272–281. https://doi.org/10.1111/j.1749-6632.1999.tb09191.x

Bremmer F, Duhamel JR, Ben Hamed S, Graf W (2002a) Heading encoding in the macaque ventral intraparietal area (VIP). Eur J Neurosci 16(8):1554–1568. https://doi.org/10.1046/j.1460-9568.2002.02207.x

Bremmer F, Klam F, Duhamel JR, Ben Hamed S, Graf W (2002b) Visual-vestibular interactive responses in the macaque ventral intraparietal area (VIP). Eur J Neurosci 16(8):1569–1586. https://doi.org/10.1046/j.1460-9568.2002.02206.x

Chandrasekaran C (2017) Computational principles and models of multisensory integration. Curr Opin Neurobiol 43:25–34. https://doi.org/10.1016/j.conb.2016.11.002

Chen A, Gu Y, Takahashi K, Angelaki DE, Deangelis GC (2008) Clustering of self-motion selectivity and visual response properties in macaque area MSTd. J Neurophysiol 100(5):2669–2683. https://doi.org/10.1152/jn.90705.2008

Chen A, DeAngelis GC, Angelaki DE (2011a) A comparison of vestibular spatiotemporal tuning in macaque parietoinsular vestibular cortex, ventral intraparietal area, and medial superior temporal area. J Neurosci 31(8):3082–3094. https://doi.org/10.1523/JNEUROSCI.4476-10.2011

Chen A, DeAngelis GC, Angelaki DE (2011b) Convergence of vestibular and visual self-motion signals in an area of the posterior sylvian fissure. J Neurosci 31(32):11617–11627. https://doi.org/10.1523/JNEUROSCI.1266-11.2011

Chen A, DeAngelis GC, Angelaki DE (2011c) Representation of vestibular and visual cues to self-motion in ventral intraparietal cortex. J Neurosci 31(33):12036–12052. https://doi.org/10.1523/JNEUROSCI.0395-11.2011

Chen A, Deangelis GC, Angelaki DE (2013) Functional specializations of the ventral intraparietal area for multisensory heading discrimination. J Neurosci 33(8):3567–3581. https://doi.org/10.1523/JNEUROSCI.4522-12.2013

Chen A, Gu Y, Liu S, DeAngelis GC, Angelaki DE (2016) Evidence for a causal contribution of macaque vestibular, but not intraparietal, cortex to heading perception. J Neurosci 36(13):3789–3798. https://doi.org/10.1523/JNEUROSCI.2485-15.2016

Churchland AK, Kiani R, Shadlen MN (2008) Decision-making with multiple alternatives. Nat Neurosci 11(6):693–702. https://doi.org/10.1038/nn.2123

Churchland AK, Kiani R, Chaudhuri R, Wang XJ, Pouget A, Shadlen MN (2011) Variance as a signature of neural computations during decision making. Neuron 69(4):818–831. https://doi.org/10.1016/j.neuron.2010.12.037

Coen P, Sit TPH, Wells MJ, Carandini M, Harris KD (2021) Mouse frontal cortex mediates additive multisensory decisions. bioRxiv, Cold Spring Harbor Laboratory. https://doi.org/10.1101/2021.04.26.441250

Dadarlat MC, O'Doherty JE, Sabes PN (2015) A learning-based approach to artificial sensory feedback leads to optimal integration. Nat Neurosci 18(1):138–144. https://doi.org/10.1038/nn.3883

Ding L, Gold JI (2010) Caudate encodes multiple computations for perceptual decisions. J Neurosci 30(47):15747–15759. https://doi.org/10.1523/JNEUROSCI.2894-10.2010

Doya K, Ishii S, Pouget A, Rao RPN (2006) Bayesian brain: probabilistic approaches to neural coding. MIT Press, Cambridge, MA

Drugowitsch J, Pouget A (2012) Probabilistic vs. non-probabilistic approaches to the neurobiology of perceptual decision-making. Curr Opin Neurobiol 22(6):963–969. https://doi.org/10.1016/j.conb.2012.07.007

Drugowitsch J, DeAngelis GC, Klier EM, Angelaki DE, Pouget A (2014) Optimal multisensory decision-making in a reaction-time task. eLife Sciences Publications, Ltd., Cambridge, UK

Duffy CJ (1998) MST neurons respond to optic flow and translational movement. J Neurophysiol 80(4):1816–1827. https://doi.org/10.1152/jn.1998.80.4.1816

Erlich JC, Brunton BW, Duan CA, Hanks TD, Brody CD (2015) Distinct effects of prefrontal and parietal cortex inactivations on an accumulation of evidence task in the rat. eLife sciences publications, Ltd., Cambridge, UK

Ernst MO, Banks MS (2002) Humans integrate visual and haptic information in a statistically optimal fashion. Nature 415(6870):429–433. https://doi.org/10.1038/415429a

Fernandez C, Goldberg JM (1976) Physiology of peripheral neurons innervating otolith organs of the squirrel monkey. II. Directional selectivity and force-response relations. J Neurophysiol 39(5):985–995. https://doi.org/10.1152/jn.1976.39.5.985

Fetsch CR, Wang S, Gu Y, Deangelis GC, Angelaki DE (2007) Spatial reference frames of visual, vestibular, and multimodal heading signals in the dorsal subdivision of the medial superior temporal area. J Neurosci 27(3):700–712. https://doi.org/10.1523/JNEUROSCI.3553-06.2007

Fetsch CR, Pouget A, DeAngelis GC, Angelaki DE (2011) Neural correlates of reliability-based cue weighting during multisensory integration. Nat Neurosci 15(1):146–154. https://doi.org/10.1038/nn.2983

Fetsch CR, DeAngelis GC, Angelaki DE (2013) Bridging the gap between theories of sensory cue integration and the physiology of multisensory neurons. Nat Rev Neurosci 14(6):429–442. https://doi.org/10.1038/nrn3503

Ghazanfar AA, Schroeder CE (2006) Is neocortex essentially multisensory? Trends Cogn Sci 10(6):278–285. https://doi.org/10.1016/j.tics.2006.04.008

Ghazanfar AA, Maier JX, Hoffman KL, Logothetis NK (2005) Multisensory integration of dynamic faces and voices in rhesus monkey auditory cortex. J Neurosci 25(20):5004–5012. https://doi.org/10.1523/JNEUROSCI.0799-05.2005

Gold JI, Shadlen MN (2007) The neural basis of decision making. Annu Rev Neurosci 30(1):535–574. https://doi.org/10.1146/annurev.neuro.29.051605.113038

Goldberg JM, Fernandez C (1971) Physiology of peripheral neurons innervating semicircular canals of the squirrel monkey. I. Resting discharge and response to constant angular accelerations. J Neurophysiol 34(4):635–660. https://doi.org/10.1152/jn.1971.34.4.635

Gu Y (2018) Vestibular signals in primate cortex for self-motion perception. Curr Opin Neurobiol 52:10–17. https://doi.org/10.1016/j.conb.2018.04.004

Gu Y, Watkins PV, Angelaki DE, DeAngelis GC (2006) Visual and nonvisual contributions to three-dimensional heading selectivity in the medial superior temporal area. J Neurosci 26(1):73–85. https://doi.org/10.1523/JNEUROSCI.2356-05.2006

Gu Y, Angelaki DE, Deangelis GC (2008) Neural correlates of multisensory cue integration in macaque MSTd. Nat Neurosci 11(10):1201–1210. https://doi.org/10.1038/nn.2191

Gu Y, Fetsch CR, Adeyemo B, Deangelis GC, Angelaki DE (2010) Decoding of MSTd population activity accounts for variations in the precision of head-

ing perception. Neuron 66(4):596–609. https://doi.org/10.1016/j.neuron.2010.04.026
Gu Y, Deangelis GC, Angelaki DE (2012) Causal links between dorsal medial superior temporal area neurons and multisensory heading perception. J Neurosci 32(7):2299–2313. https://doi.org/10.1523/JNEUROSCI.5154-11.2012
Gu Y, Cheng Z, Yang L, DeAngelis GC, Angelaki DE (2016) Multisensory convergence of visual and vestibular heading cues in the pursuit area of the frontal eye field. Cereb Cortex 26(9):3785–3801. https://doi.org/10.1093/cercor/bhv183
Hanks TD, Summerfield C (2017) Perceptual decision making in rodents, monkeys, and humans. Neuron 93(1):15–31. https://doi.org/10.1016/j.neuron.2016.12.003
Hanks TD, Ditterich J, Shadlen MN (2006) Microstimulation of macaque area LIP affects decision-making in a motion discrimination task. Nat Neurosci 9(5):682–689. https://doi.org/10.1038/nn1683
Horwitz GD, Newsome WT (1999) Separate signals for target selection and movement specification in the superior colliculus. Science 284(5417):1158–1161. https://doi.org/10.1126/science.284.5417.1158
Horwitz GD, Newsome WT (2001) Target selection for saccadic eye movements: prelude activity in the superior colliculus during a direction-discrimination task. J Neurophysiol 86(5):2543–2558. https://doi.org/10.1152/jn.2001.86.5.2543
Hou H, Gu Y (2020) Multisensory integration for self-motion perception. In: The senses: a comprehensive reference. Elsevier, pp 458–482, Amsterdam, Netherlands
Hou H, Zheng Q, Zhao Y, Pouget A, Gu Y (2019) Neural correlates of optimal multisensory decision making under time-varying reliabilities with an invariant linear probabilistic population code. Neuron 104(5):1010–1021 e1010. https://doi.org/10.1016/j.neuron.2019.08.038
Huk AC, Shadlen MN (2005) Neural activity in macaque parietal cortex reflects temporal integration of visual motion signals during perceptual decision making. J Neurosci 25(45):10420–10436. https://doi.org/10.1523/JNEUROSCI.4684-04.2005
Ibos G, Freedman DJ (2014) Dynamic integration of task-relevant visual features in posterior parietal cortex. Neuron 83(6):1468–1480. https://doi.org/10.1016/j.neuron.2014.08.020
Katz LN, Yates JL, Pillow JW, Huk AC (2016) Dissociated functional significance of decision-related activity in the primate dorsal stream. Nature 535(7611):285–288. https://doi.org/10.1038/nature18617
Kim JN, Shadlen MN (1999) Neural correlates of a decision in the dorsolateral prefrontal cortex of the macaque. Nat Neurosci 2(2):176–185. https://doi.org/10.1038/5739
Kim SS, Gomez-Ramirez M, Thakur PH, Hsiao SS (2015) Multimodal interactions between proprioceptive and cutaneous signals in primary somatosensory cortex. Neuron 86(2):555–566. https://doi.org/10.1016/j.neuron.2015.03.020
Knill DC, Pouget A (2004) The Bayesian brain: the role of uncertainty in neural coding and computation. Trends Neurosci 27(12):712–719. https://doi.org/10.1016/j.tins.2004.10.007
Knill DC, Richards W (1996) Perception as Bayesian inference. Cambridge University Press, Cambridge, UK
Knill DC, Saunders JA (2003) Do humans optimally integrate stereo and texture information for judgments of surface slant? Vis Res 43(24):2539–2558. https://doi.org/10.1016/s0042-6989(03)00458-9
Kravitz DJ, Saleem KS, Baker CI, Mishkin M (2011) A new neural framework for visuospatial processing. Nat Rev Neurosci 12(4):217–230. https://doi.org/10.1038/nrn3008
Landy MS, Banks MS, Knill DC (2011) Ideal-observer models of Cue integration. In: Sensory Cue integration. Oxford University Press, pp 5–29, Oxford, UK
Laurens J, Liu S, Yu XJ, Chan R, Dickman D, DeAngelis GC, Angelaki DE (2017) Transformation of spatiotemporal dynamics in the macaque vestibular system from otolith afferents to cortex. elife 6:e20787. https://doi.org/10.7554/eLife.20787
Lemus L, Hernandez A, Luna R, Zainos A, Romo R (2010) Do sensory cortices process more than one sensory modality during perceptual judgments? Neuron 67(2):335–348. https://doi.org/10.1016/j.neuron.2010.06.015
Li N, Daie K, Svoboda K, Druckmann S (2016) Robust neuronal dynamics in premotor cortex during motor planning. Nature 532(7600):459–464. https://doi.org/10.1038/nature17643
Lisberger SG, Movshon JA (1999) Visual motion analysis for pursuit eye movements in area MT of macaque monkeys. J Neurosci 19(6):2224–2246. https://doi.org/10.1523/jneurosci.19-06-02224.1999
Liu J, Newsome WT (2005) Correlation between speed perception and neural activity in the middle temporal visual area. J Neurosci 25:711–722
Ma WJ (2019) Bayesian decision models: a primer. Neuron 104(1):164–175. https://doi.org/10.1016/j.neuron.2019.09.037
Maciokas JB, Britten KH (2010) Extrastriate area MST and parietal area VIP similarly represent forward headings. J Neurophysiol 104(1):239–247. https://doi.org/10.1152/jn.01083.2009
Mante V, Sussillo D, Shenoy KV, Newsome WT (2013) Context-dependent computation by recurrent dynamics in prefrontal cortex. Nature 503(7474):78–84. https://doi.org/10.1038/nature12742
Maunsell JH, Van Essen DC (1983) Functional properties of neurons in middle temporal visual area of the macaque monkey. I. Selectivity for stimulus direction, speed, and orientation. J Neurophysiol 49(5):1127–1147. https://doi.org/10.1152/jn.1983.49.5.1127
Meredith AM, Stein BE (1986a) Spatial factors determine the activity of multisensory neurons in cat superior

colliculus. Brain Res 365(2):350–354. https://doi.org/10.1016/0006-8993(86)91648-3

Meredith MA, Stein BE (1986b) Visual, auditory, and somatosensory convergence on cells in superior colliculus results in multisensory integration. J Neurophysiol 56(3):640–662. https://doi.org/10.1152/jn.1986.56.3.640

Meredith MA, Nemitz JW, Stein BE (1987) Determinants of multisensory integration in superior colliculus neurons. I Temporal factors J Neurosci 7(10):3215–3229. https://doi.org/10.1523/jneurosci.07-10-03215.1987

Morgan ML, Deangelis GC, Angelaki DE (2008) Multisensory integration in macaque visual cortex depends on cue reliability. Neuron 59(4):662–673. https://doi.org/10.1016/j.neuron.2008.06.024

Najafi F, Churchland AK (2018) Perceptual decision-making: a field in the midst of a transformation. Neuron 100(2):453–462. https://doi.org/10.1016/j.neuron.2018.10.017

Nikbakht N, Tafreshiha A, Zoccolan D, Diamond ME (2018) Supralinear and Supramodal integration of visual and tactile signals in rats: psychophysics and neuronal mechanisms. Neuron 97(3):626–639 e628. https://doi.org/10.1016/j.neuron.2018.01.003

Page WK, Duffy CJ (2003) Heading representation in MST: sensory interactions and population encoding. J Neurophysiol 89(4):1994–2013. https://doi.org/10.1152/jn.00493.2002

Raposo D (2016) The role of posterior parietal cortex in multisensory decision-making. Doctoral dissertation

Raposo D, Sheppard JP, Schrater PR, Churchland AK (2012) Multisensory decision-making in rats and humans. J Neurosci 32(11):3726–3735. https://doi.org/10.1523/JNEUROSCI.4998-11.2012

Raposo D, Kaufman MT, Churchland AK (2014) A category-free neural population supports evolving demands during decision-making. Nat Neurosci 17(12):1784–1792. https://doi.org/10.1038/nn.3865

Ratcliff R (1978) A theory of memory retrieval. Psychol Rev 85(2):59–108. https://doi.org/10.1037/0033-295x.85.2.59

Ratcliff R, McKoon G (2008) The diffusion decision model: theory and data for two-choice decision tasks. Neural Comput 20(4):873–922. https://doi.org/10.1162/neco.2008.12-06-420

Ratcliff R, Rouder JNJPS (1998) Modeling response times for two-choice. Decisions 9:347–356

Ratcliff R, Smith PL (2004) A comparison of sequential sampling models for two-choice reaction time. Psychol Rev 111(2):333–367. https://doi.org/10.1037/0033-295X.111.2.333

Reig R, Silberberg G (2014) Multisensory integration in the mouse striatum. Neuron 83(5):1200–1212. https://doi.org/10.1016/j.neuron.2014.07.033

Rigotti M, Barak O, Warden MR, Wang XJ, Daw ND, Miller EK, Fusi S (2013) The importance of mixed selectivity in complex cognitive tasks. Nature 497(7451):585–590. https://doi.org/10.1038/nature12160

Rodman HR, Albright TD (1987) Coding of visual stimulus velocity in area MT of the macaque. Vis Res 27(12):2035–2048. https://doi.org/10.1016/0042-6989(87)90118-0

Roitman JD, Shadlen MN (2002) Response of neurons in the lateral intraparietal area during a combined visual discrimination reaction time task. J Neurosci 22(21):9475–9489. https://doi.org/10.1523/jneurosci.22-21-09475.2002

Schlack A, Hoffmann KP, Bremmer F (2002) Interaction of linear vestibular and visual stimulation in the macaque ventral intraparietal area (VIP). Eur J Neurosci 16(10):1877–1886. https://doi.org/10.1046/j.1460-9568.2002.02251.x

Seilheimer RL, Rosenberg A, Angelaki DE (2014) Models and processes of multisensory cue combination. Curr Opin Neurobiol 25:38–46. https://doi.org/10.1016/j.conb.2013.11.008

Shadlen MN, Kiani R (2013) Decision making as a window on cognition. Neuron 80(3):791–806. https://doi.org/10.1016/j.neuron.2013.10.047

Shadlen MN, Newsome WT (1996) Motion perception: seeing and deciding. Proc Natl Acad Sci U S A 93(2):628–633. https://doi.org/10.1073/pnas.93.2.628

Shadlen MN, Newsome WT (2001) Neural basis of a perceptual decision in the parietal cortex (area LIP) of the rhesus monkey. J Neurophysiol 86(4):1916–1936. https://doi.org/10.1152/jn.2001.86.4.1916

Shao M, DeAngelis GC, Angelaki DE, Chen A (2018) Clustering of heading selectivity and perception-related activity in the ventral intraparietal area. J Neurophysiol 119(3):1113–1126. https://doi.org/10.1152/jn.00556.2017

Sheppard JP, Raposo D, Churchland AK (2013) Dynamic weighting of multisensory stimuli shapes decision-making in rats and humans. J Vis 13(6):4–4. https://doi.org/10.1167/13.6.4

Siegel M, Buschman TJ, Miller EK (2015) Cortical information flow during flexible sensorimotor decisions. Science 348(6241):1352–1355. https://doi.org/10.1126/science.aab0551

Smith AT, Greenlee MW, DeAngelis GC, Angelaki DE (2017) Distributed visual–vestibular processing in the cerebral cortex of man and macaque. Multisens Res 30(2):91–120. https://doi.org/10.1163/22134808-00002568

Sober SJ, Sabes PN (2005) Flexible strategies for sensory integration during motor planning. Nat Neurosci 8(4):490–497. https://doi.org/10.1038/nn1427

Stein BE, Meredith MA (1993) The merging of the senses. MIT Press, Cambridge, MA

Stein BE, Stanford TR (2008) Multisensory integration: current issues from the perspective of the single neuron. Nat Rev Neurosci 9(4):255–266. https://doi.org/10.1038/nrn2331

Stein BE, Stanford TR, Rowland BA (2020) Multisensory integration and the Society for Neuroscience: then and now. J Neurosci 40(1):3–11. https://doi.org/10.1523/JNEUROSCI.0737-19.2019

Summerfield C, de Lange FP (2014) Expectation in perceptual decision making: neural and computational mechanisms. Nat Rev Neurosci 15(11):745–756. https://doi.org/10.1038/nrn3838

Takahashi K, Gu Y, May PJ, Newlands SD, DeAngelis GC, Angelaki DE (2007) Multimodal coding of three-dimensional rotation and translation in area MSTd: comparison of visual and vestibular selectivity. J Neurosci 27(36):9742–9756. https://doi.org/10.1523/JNEUROSCI.0817-07.2007

Ursino M, Cuppini C, Magosso E (2014) Neurocomputational approaches to modelling multisensory integration in the brain: a review. Neural Netw 60:141–165. https://doi.org/10.1016/j.neunet.2014.08.003

van Beers RJ, Sittig AC, Gon JJ (1999) Integration of proprioceptive and visual position-information: an experimentally supported model. J Neurophysiol 81(3):1355–1364. https://doi.org/10.1152/jn.1999.81.3.1355

Yang L, Gu Y (2017) Distinct spatial coordinate of visual and vestibular heading signals in macaque FEFsem and MSTd. eLife Sciences Publications, Ltd., Cambridge, UK

Yau JM, DeAngelis GC, Angelaki DE (2015) Dissecting neural circuits for multisensory integration and crossmodal processing. Philos Trans R Soc Lond Ser B Biol Sci 370(1677):20140203. https://doi.org/10.1098/rstb.2014.0203

Yu X, Gu Y (2018) Probing sensory readout via combined choice-correlation measures and microstimulation perturbation. Neuron 100(3):715–727 e715. https://doi.org/10.1016/j.neuron.2018.08.034

Zhang T, Britten KH (2004) Clustering of selectivity for optic flow in the ventral intraparietal area. Neuroreport 15(12):1941–1945. https://doi.org/10.1097/00001756-200408260-00022

Zhang T, Britten KH (2011) Parietal area VIP causally influences heading perception during pursuit eye movements. J Neurosci 31(7):2569–2575. https://doi.org/10.1523/JNEUROSCI.5520-10.2011

Zhang T, Heuer HW, Britten KH (2004) Parietal area VIP neuronal responses to heading stimuli are encoded in head-centered coordinates. Neuron 42(6):993–1001. https://doi.org/10.1016/j.neuron.2004.06.008

Zhang W-H, Wang H, Chen A, Gu Y, Lee TS, Wong KYM, Wu S (2019) Complementary congruent and opposite neurons achieve concurrent multisensory integration and segregation. eLife Sciences Publications, Ltd., Cambridge, UK

Zheng Q, Zhou L, Gu Y (2021) Temporal synchrony effects of optic flow and vestibular inputs on multisensory heading perception. Cell Rep 37(7):109999. https://doi.org/10.1016/j.celrep.2021.109999

Zhong L, Zhang Y, Duan CA, Deng J, Pan J, Xu NL (2019) Causal contributions of parietal cortex to perceptual decision-making during stimulus categorization. Nat Neurosci 22(6):963–973. https://doi.org/10.1038/s41593-019-0383-6

3 More Than the Sum of Its Parts: Visual–Tactile Integration in the Behaving Rat

Nader Nikbakht

Abstract

We experience the world by constantly integrating cues from multiple modalities to form unified sensory percepts. Once familiar with multimodal properties of an object, we can recognize it regardless of the modality involved. In this chapter we will examine the case of a visual–tactile orientation categorization experiment in rats. We will explore the involvement of the cerebral cortex in recognizing objects through multiple sensory modalities. In the orientation categorization task, rats learned to examine and judge the orientation of a raised, black and white grating using touch, vision, or both. Their multisensory performance was better than the predictions of linear models for cue combination, indicating synergy between the two sensory channels. Neural recordings made from a candidate associative cortical area, the posterior parietal cortex (PPC), reflected the principal neuronal correlates of the behavioral results: PPC neurons encoded both graded information about the object and categorical information about the animal's decision. Intriguingly single neurons showed identical responses under each of the three modality conditions providing a substrate for a neural circuit in the cortex that is involved in modality-invariant processing of objects.

Keywords

Multimodal integration · Psychophysics · Vision · Touch · Rat · Linearity · Bayesian · Mutual information · Posterior parietal cortex Neuronal coding

3.1 Introduction

A central question in neuroscience is how the brain "constructs" perception from sensory data collected from the outside world through sensory receptors. Consequently, the study of sensory systems is among the oldest endeavors undertaken to understand how our brains function. To explain perception, Democritus (430–420 B.C.) hypothesized that our senses receive myriad small images (eidola) emanating from the world. Then these images are processed and integrated into thinking. He postulated that this process provided the building blocks of sensation, perception, and action (Jung 1984). Centuries later, essentially, the same question is asked by modern systems neuroscientists: How does the brain

N. Nikbakht (✉)
Massachusetts Institute of Technology, Cambridge, MA, USA
e-mail: nikbakht@mit.edu

Y. Gu, A. Zaidel (eds.), *Advances of Multisensory Integration in the Brain*, Advances in Experimental Medicine and Biology 1437, https://doi.org/10.1007/978-981-99-7611-9_3

transform a sensory input into perception? Where is this perception stored in memory and how does it affect our actions? Through answering these timeless questions, we may begin to understand how our experience of the physical world makes us who we are.

Historically, theoretical and experimental studies in neuroscience have often focused on the study of individual sensory systems in isolation. These investigations have laid the foundation for new questions about cognition: where and how in the brain are the sensory information represented and stored? What aspects of the neural responses to sensory stimuli, together with sensory information representation, underlie perceptual decision-making? How are all these processes learned? How does the brain construct a conceptual knowledge of the world to guide adaptive behavior?

In this chapter we will review a set of experiments performed using rat as a model organism, with the specific aim of describing some of the neural mechanisms by which the brain combines information from different sensory aspects of an object to form a unified percept of it.

Interest in multisensory research, in particular, using rodent models, has peaked in the past decade. Most of the multisensory studies have focused on either neurophysiological observations, or on mathematical quantification of the multisensory behavior. However, a systemic study of the problem is one that links these approaches by: (1) assessing the sensory integration process through measurable changes in behavior and (2) probing neural processing in freely moving animals to examine the neural computations involved.

In the first step, well-defined sensory stimuli are introduced, and the behavioral choices of the subjects are measured to evaluate the trial-by-trial percept. This is the "*psychophysical*" step where the relationship between the stimuli and the subjects' percept is characterized along some physical dimension. Thanks to this quantification, as part of the psychophysical modeling, one could benchmark the behavioral findings against the predictions of a mathematical model and uncover the predictive power of the model and its possible pitfalls.

In step 2, we measure the relationship between neuronal activity and the stimuli presented on each trial. This allows us to discover the "*neuronal coding*"—i.e., how physical events are transformed into neuronal signals by the relevant neural circuits. Then by measuring the relationship between the neural activity and the subjects' behavioral choice, we learn about "*decoding*"—i.e., how the neural representation of a stimulus is converted into a perceptual decision.

To motivate an elegant experimental design, it is useful to study the evolutionary history through which multisensory perception (or any other biological phenomenon) has emerged. Ideally, we want to bring ethologically and ecologically relevant behavioral testing into the laboratory inasmuch as the precision of our measurements will not be affected.

3.2 Evolution of Multisensory Perception

In the early evolution of mammals, a magnificent step was the expansion and re-organization of the dorsal cortex, a region in the roof of the reptilian brain, into what we now call the mammalian neocortex (Kaas 2010; Northcutt and Kaas 1995). From a single-layer mixture of excitatory and inhibitory cells, this area changed into the hallmark of all existing mammalian species: a six-layer structure with parcellated brain regions and complex microcircuitry that allows multimodal representations of the world.

According to one hypothesis, what drove the evolution of neocortex is survival in a world with limited and ambiguous sensory input. It has been proposed that early mammals were either nocturnal or, more likely, crepuscular. The evidence suggests that they might have been whiskered. So it is conceivable that they combined noisy and incomplete sensory signals (e.g., tactile signals with visual ones under low light). In this scenario, survival of the animal in a complex, dynamic, and hazardous environment would drive the neocortical evolution where cortical maps for multiple sensory modalities emerged and enabled mechanisms for speedy integration across these maps. As a result, it would become possible to make quick yet accurate inferences based on noisy, scarce, or ambiguous sensory input (Jerison 2012).

According to another hypothesis, the evolutionary pressure on neocortex is linked to endothermy: endothermic mammals have a much greater need for energy compared to reptiles of similar size (Allman 2000). This increased demand in turn meant that search for food had to become much more efficient. As a result, neocortex evolved to have a rich multimodal representational capability that would allow mammals to make better and faster decisions about when, where, and how to forage.

The significance of multisensory perception is highlighted by its universality among the animal species (Naumer and Kaiser 2010; van Hemmen et al. 2012). Each sensory system enables animals to respond to signals through different energy sources: chemical gradients (olfaction), kinetic energy (haptics), electromagnetic (vision), pressure waves (hearing), gravity and acceleration (vestibular system), etc. Since these signals originate from real-world objects possessing multiple physical attributes, one could easily posit that sensory systems have evolved to function together as well. For example, it would be non-optimal to have a neural circuit dedicated to each individual sensory stimuli, such as one circuit to move our gaze toward the source of a familiar voice in the crowd, another one for eye movements in response to a touch on the skin, and yet another system for eye movements in response to the vestibular inputs while walking.

Therefore, the mammalian nervous system evolved to develop, alongside single sensory systems, multimodal circuits that integrate different modalities. By doing so, it enabled the detection of sensory signals rapidly and with less ambiguity than what could be achieved with each single sensory system. This fusion of information among different senses is called "*multisensory integration*," and the response gain to the cross-modal stimulus compared with the response to the most effective of its component stimuli is termed "*multisensory enhancement*" (Stein and Stanford 2008).

Another significant evolutionary adaptation is *modality-shared* or *supramodal* representation of percepts. Modality sharing is a natural consequence of the fact that real-life objects have features that span multiple sensory modalities. Recognizing things despite the change of conditions surrounding them is an adaptive advantage, e.g., hearing and touch might substitute vision when the animal is in the dark, allowing it to avoid danger or forage successfully.

3.3 Models of Multisensory Cue Combination

A common way to study neuronal circuitry behind a particular cognitive task is to propose a biologically plausible computational model for the neuronal computations that are involved. With such model at hand, one can compare the performance of subjects in a psychophysical task with the model predictions and attempt to gain insights about the underlying neural computations.

These models are generally *ideal observer* models where the *optimal* performance in a given task is quantified. Here, optimality is defined according to a mathematical function (for example, maximizing a performance metric or minimizing a cost function) (Fetsch et al. 2013), and optimal performance indicates the best possible performance given the conditions of the task and the noise of the sensory–motor system. In order to assess optimality, the experimenter has to explicitly consider all specifications of the stimuli and the task that may affect the performance.

When it comes to multisensory cue combination, whether this process is optimal is an open research question. It is common to develop models that assume it to be so, as it seems plausible that animals have evolved to optimally combine sensory information (Landy et al. 2011). Historically, ideal observer cue-combination models were first used to describe problems in machine vision (Jacobs 1999) and later in other sensory system studies (Alais and Burr 2004; Körding and Wolpert 2006; Ernst and Banks 2002), for both combinations of cues within the same modality or multimodal cues.

Before diving into the central case of this chapter, the visual–tactile orientation categorization task, we first review a simple linear cue integration model and then jump to one of the best studied ideal observer models for cue integration: Bayesian cue combination.

3.3.1 Linear Models for Maximum Reliability

Before planning an action or making a decision, an animal estimates some relevant environmental parameters, such as how far it needs to jump to reach the fruit or whether it is ripe enough to worth any effort. In order to make reliable estimates, it often takes advantage of multiple sources of information or sensory *cues*—i.e., any piece of information or signal about the state of the environment. It can be reasonably expected that these estimates come with some noise, which leads to an associated perceptual uncertainty that is proportional to the variance of the noise. In the simple case of Gaussian noise, an ideal observer could perform linear cue integration in order to minimize the uncertainty (maximize the reliability) of its estimate. This would be an optimal strategy while remaining unbiased (i.e., the estimate would be correct on average because the mean of the sampling distribution of the estimate can be shown to be equal to the parameter being estimated). Optimality reflects an estimate with the lowest possible degree of uncertainty (i.e., variance).

More formally, suppose cues are n independent, Gaussian random variables, $s_i, i = 1, 2, \ldots n$, with a common mean μ and variances σ_i^2. The minimum-variance unbiased estimator of μ is a weighted average:

$$\hat{s} = \sum_{i=1}^{n} w_i s_i, \tag{3.1}$$

where w_i is the perceptual weight of the ith cue, which is proportional to that cue's relative *reliability*, η_i:

$$w_i = \frac{\eta_i}{\sum_{j=1}^{n} \eta_j}, \quad \eta_i = \frac{1}{\sigma_i^2}. \tag{3.2}$$

The reliability of the optimal integrated estimate, η_{opt}, is therefore greater than or equal to the reliability of the single cue estimates:

$$\eta_{opt} = \sum_{i=1}^{n} \eta_i. \tag{3.3}$$

Therefore, the variance of the integrated estimate is generally lower than the variance of each individual estimate or at most it is equal to the lowest variance. This shows how cue integration results in better estimates and how linear models of cue integration maximize reliability. It should be emphasized that this is only true for estimates that are Gaussian distributed and independent of each other (estimate noise and errors are independent). However, if the noise of different cue estimates is correlated, the minimum-variance unbiased estimator will not necessarily be a weighted average as above but could be a nonlinear function of the individual estimates or one that incorporates the covariance of the cues in its expression (Landy et al. 2011; Fetsch et al. 2013).

3.3.2 Bayesian Cue Integration

Linear models are frequently used in cue-combination literature. However, limitations of these models become apparent as the task complexity increases. As we will see in the case of visual–tactile orientation experiment, Bayesian decision theory provides a more suitable computational framework to address the complexities that the observer faces.

According to the Bayes' rule:

$$P(s \mid d) = \frac{P(d \mid s)P(s)}{P(d)}, \tag{3.4}$$

where s is the object property to be estimated and d is the sensory data observed by the animal (provided by a noisy sensor). $P(s \mid d)$ is the probability distribution representing the probability of s being true given the observed data d, also known as the posterior probability of s given d. If this distribution is a narrow one, it means the sensory observation is closely linked to the actual object property and estimate is reliable. $P(d \mid s)$ is the probability of obtaining a sensory input (evidence) given each possible value of s, also known as the likelihood function of each sensory cue. Likelihood does not necessarily behave like a probability distribution and does not need to integrate to one. $P(d)$ does not depend on s and can be treated as a constant normalizing factor so that the entire expression integrates to one; it can be generally ignored in the estimation process.

Given the posterior distribution $P(s \mid d)$, the Bayesian observer can make an optimal decision where a loss function is minimized. The loss function is based on the estimation error, i.e., the difference between the perceived estimate and the actual value of the environmental property. The optimal decision can also be expressed in terms of the expected gain function *EG*.

In many psychophysical tasks, the observer makes a categorical decision based on the sensory data: $a(d)$. The gain function $g(a(d), s)$ can formally be defined as the negative of the loss function. The optimal decision is then as follows:

$$a_{opt} = arg\ max\ EG(a), \qquad (3.5)$$

where

$$EG(a) = \iint g(a(d), s) P(d|s) P(s)\, dd\, ds. \qquad (3.6)$$

In the case of the uniform prior distribution, $P(s)$ where all values of s are equally probable before the observation is made, the above Eq. (3.5) is equivalent to a *maximum likelihood estimation*. If the prior distribution is not uniform, a_{opt} is estimated optimally using *maximum a posteriori* (MAP) estimation.

In the case of the multisensory integration, let us start by assuming that the sensory data associated with each modality are independent. Then the likelihood function can be written as the following:

$$P(d_1, \ldots, d_n \mid s) = \prod_{i=1}^{n} P(d_i \mid s). \qquad (3.7)$$

Therefore, for the posterior distribution, using Bayes' rule in Eq. (3.4), we can write

$$P(s \mid d_1, \ldots, d_n) \propto \prod_{i=1}^{n} P(d_i \mid s). \qquad (3.8)$$

Thus, with a prior distribution $P(s)$ that is uniform or significantly broader than the Gaussian likelihood functions, the proportionality in (3.8) approaches to equality; hence, the posterior distribution is simply the product of the sensory likelihoods for different modalities that is another Gaussian with mean corresponding to Eq. (3.1) and a variance that corresponds to the inverse of Eq. (3.3) (Knill and Saunders 2003; Angelaki et al. 2009; Landy et al. 2011; Fetsch et al. 2013).

3.3.3 Bayesian Modeling of Multisensory Behavioral Experiments

In the past three decades, a large number of behavioral experiments have put the validity of optimal cue-combination framework to test in psychophysical discrimination tasks. In these experiments, usually the subjects are presented with stimuli of one or two sensory modalities, either separately or together. Depending on the modalities involved, these stimulus presentations were either simultaneous or temporally disjointed. In these experiments, human subjects, non-human primates, or rodents decreased perceptual uncertainty (made fewer errors) by combining multiple cues. The vast majority of these experimental results were well explained by statistically optimal observer models (Jacobs 1999; Alais and Burr 2004; Battaglia et al. 2003; Gu et al. 2008; Ernst and Banks 2002; Fetsch et al. 2009) albeit with important exceptions (Raposo et al. 2012; Nikbakht et al. 2018).

Generally, in these experiments, the noise of the sensory cues (i.e., the inverse of their reliability) is measured as the variance of the psychophysical performance under unimodal conditions. Then the optimal combined performance could be computed based on a given theoretical model and compared to the experimental multisensory performance.

Consider a multisensory two-alternative forced-choice (2AFC) task involving two modalities. With the assumption that sensory noise is Gaussian distributed, the psychometric data could be described with a cumulative Gaussian function through fitting. This would yield two parameters: (1) the mean of the fitted cumulative Gaussian function (as the point of subjective equality, PSE) and its standard deviation, σ (as psychophysical threshold). From these unimodal threshold values, one could estimate the reliability of each of the two cues. These reliability values specify the weights that an optimal observer should apply to each cue (Eq. (3.1)).

To better understand the rationale behind these formulations, we should first understand the logic of signal detection theory to describe behavioral performance. An experimental subject estimates some features of a presented stimulus (s). The subject's internal estimate of that feature ($\hat{s}$) is noisy. For simplicity, we can assume that this internal noisy representation, $\hat{s}$, is Gaussian distributed. If the representation is unbiased:

$$\hat{s} \sim N(s, \sigma^2). \tag{3.9}$$

If $\hat{s}$ is a biased estimate, then its mean will not be equal to s. In many parameterized discrimination or categorization tasks, the subjects are asked to compare either two cues or one cue and a fixed reference, r. The subjects perform their calculations on the $\hat{s}$ values so considering comparison between $\hat{s}$ and a fixed reference, r, and then the probability that subjects estimate $s > r$ depends on the cumulative distribution of $\hat{s}$. This psychometric function is a Gaussian cumulative distribution function.

Binary response models can describe the behavior of most subjects in a given 2AFC task. The subject's responses on the ith trial can be denoted by the binary variable y_i. These responses can only take two possible values which we denote as 1 and 0. For example, y_i may represent the presence or absence of a certain stimulus feature, or left versus right response in a two-alternative task:

$$y_i = \begin{cases} 1 & \text{with probability } P_i \\ 0 & \text{with probability } 1-P_i \end{cases},$$

where P_i is given by cumulative distribution function of the standard normal distribution, Φ, evaluated at s_i', where s_i' is the stimulus parameter taken from a normal distribution defined by μ and σ:

$$P(y=1 \mid s') = \Phi(s'; \mu, \sigma). \tag{3.10}$$

According to Eq. (3.10), the probability that a subject answers one of the two response categories is given by an inverse CDF function evaluated at the subject's internal representation of the stimulus. For N trials, we can write the likelihood function as

$$\mathcal{L} = \prod_{i=1}^{N} P_i^{y_i} (1-P_i)^{1-y_i}. \tag{3.11}$$

From Eq. (3.10), we have

$$\mathcal{L}(\beta) = \prod_{i=1}^{N} \Phi(s_i'\beta)^{y_i} (1-\Phi(s_i'\beta))^{1-y_i}. \tag{3.12}$$

We could then derive the log-likelihood expression for the responses:

$$\ln \mathcal{L}(\beta) = \sum_{i=1}^{N} \left(y_i \ln \Phi(s_i'\beta) + (1-y_i) \ln(1-\Phi(s_i'\beta)) \right). \tag{3.13}$$

From Eqs. (3.7) and (3.8), we know that in the case of multiple cues, the joint (multimodal) likelihood is the product of each individual cue's likelihood function. In the case of Gaussian likelihood functions, the joint likelihood is also a Gaussian. Mathematically, the mean (μ) of the joint likelihood function will be the weighted sum of the means of the individual likelihood functions (Knill and Saunders 2003).

3.4 Rodent Models for the Study of Perception

Non-human primates have been the traditional animal models for the laboratory testing of higher cognitive functions such as decision-making. But working with primates has major limitations. These include costly and complicated animal maintenance and welfare in primate facilities. The number of available subjects for each study is also usually very low, which limits the statistical strength and generalizability of findings across individuals. Ethical issues regarding primate welfare are particularly abundant. The availability of molecular biology toolkits and genetic lines and much lower reproductive rates are also limiting the type of neuroscience experiments that could be done using primate models. Rodents, on the other hand, have served as ideal animal models for neuroscience research on spatial navigation and reward processing but are they also good candidates for the study of cognitive mechanisms? Until recently, "perception" was mainly attributed to primates but not to rodents. This view has now largely changed, and rodents have become popular and familiar model animals for the study of higher cognitive functions such as perception and decision-making (Diamond and Arabzadeh 2013; Zoccolan 2015; Carandini and Churchland 2013).

Three main features of rodents have led to this popularity. First, emerging technologies are far more advanced in rodents than in other species. Some of these techniques include transgenic lines for specific cell type targeting, transcriptomics, optogenetics and connectivity maps in brain areas, two-photon calcium imaging, and large-scale simultaneous neural ensemble recordings from multiple brain regions make the rodent a desirable model for systems neuroscience research—particularly for addressing questions at the neural circuit level. Second, evidence for rodent cognitive capacities is mounting (Abbott 2010; Nikbakht and Diamond 2021), such as working memory (Fassihi et al. 2014; Akrami et al. 2018), decision-making (Brunton et al. 2013; Toso et al. 2021), rule learning (Murphy et al. 2008), and the ability to weigh sensory evidence (Kepecs et al. 2008). Finally, since rodents are evolutionary close relatives of humans (Dawkins 2010), they bear fundamental similarities in brain organization to humans, including a basic common plan for the cortex (Krubitzer 2007; Carandini and Churchland 2013).

In the following section, we will give a brief overview of the two relevant sensory systems in rats in the context of visual–tactile orientation categorization experiment: rats' whisker sensory system as an "expert" active sensory system and then rats' visual system.

3.4.1 Rat Tactile Sensory System

Examining the so-called expert sensory systems is an instructive approach in the study of perception. Expert cortical processing systems perform reliable and fast transformations of complex physical stimuli into meaningful percepts (Diamond and Arabzadeh 2013).

One of the best studied examples of such systems is the primate visual system, well-known for its efficiency in extracting meaning from the visual scenes. In rats and mice, an equally well-studied expert system is the whisker-mediated somatosensation. In their natural habitats, rats are active in environments with dim or no light where their survival critically depends on their sense of touch. Vincent (1912) in a classic study demonstrated that a rat's ability to navigate through a raised labyrinth is whisker-mediated. Whisker touch (along with olfaction) represents a major channel through which rodents collect information from their surroundings (Diamond et al. 2008b). They use their whiskers to recognize the position and identity of objects, particularly in the absence of light.

Mammalian whiskers (vibrissae) are specialized types of hair, identifiable by their large size and large and heavily innervated follicles. The follicles, in addition, contain blood-filled sinus tissues and have an intricate muscle matrix around the base and a network of motor control centers for their active movement. Most importantly, the whiskers have an identifiable representation in the somatosensory cortex (Prescott et al. 2011b).

Whiskers are arranged in the form of a grid layout on the rat's snout and include around 35 vibrissae. In addition, on their chins and around their lips, rats have an array of shorter, non-actuated microvibrissae[1] (Fig. 3.1). These two types of whiskers make up an array of highly sensitive hap-

[1] Microvibrissae are short (a few mm), densely spaced (87 per cm^2) whiskers located on the anterior part of a rat's snout. They are not ordered in a regular grid and show little or no whisking motion (Diamond et al. 2008b).

Fig. 3.1 Macrovibrissae are often used to locate objects that are then investigated further with the shorter, non-actuated microvibrissae on the chin and lips. Pictured are high-speed video frames of a rat locating a coin with its macrovibrissae (top) and next brushes the microvibrissae against the coin surface (bottom). (Reprint with permission from Fox et al. 2009)

tic detectors that generate and collect tactile information (Diamond and Arabzadeh 2013).

Somatosensory cortex of the rat is comprised of a primary field (SI) and a secondary field (SII). In the SI, macrovibrissae have a distinct representation as revealed by histological, connectivity maps and electrophysiological studies that resemble the shape of barrels—hence, the name barrel cortex. We know that there is a one-to-one correspondence between macrovibrissae and barrels (Diamond and Arabzadeh 2013).[2]

As mentioned earlier, the rat vibrissae system is a good example of an *active sensory system*. A distinct feature of active sensation is the controlled movement of the sensor in a manner that is optimized for the task, so as to maximize information gain (Nelson and MacIver 2006). Studies on the rat's whisker system have demonstrated that self-generated whisker motion is critical for following a wall in darkness (Jenks et al. 2010), identifying object properties such as texture (Diamond et al. 2008a), shape, and size (Brecht et al. 1997; Harvey et al. 2001; Polley et al. 2005), and estimating distances (Harris et al. 1999).

The whisker-mediated perception can be constructed through two distinct modes of active sensing: (1) generative mode and (2) receptive mode (Diamond and Arabzadeh 2013). The generative mode is the process in which rats sweep their whiskers in cycles of retraction and protraction at a frequency of about 10 Hz. Generative whisking actively makes contact with the objects in the environment. Whisking is the starting point for the texture perception (Diamond and Arabzadeh 2013) in which the rat actively creates the percept by its own movement under precisely controlled sensory–motor coordination. Observing rats making discrimination of textures and localization of objects reveals active whisking (Diamond 2010; Diamond et al. 2008a; Zuo et al. 2011). In the receptive mode, however, rats would "actively" immobilize their whiskers to optimize the collection of vibro-tactile stimuli from an object that is moving on its own such as a vibrating surface (Hachen et al. 2021). Although it is extremely challenging to measure rodents' whisker use in natural settings, one could reasonably assume that some forms of whisker-mediated perception rely on blocking motor output to the whisker muscles. A familiar experience for us is when we are trying to feel the vibrations of an object. We would place our fingers on the object surface without moving them and attend to the movements; rats do the same with their whiskers to detect the vibrations made by a potential predator approaching their underground borrow (Fig. 3.2). High-speed video recordings revealed that rats use a similar sensory strategy while judging the orientation of a surface or an object

[2] Specifically, the barrel cortex is revealed by the dark-staining regions of the layer four of the somatosensory cortex where inputs from the contralateral side of the body come in from the thalamus (Woolsey et al. 1975).

Fig. 3.2 Whisker-mediated touch in the receptive mode. As a cat is walking in proximity of the rat, the vibrations might be transferred to the whiskers through their contact with the walls and floor of the burrow. (Illustration by Marco Gigante)

feature. Rats would likely rely on the contacts made by their immobilized whiskers as they move their head and body with respect to the surface of the object.

3.4.2 Rat Vision

Traditionally, it was thought that mice and rats rely on a different combination of senses—somatosensation, olfaction, and hearing—compared to visually dominant primates. In fact, rats and mice have an excellent olfactory system with nose, as well as vomeronasal organ, and large olfactory bulbs that enables them to make incredibly acute and reliable decisions (Kepecs et al. 2008; Poo et al. 2022). Rats have 500–1,000 types of olfactory receptors that are coded for by a similar number of genes. This library of olfactory genes occupies about 1% of the rat's entire DNA and underscores the evolutionary significance of this sense for them.

Nevertheless, despite the vision being a main research topic in neuroscience, and despite the universality of the use of rats and mice as the most widespread animal models, until recently, rodents had been largely overlooked by the visual neuroscience community. One contributing factor was because their brains were assumed to lack high-level visual processing capability (Zoccolan 2015). A generic misconception among neuroscientists is that rodent behavior is only weakly influenced by vision. This assumption is based on several observations—mice and rats are nocturnal/crepuscular animals (Burn 2008),[3] and their visual acuity is relatively low (e.g., 1 cycle/deg in pigmented rats (Prusky et al. 2002, 2000; Birch and Jacobs 1979; Keller et al. 2000; Dean 1981), compared to 30–60 cycles/deg in the human and macaque fovea (Campbell and Gubisch 1966; Merigan and Katz 1990)). Although neurons in the rodent primary visual cortex share many of the tuning properties found in higher primates, such as orientation tuning (Girman et al. 1999; Bonin et al. 2011; Shaw et al. 1975), they are not arranged into the functional cortical modules, such as the orientation columns (Van Hooser et al. 2005).

However, evidence is mounting that vision is indeed a crucial source of sensory information for rodents. Particularly, they may rely on vision to navigate the environment—typically, when foraging over large distances around their nests (Chen et al. 2013; Barnett 2007). Additionally, rats have been shown to make excellent use of the available visual cues; when tested in spatial navigation tasks, they mostly rely on the available visual cues as opposed to the olfactory or auditory cues or path integration (Morris 1981; Suzuki et al. 1980). A signature of this visual dependence is hippocampal place field locking to the environmental visual cues and its remapping when these cues are altered (O'Keefe and Speakman 1987; Jezek et al. 2011; Muller and Kubie 1987). In addition, several cortical areas in the rodents are devoted to vision and provide possibility of complex visual processing such as shape-based visual object recognition (Gaffan et al. 2004) and transformation-tolerant (invariant) visual object recognition of three-dimensional shapes from two-dimensional images (Zoccolan et al. 2009; Tafazoli et al. 2012).

Although rats and mice are considered nocturnal, it has been shown that mice are consistently nocturnal only if they have unlimited food access (e.g., in laboratory animal facilities); however, when they are food restricted, they spend parts of the night sleeping, likely to lower their body

[3]Crepuscular animals are those that are active primarily during twilight (dawn and dusk).

temperature and minimize their energy expenditure (Hut et al. 2011). In nature they can even become entirely diurnal (Daan et al. 2011). Furthermore, the absence of reflective tapetum in the eyes, a hallmark of nocturnal animals, contradicts the view that mice and rats are entirely nocturnal (Carandini and Churchland 2013).

3.5 Visual–Tactile Orientation Categorization Task

The experiment in Fig. 3.3 was designed to examine the engagement of the cerebral cortex in recognizing object features by combining multiple sensory modalities (Nikbakht et al. 2018). The stimulus object, shown in Fig. 3.3a, was a disc with a diameter of 10 cm consisting of raised alternating black and white parallel bars. It was thus accessible to the rats as a visual as well as a tactile square-wave grating. In the task, rats were trained to face the disc to judge the grating orientation on each trial. They learned to categorize orientations in the range of 0°–45° as "horizontal," and those in the range of 45°–90° as "vertical" (Fig. 3.3b). Grating angles randomly varied in 5° steps between 0° and 90°. The sequence of events in the behavioral task was controlled by a computer program. A given trial started when the rats poked their head into the stimulus delivery port and interrupted an optical sensor that triggered the opening of an opaque gate. The gate opening followed by either a tactile (T) trial (in the absence of visible light, with the rat allowed to touch the object with its snout and whiskers), a visual (V) trial (illuminated under visible light but with a transparent screen in front of the object to prevent informative touch), or a visual–tactile (VT) trial (with the object illuminated and also accessible by touch). After examining the stimulus, the rat turned its head toward one of the two reward spouts (L or R) and licked—if it made a correct choice a drop of pear juice was given to it as reward. The boundary angle, 45°, was randomly rewarded on left or right. The three stimulus modality conditions (V, T, and VT) were also randomly interleaved with 32% likelihood on each trial. The remaining 4% of trials were control (catch) trials (on which neither visual nor tactile access to the stimulus was available).[4]

In order to examine the combination of visual and tactile cues, the rats' performance on unimodal trials was first measured. Next, the rats' performance data in the three conditions were analyzed to determine whether on visual–tactile trials the rats combined or else kept the visual and tactile cues separate.

The psychometric curves in Fig. 3.4a, b show the proportion of trials in which the rats judged the grating orientation as vertical in the three different modalities. Figure 3.4a demonstrates that rats had different modality-specific acuities: for example, rat 3 performed better in the T condition (left), while rats 12 and 1 performed better in the V condition (middle and right). However, all three representative rats performed more accurately in the VT than in V or T alone. The boost in the bimodal performance indicates that, when both modalities were accessible, rats did not select a single, preferred sensory channel; but instead, they benefited from both sensory input channels through multisensory combination.

In Fig. 3.4b the psychometric fits for 12 rats are summarized. In this plot the pale curves show individual rats' data, and the average performance is shown in thick curves. The curves are colored according to modality. The curve fits are cumulative Gaussian functions with their underlying standard deviation, σ, as a robust measure of discrimination acuity. A smaller σ implies a narrow Gaussian noise distribution (less trial-to-trial variability) and steeper psychometric curve and therefore higher acuity. As is evident from the slope of the curves, individual rats performed best on bimodal trials, independently of the individual's preferred modality: the decrease in the bimodal versus V and T conditions was significant for all rats. On average, rats had good tactile

[4] In these trials, the transparent screen remained in front of the stimulus and the visible light was switched off. Performance was not significantly different from the expected 50% chance performance. These catch trials were important control for ruling out the possibility of rats utilizing confounding signals such as the noise of the motor used to rotate the grating object, olfactory cues, etc.

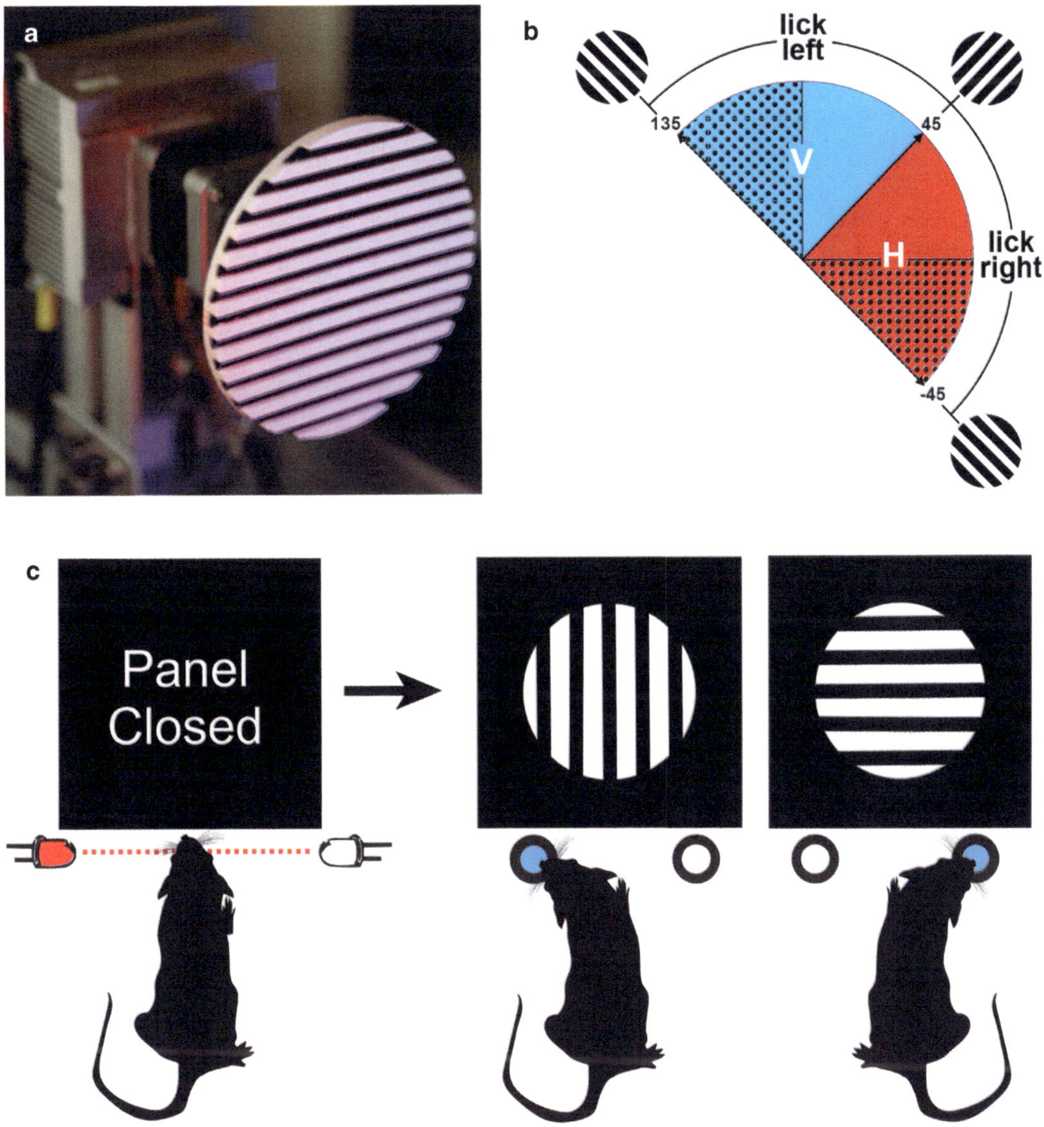

Fig. 3.3 Visual–tactile orientation categorization task. (**a**) The grating object was a 3D-printed disc of square-wave gratings with salient visual and tactile features. (**b**) Stimulus orientations and the categorization rule of the task. Rats learned to categorize orientations from 0° to 45° (solid red) in one category and orientations from 45° to 90° (solid blue) into another. When tested with -45° to 0° and 90°–135° (shown in red and blue stippling), they immediately generalized the categorization rule that suggests that the categories corresponded to horizontal (H) and vertical (V). (**c**) Sequence of stages in a trial. The trial was initiated after a head poke by the rat interrupted a IR-light beam. This triggered the opening of an opaque gate. The stimulus was then presented to the rat by either visual, tactile, or visual–tactile access. After sampling the stimulus, the rat turned toward one of the two spouts to collect a reward if correct. (Reprint with permission from Nikbakht et al. 2018)

orientation acuity (T) but on average showed slightly better visual performance (V). Regardless of individual variability in the visual or tactile performance, rats' accuracy was significantly superior in the bimodal condition, indicating multisensory enhancement.

How does the observed performance in bimodal trials compare to the prediction of the

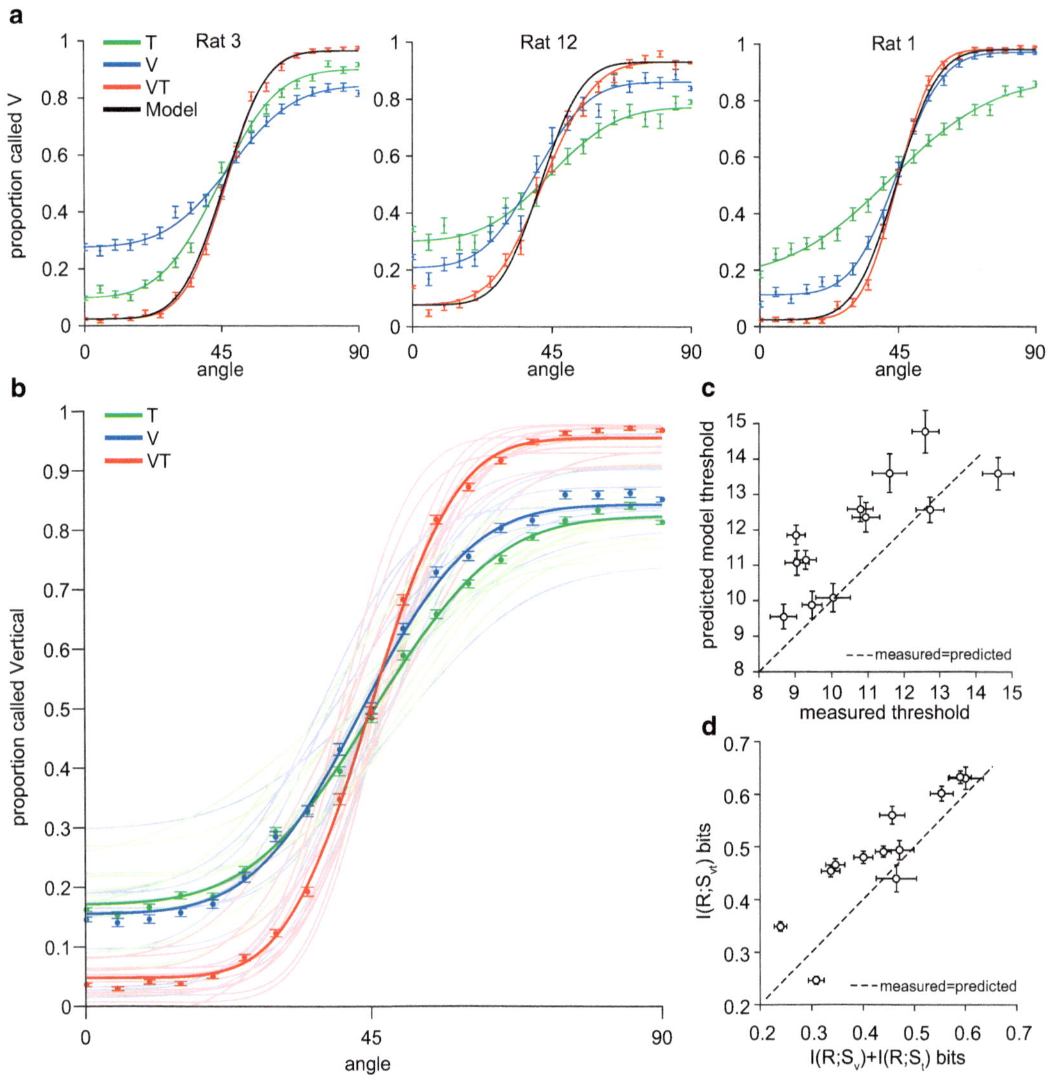

Fig. 3.4 Multisensory visual–tactile convergence. (**a**) Psychometric performance from three representative rats. Despite their differences in modality-specific performance, all of the rats showed better performance on bimodal trials compared to the V and T trials. The green curves correspond to the T condition, blue curves to V, and red to VT. Black traces show the predicted psychometric curves from the Bayesian cue-combination model. Error bars are 95% binomial confidence intervals. (**b**) Summary of psychometric performance of 12 rats in the multimodal task. Bold data points and curves are the performance average over all rats in each modality. Pale curves show individual rat's performance. Error bars are 95% binomial confidence intervals. (**c**) Comparing psychometric thresholds as predicted by the Bayesian cue combination of V and T (ordinate) with behaviorally measured thresholds on VT trials (abscissa). Each point corresponds to one rat. Error bars indicate 95% confidence intervals by bootstrapping. (**d**) Mutual information between behavioral choice and stimulus category in VT trials (ordinate) vs. the linear summation of mutual information between behavioral choice and stimulus category from V and T trials (abscissa). Each data point shows one rat. Error bars indicate 95% confidence intervals by bootstrapping. (Reprint with permission from Nikbakht et al. 2018)

optimal linear model? As discussed at the beginning of this chapter, optimal performance can be defined as the linear combination of the two independent signals, based on the framework of Bayesian decision theory. Fitting the psychometric data with a cumulative Gaussian function yields two parameters: the point of subjective equality (the orientation aligned to the mean of the best-fitting cumulative Gaussian function) and the psychophysical threshold (the Gaussian function standard deviation, σ). According to the Bayesian framework, the standard deviation in the bimodal condition is related to the standard deviations of the unimodal conditions as described by Eq. (3.2), in the case of vision and touch:

$$\frac{1}{\sigma_{vt}^2} = \frac{1}{\sigma_v^2} + \frac{1}{\sigma_t^2}. \tag{3.14}$$

Solving Eq. (3.14) for σ_{vt} yields the optimal bimodal threshold from unimodal estimates. Psychometric curves from each rat computed based on this optimal cue-combination model are drawn as black sigmoid curves in Fig. 3.4a. Figure 3.4c compares the measured threshold (σ) on bimodal trials versus the threshold predicted by a linear combination of V and T signals for all rats: Out of 12 rats, 2 rats combined V and T signals in a linear manner (e.g., rat 3 in Fig. 3.4a), 9 in a supra-linear manner (similar to rat 1 in Fig. 3.4a), and 1 sublinearly (rat 12 in Fig. 3.4a).

A different approach for comparing behavioral performance with the predictions of a linear combination model is based on information theory (MI). Here one could treat V and T signals as two information channels. Using Shannon's mutual information formula, we can quantify the amount of information that the behavioral response of rats (left or right) conveys about the stimulus category (horizontal or vertical) (Shannon 1948; Cover and Thomas 2012):

$$I(R;S) = \sum_{r,s} P(s)P(r \mid s)\log_2 \frac{P(r \mid s)}{P(r)}, \tag{3.15}$$

where $P(s)$ is the probability that a given stimulus category (horizontal or vertical) is presented, $P(r \mid s)$ is the conditional probability of the rat's response (right or left choice) given the stimulus category, and $P(r)$ is the marginal probability of rats' responses r (left or right response) unconditioned on the stimulus category.

Similar to the Bayesian model comparison, using MI, one could compare the total quantity of information available to the rat on bimodal trials with the linear sum of the information available on V and T trials. The assumption behind this approach is that the information available within each sensory modality is directly converted into a choice (Adibi et al. 2012). Thus, we can calculate this amount of information based on the behavioral performance, for example, 1.0 bits of information corresponds to 100% accuracy, while 0 bits corresponds to 50% accuracy.

In theory, the observed combination of touch and vision information could be linear, in case of the linear combination of two independent cues, sublinear, in case of redundancy between the available information in the two signals, or supralinear, which would reflect synergy. If the summed quantities information from V plus T were to exceed 1.0 bits, then the rat would have more information than it could express by its choices and the model would fail.

The results of mutual information approach are summarized in Fig. 3.4d. 10 out of 12 rats combined V and T signals supralinearly, 1 rat in a linear manner, and 1 in a sublinear manner.

Both models—Bayesian cue combination and mutual information—suggest that in most rats the sensory signals carried by the two channels were not combined independently but were merged synergistically.

The results of the visual–tactile orientation categorization behavior offer important insights into how information from multiple sensory channels is combined. The majority of rats exhibited bimodal performance that was significantly better than predictions of the standard linear Bayesian ideal observer model. We refer to this deviation from the model as "supralinearity." The mutual information analysis also showed that in the bimodal condition the information that rats carried about stimulus orientation was greater than the sum of information quantities they had in the V and T conditions.

Supralinear behavioral combination of sensory modalities has been previously reported (Fetsch et al. 2009, 2012; Kiani et al. 2013; Raposo et al. 2012). One hypothesis is that supralinear combination could occur if the available information in unimodal trials were exploited imperfectly, while the information available on bimodal trials were fully exploited. In this scenario, the model would have included an underestimated measure of unimodal signals, and as a result the bimodal accuracy would surpass the linear combination of (underestimated) unimodal signals. But what could lead to underestimation of the unimodal performance? Incomplete exploitation of information on unimodal trials could be due to reduced motivation. Rats are extremely sensitive to the rate of reward they cumulatively receive in the task. Since they make more errors on unimodal trials, they naturally obtain lower average reward rates in unimodal trials than in bimodal trials. Thus, they may pay less attention to the task in unimodal trials compared to the bimodal trials, where their efforts would be more frequently rewarded. With lower expected net reward on unimodal trials, rats may prioritize speed over accuracy (Drugowitsch et al. 2014). According to the bounded evidence accumulation framework, faster decisions could be made by lowering the decision threshold, thus prematurely stopping the processing of the unimodal stimuli (Kiani et al. 2008; Zuo and Diamond 2019).

However, the results of the orientation categorization task do not support the hypothesis of non-optimal information exploitation on unimodal trials. An indirect measure of the strength of evidence accumulation for decision-making tasks is trial response time (RT). In the orientation categorization task, regardless of the modality condition tested, rats' RT was nearly 150 ms longer on trials with orientation near the 45° category boundary. This indicates that rats did not truncate collecting evidence on more difficult (less frequently rewarded) trials. Moreover, similar RT distributions among all three modality conditions suggest more or less equal effort in sensory evidence accumulation on both unimodal and multimodal trials.

Another way to mitigate the motivational effects is using a block experimental design. Rats were tested with blocks of sessions comprised of a single modality condition. In V-only or T-only sessions, rats could not choose to attend preferentially to bimodal trials, so unimodal performance should come close to its maximal possible value. Yet, comparing across sessions, rats still combined V and T signals in a supralinear fashion to guide bimodal performance.

If we cannot fully explain the supralinear combination by increased motivation on bimodal trials, what could better explain these intriguing results? Both of the linear combination models analyzed here (Bayesian cue combination and mutual information) operate under the basic assumption that sensory channels involved operate independently of each other. Supralinearity could arise from violation of this assumption—here the visual and tactile signals might not be processed independently of each other in the bimodal trials. One possible mechanism could be that the co-occurrence of the cues from one sensory modality affects how the cues from the other modality are acquired. For instance, the presence of visual cues might assist the rat to palpate the stimulus with its whiskers more efficaciously, thereby enhancing the touch signal (Prescott et al. 2011a; Schroeder et al. 2010). This mechanism is consistent with active sensing. To test this hypothesis, head and whisker kinematics analysis is necessary. A different (non-mutually non-exclusive) mechanism of interaction is formation of synergy between the two processing pathways within the brain. Intra-cortical inhibition could enable a sensory region whose neural activity is evoked by one modality suppress the non-specific noise in the other modality's responses. Evidence for direct functional connectivity between different sensory cortices is emerging. In an experiment involving activation of the mouse auditory cortex by a noise burst, the auditory cortex could drive local inhibition in the primary visual cortex via corticocortical GABAergic connections (Iurilli et al. 2012). Similarly, short auditory noise pulses could sharpen the orientation tuning of L2/L3 pyramidal neurons in the primary visual cortex (Ibrahim et al. 2016).

An alternative hypothesis is the existence of noise from other sensory modalities in the unimodal conditions. For instance during the visual trials, the somatosensory system receives stimulus-irrelevant (noisy) signals. This noise could corrupt the unimodal estimates and lead to their underestimation. According to the Bayesian decision theory, when an informative cue is combined with a non-informative (uncorrelated noise) cue, the animal will completely rely on the informative cue and ignore the noisy input. However, in some cases, adding a completely noisy input to an informative stimulus, human subjects sub-optimally integrate some of the noise. This sub-optimal integration in unimodal conditions would lead to an apparent supralinearity in the multisensory condition (Zaidel et al 2015; Shalom and Zaidel 2018).

3.6 Convergence of Sensory Pathways in the Posterior Parietal Cortex

Our knowledge of a sensory object is not limited to a single modality, e.g., we may know the taste and smell of an apple as well as how it feels to our touch. Since multiple modalities are connected to our knowledge of an object, we can hypothesize that in some region in the brain, various sensory modalities must converge to form a multisensory representation of the world.

Where in the brain might the sensory channels of touch and vision merge together? One could safely hypothesize that the most likely region would be a cortical region, where input from both the tactile and visual senses are received. A good candidate for this is the posterior parietal cortex (PPC), which lies in between the somatosensory and the visual cortices and projects back to both of these areas (Cavada and Goldman-Rakic 1989; Miller and Vogt 1984; Leichnetz 2001) (Fig. 3.6a). PPC is also mutually connected to almost all of the other multimodal association cortices, for example, frontal, prefrontal, temporal, and limbic association cortex as well as motor and premotor areas (Akers and Killackey 1978; Reep et al. 1994) (Fig. 3.5).

To define the location of PPC, one can rely on several criteria, including connectivity to other cortical areas, myeloarchitecture, cytoarchitecture, and expression patterns of various neurotransmitters as well as neuroactive substances. It is suggested that PPC can be defined satisfactorily in most species focusing on the thalamocortical and/or corticocortical connectivity together with electrophysiological and behavioral comparisons (Whitlock et al. 2008).

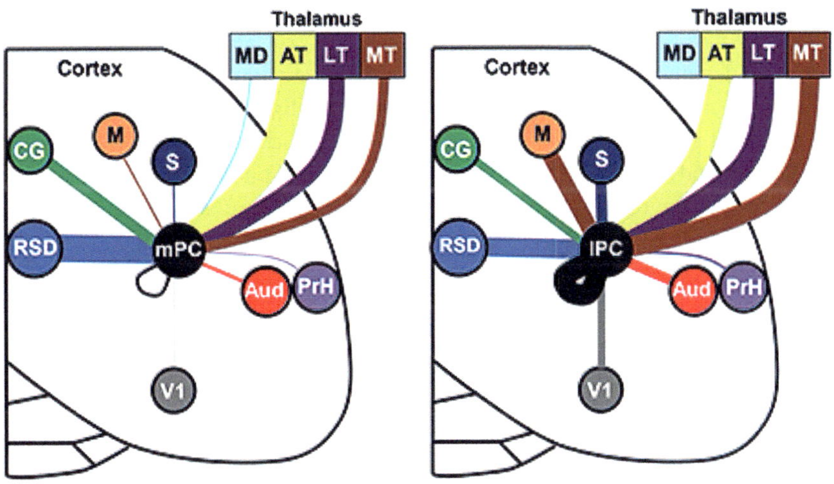

Fig. 3.5 Distinct input patterns for medial vs. lateral parietal cortex of the rat. Brain regions are shown in different colors, and the thickness of the connecting lines represents the strength of thalamic and cortical projections to the medial PC (left panel) and lateral PC (right panel), where the thalamic data are normalized separately from cortical data. This analysis shows projections to lateral PC (lPC) involve stronger projections from the visual (V1), somatosensory (S), auditory (AUD), motor (M) cortex, and motor thalamus (MT), while the medial PC (mPC) receives stronger inputs from the dorsal retrosplenial cortex (RSD) and cingulate region (CG). The remaining regions are shown as the following: mediodorsal thalamus (MD), anterior thalamus (AT), lateral thalamus (LT), and perirhinal cortex (PrH). (Figure adapted from Wilber et al. 2014)

The PPC in rats lies between somatosensory cortex anteriorly and visual cortex posteriorly (Krubitzer 1995). It is distinguished from the adjacent somatosensory cortex by that only the somatosensory cortex exhibits strong callosal connections (Akers and Killackey 1978).

In various species, connection of the PPC with the higher-order associative posterior thalamus is distinct. In non-human primates, it features a thalamic input from the posterior domain of the thalamus, particularly the pulvinar. For the rat and other mammals, the lateral posterior nucleus and adjacent nuclei predominantly constitute the thalamic input, including laterodorsal and posterior nuclei that are considered to be homologous to the primate posterior thalamic domain and pulvinar. Similar to primates, in rats the thalamic projections from the ventral nuclear complex and lateral geniculate do not seem to innervate the PPC (Reep et al. 1994). These features of the thalamic inputs to PPC help define the boundaries of it with respect to neighboring somatosensory and visual cortices (Wilber et al. 2014).

In the case of the orientation categorization experiment, extracellular recordings from PPC neurons recorded during the task revealed two broad response categories: One group of neurons were modulated by the stimulus orientation in a graded manner, as demonstrated in Fig. 3.6c. In contrast, another group of neurons showed a significantly large signal pertaining to the category of the stimulus (Fig. 3.6d). The firing rate of neurons in the second category was modulated according to the upcoming action (left or right). In line with this, the orientation of the stimulus in a given choice category poorly modulated the neurons of the second category. Both groups of neurons in PPC, whether they encoded stimulus orientation by a graded code or stimulus category by a step-like code, did so regardless of the sensory channel(s) involved (Fig. 3.6b).

In summary, Fig. 3.7a shows that each of the hundreds of recorded PPC neurons showed this same property: they were all equally excited by the two sensory channels, separately (V or T) or combined (VT). Figure 3.7b shows that this neural pattern was consistent independently of the stimulus orientation. The fact that nearly all data points lie close to the diagonal evidences supramodality, showing that a given neuron's overall engagement in the task was nearly equal under V, T, V-T modality conditions. In other words, this neural population does not distinguish the modality in which the object was presented.

What accounts for the differences observed across studies involving the PPC in different tasks? In a cortical circuit, capable of flexible computations, the functional properties emerge in the context of behavior. Considering the many different inputs to PPC—which PPC could "sample" in order to construct any of a large number of potential representations of the world—it would seem unlikely that as large a proportion of neurons as found in the rat visual–tactile orientation experiment would be tuned to orientation or would exhibit a significant categorical separation between horizontal and vertical, if not for this specific behavioral training.

Vertical and horizontal are cardinal, behaviorally significant orientations in natural settings. As humans, we are sensitive to small deviations from these cardinal angles. Figure 3.8a shows a modified version of the orientation categorization task. Was vertical versus horizontal categorization a special case? To answer this question a group of rats were trained to categorize the stimulus orientation in relation to boundaries other than 45°. Specifically, rats began a test session with categorization boundary set at either 22.5° or else 67.5°. At a later point during the testing session, the reward boundary was switched from 22.5 to 67.5 or vice versa. Figure 3.8b shows the behavioral result. The average rat could set its choices to a boundary other than 45° and readily switch its criterion partway through the test session.

The results of the shifted boundary experiment demonstrate rats' behavioral flexibility in learning to adapt to a new decision rule in the categorization task. Neural circuits underlying the ability to learn stimulus–outcome associations in a dynamic environment are distributed across multiple brain networks involving sensory, associative, and motor areas. PPC is posed as a good candidate for transformation of sensory information into categorical choice information.

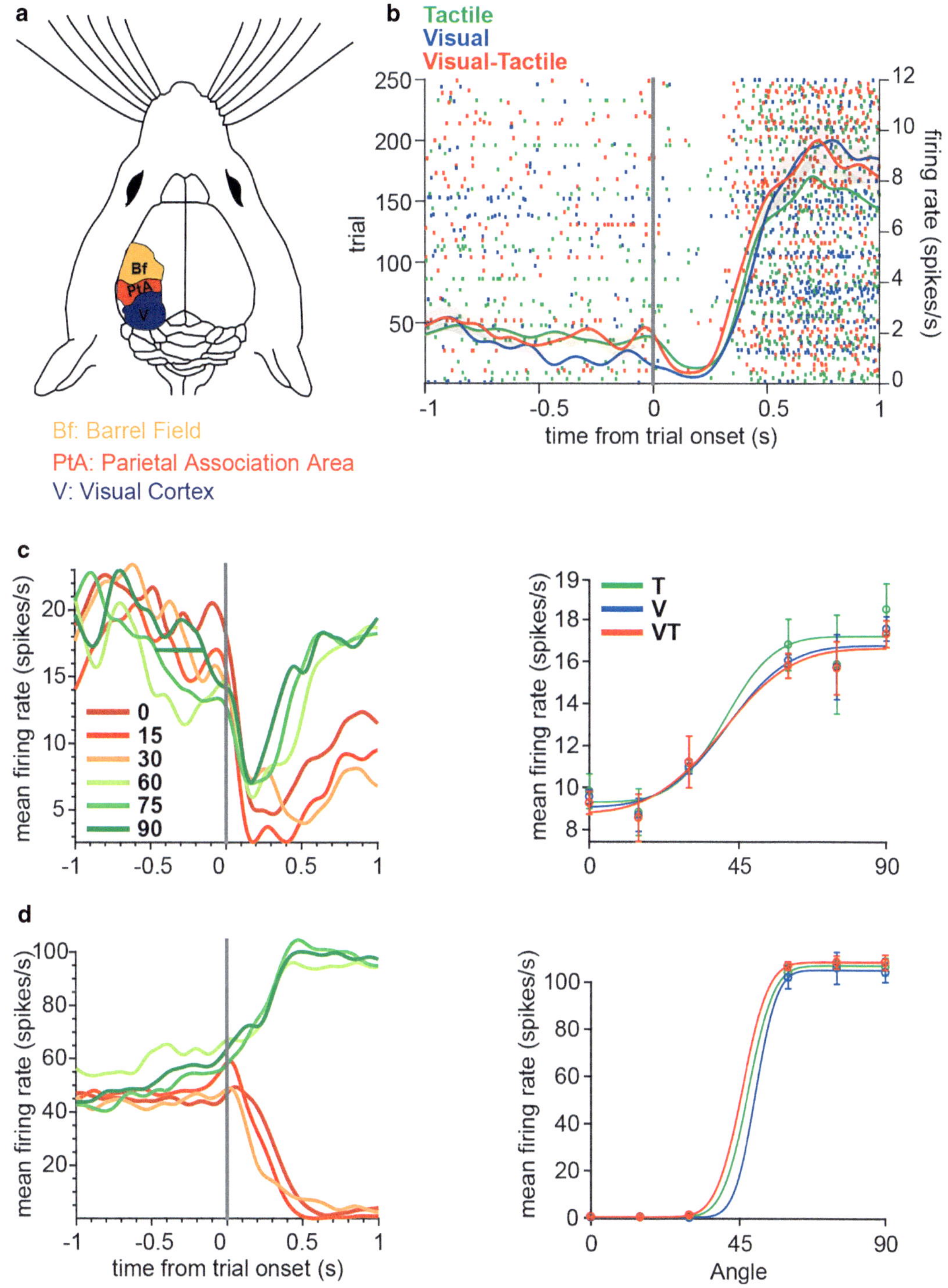

Fig. 3.6 Representation of orientation and stimulus category in posterior parietal cortex. (**a**) The recording target area, PPC (red), is located between visual cortex (blue) and the vibrissal region of somatosensory cortex, Bf (orange). (**b**) A raster plot showing an example PPC neuron firing on 250 randomly selected trials. Modality conditions (V, T, and VT) are labeled by color. Neural responses in all three modality conditions were similar. (**c** and **d**) Peri-event time histograms (PETHs, left plots, smoothed with a Gaussian kernel = 50 ms), and angular tuning curves (right plots, separated by modality) of two example PPC neurons carrying graded vs. categorical stimulus information. Trials are grouped by stimulus orientation (see color key). (Reprint with permission from Nikbakht et al. 2018)

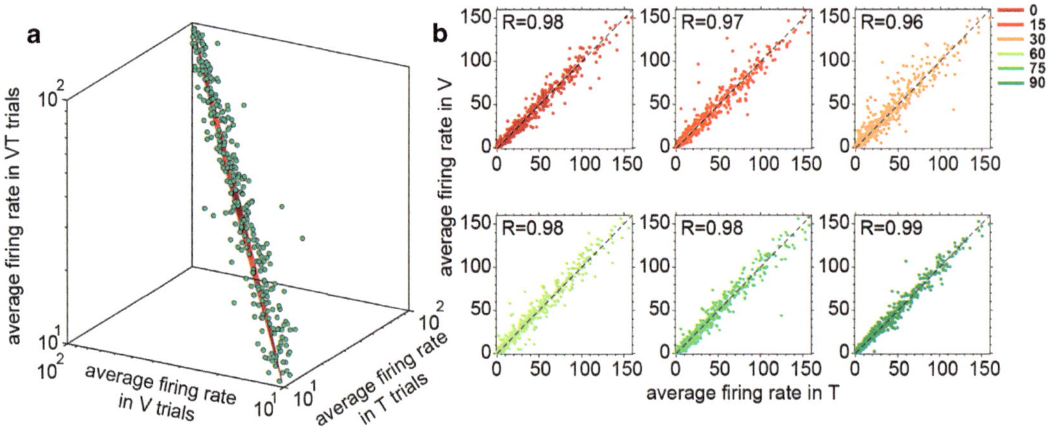

Fig. 3.7 Supramodal neural responses in the PPC. (**a**) Data points demonstrate average firing rates of 622 PPC neurons in window 300 ms before the response of the rat in V, T, and VT trials. All neurons are clustered about a line (orange) indicating equal responses in all three modality conditions. (**b**) Points depict the firing rate of all neurons, again in the 300 ms window preceding the response, in T (abscissa) versus V (ordinate) trials grouped by orientation (colors). R is the Pearson linear correlation coefficient. (Reprint with permission from Nikbakht et al. 2018)

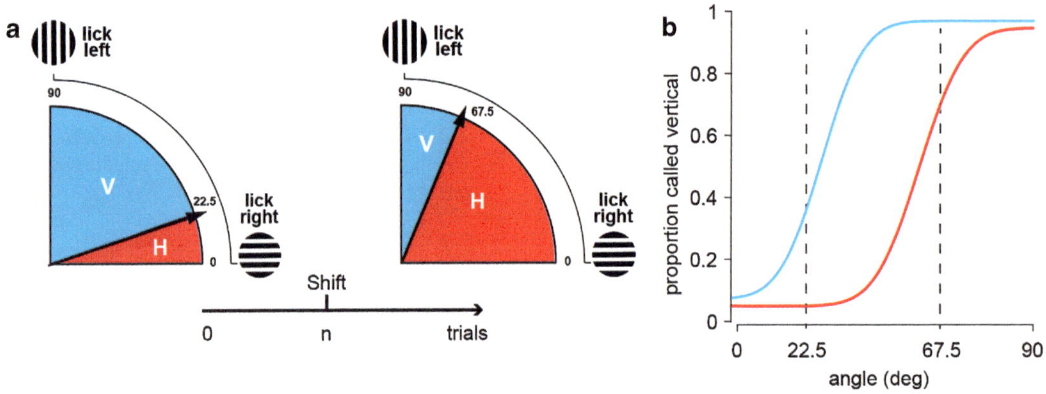

Fig. 3.8 The modified version of the orientation categorization experiment with shifted categorization boundaries. (**a**) Rats randomly started the session with category boundary set at either 22.5° or 67.5°. After *n* trials the boundary was switched. (**b**) The shift in the mean of the psychometric curves shows that on average rats could adjust their choices to a boundary other than 45° and readily switch their categorization rule in the middle of the test session

Taken together these results demonstrate that PPC contains a high degree of task-dependent plasticity. While it is possible that PPC receives orientation-tuned signals from earlier stages of visual cortical processing (Wang et al. 2012), tactile orientation tuning has never been reported in the somatosensory cortex, and therefore, it is possible that it may be generated in PPC to meet specific behavioral demands. Curiously, the categorization boundary of 45° was set completely arbitrarily; therefore, its neural correlates must also be formed to meet behavioral demands. These findings point to the conclusion that the circuitry in PPC is capable of producing functional properties that are not hardwired but emerge through learning to enable the animal to adapt to the requirements of the ongoing task.

The modality-free encoding of stimuli in the rat PPC contrasts the finding of a previous study in which rats judged a train of auditory/visual pulses. In this chapter most neurons in PPC were shown to respond differentially on trials from the two modalities such that the stimulus modality itself could be decoded from neuronal firing (Raposo et al. 2014). A potential mechanism explaining this difference is that modality-invariant coding in PPC arises from the simultaneous arrival of congruous signals, through two sensory channels, about a real object—rather than arbitrary pairings of computer-generated stimulus events. Coherence between modalities might better reflect the statistics of the real world. Indeed, vision and touch both evolved to extract shape, form, and spatial properties from the environment. A rat might need to maneuver through oriented edges (think of inside a burrow), whether those edges are seen or felt. In general, the brain's capacity to call up knowledge about things independently of sensory input channel (Quiroga et al. 2005) requires a stage of processing in which the same information is encoded by both channels, and the responses of neurons in PPC highlight the core function of an associative cortical region whose job is to create representations of the real world using sensory messages as the building blocks.

In summary, the results of the visual–tactile orientation categorization task suggest multisensory synergy and inspire important questions about the neuronal and behavioral mechanisms of supralinearity, such as what circumstances lead animals to combine sensory signals synergistically. A striking result of the physiological recordings was the invariance of neuronal activity in the PPC to modality. This indicates that PPC circuitry might harbor the neural substrate for abstraction of stimuli from sensory domains—an essential feature of conceptual knowledge. Further studies examining this circuitry will shed light on the neural mechanisms enabling abstraction of sensory data into a meaningful percept.

References

Abbott A (2010) Neuroscience: the rat pack. Nat News 465(7296):282–283

Adibi M, Diamond ME, Arabzadeh E (2012) Behavioral study of whisker-mediated vibration sensation in rats. Proc Natl Acad Sci 109(3):971–976

Akers RM, Killackey HP (1978) Organization of cortico-cortical connections in the parietal cortex of the rat. J Comp Neurol 181(3):513–537

Akrami A, Kopec CD, Diamond ME, Brody CD (2018) Posterior parietal cortex represents sensory history and mediates its effects on behaviour. Nature 554(7692):368–372.

Alais D, Burr D (2004) The ventriloquist effect results from near-optimal bimodal integration. Curr Biol 14(3):257–262

Allman JM (2000) Evolving brains. Scientific American Library, New York.

Angelaki DE, Gu Y, DeAngelis GC (2009) Multisensory integration: psychophysics, neurophysiology, and computation. Curr Opin Neurobiol 19(4):452–458

Barnett SA (2007) The rat: a study in behavior. Transaction Publishers, Piscataway

Battaglia PW, Jacobs RA, Aslin RN (2003) Bayesian integration of visual and auditory signals for spatial localization. J Opt Soc Am A 20(7):1391–1397

Birch D, Jacobs GH (1979) Spatial contrast sensitivity in albino and pigmented rats. Vision Res 19(8):933–937

Bonin V, Histed MH, Yurgenson S, Reid RC (2011) Local diversity and fine-scale organization of receptive fields in mouse visual cortex. J Neurosci 31(50):18506–18521

Brecht M, Preilowski B, Merzenich MM (1997) Functional architecture of the mystacial vibrissae. Behav Brain Res 84(1):81–97

Brunton BW, Botvinick MM, Brody CD (2013) Rats and humans can optimally accumulate evidence for decision-making. Science 340(6128):95–98

Burn CC (2008) What is it like to be a rat? rat sensory perception and its implications for experimental design and rat welfare. Appl Anim Behav Sci 112(1):1–32

Campbell F, Gubisch R (1966) Optical quality of the human eye. J Physiol 186(3):558

Carandini M, Churchland AK (2013) Probing perceptual decisions in rodents. Nat Neurosci 16(7):824–831

Cavada C, Goldman-Rakic PS (1989) Posterior parietal cortex in rhesus monkey: II. evidence for segregated corticocortical networks linking sensory and limbic areas with the frontal lobe. J Comp Neurol 287(4):422–445

Chen G, King JA, Burgess N, O'Keefe J (2013) How vision and movement combine in the hippocampal place code. Proc Natl Acad Sci 110(1):378–383

Cover TM, Thomas JA (2012) Elements of information theory. Wiley, New York

Daan S, Spoelstra K, Albrecht U, Schmutz I, Daan M, Daan B, Rienks F, Poletaeva I, Dell' G, Vyssotski A, et al (2011) Lab mice in the field: unorthodox daily activity and effects of a dysfunctional circadian clock allele. J Biol Rhythms 26(2):118–129

Dawkins R (2010) The ancestor's tale: a pilgrimage to the dawn of life. Weidenfeld & Nicolson, London

Dean P (1981) Grating detection and visual acuity after lesions of striate cortex in hooded rats. Exp Brain Res 43(2):145–153

Diamond ME (2010) Texture sensation through the fingertips and the whiskers. Curr Opin Neurobiol 20(3):319–327

Diamond ME, Arabzadeh E (2013) Whisker sensory system–from receptor to decision. Prog Neurobiol 103:28–40

Diamond ME, von Heimendahl M, Arabzadeh E (2008a) Whisker-mediated texture discrimination. PLoS Biol 6(8):e220

Diamond ME, von Heimendahl M, Knutsen PM, Kleinfeld D, Ahissar E (2008b) 'Where' and 'what' in the whisker sensorimotor system. Nat Rev Neurosci 9(8):601–612

Drugowitsch J, DeAngelis GC, Klier EM, Angelaki DE, Pouget A (2014) Optimal multisensory decision-making in a reaction-time task. eLife 3:e03005

Ernst MO, Banks MS (2002) Humans integrate visual and haptic information in a statistically optimal fashion. Nature 415(6870):429–433

Fassihi A, Akrami A, Esmaeili V, Diamond ME (2014) Tactile perception and working memory in rats and humans. Proc Natl Acad Sci 111(6):2331–2336

Fetsch CR, DeAngelis GC, Angelaki DE (2013) Bridging the gap between theories of sensory cue integration and the physiology of multisensory neurons. Nat Rev Neurosci 14(6):429–442

Fetsch CR, Pouget A, DeAngelis GC, Angelaki DE (2012) Neural correlates of reliability-based cue weighting during multisensory integration. Nat Neurosci 15(1):146–154

Fetsch CR, Turner AH, DeAngelis GC, Angelaki DE (2009) Dynamic reweighting of visual and vestibular cues during self-motion perception. J Neurosci 29(49):15601–15612

Fox CW, Mitchinson B, Pearson MJ, Pipe AG, Prescott TJ (2009) Contact type dependency of texture classification in a whiskered mobile robot. Auton Robot 26:223–239

Gaffan E, Healey A, Eacott M (2004) Objects and positions in visual scenes: effects of perirhinal and postrhinal cortex lesions in the rat. Behav Neurosci 118(5):992

Girman SV, Sauvé Y, Lund RD (1999) Receptive field properties of single neurons in rat primary visual cortex. J Neurophysiol 82(1):301–311

Gu Y, Angelaki DE, DeAngelis GC (2008) Neural correlates of multisensory cue integration in macaque MSTd. Nat Neurosci 11(10):1201–1210

Hachen I, Reinartz S, Brasselet R, Stroligo A, Diamond M (2021) Dynamics of history-dependent perceptual judgment. Nat Commun 12(1):1–15

Harris JA, Petersen RS, Diamond ME (1999) Distribution of tactile learning and its neural basis. Proc Natl Acad Sci 96(13):7587–7591

Harvey M, Roberto B, Zeigler HP (2001) Discriminative whisking in the head-fixed rat: optoelectronic monitoring during tactile detection and discrimination tasks. Somatosens Motor Res 18(3):211–222

Hut RA, Pilorz V, Boerema AS, Strijkstra AM, Daan S (2011) Working for food shifts nocturnal mouse activity into the day. PLoS One 6(3):e17527

Ibrahim LA, Mesik L, Ji X-Y, Fang Q, Li, H-F, Li, Y-T, Zingg, B, Zhang, LI, Tao, HW (2016) Cross-modality sharpening of visual cortical processing through layer-1-mediated inhibition and disinhibition. Neuron 89(5):1031–1045

Iurilli G, Ghezzi D, Olcese U, Lassi G, Nazzaro C, Tonini R, Tucci V, Benfenati F, Medini P (2012) Sound-driven synaptic inhibition in primary visual cortex. Neuron 73(4):814–828

Jacobs RA (1999) Optimal integration of texture and motion cues to depth. Vision Res 39(21):3621–3629

Jenks RA, Vaziri A, Boloori A-R, Stanley GB (2010) Self-motion and the shaping of sensory signals. J Neurophysiol 103(4):2195–2207

Jerison H (2012) Evolution of the brain and intelligence. Elsevier, Amsterdam

Jezek K, Henriksen EJ, Treves A, Moser EI, Moser M-B (2011) Theta-paced flickering between place-cell maps in the hippocampus. Nature 478(7368):246–249

Jung R (1984) Sensory research in historical perspective: some philosophical foundations of perception. Wiley Online Library.

Kaas J (2010) The evolution of neocortex from early mammals to humans. Int J Dev Neurosci 28(8):648

Keller J, Strasburger H, Cerutti DT, Sabel BA (2000) Assessing spatial vision—automated measurement of the contrast-sensitivity function in the hooded rat. J Neurosci Methods 97(2):103–110

Kepecs A, Uchida N, Zariwala HA, Mainen ZF (2008) Neural correlates, computation and behavioural impact of decision confidence. Nature 455(7210):227–231

Kiani R, Churchland AK, Shadlen MN (2013) Integration of direction cues is invariant to the temporal gap between them. J Neurosci 33(42):16483–16489

Kiani R, Hanks TD, Shadlen MN (2008) Bounded integration in parietal cortex underlies decisions even when viewing duration is dictated by the environment. J Neurosci 28(12):3017–3029

Knill DC, Saunders JA (2003) Do humans optimally integrate stereo and texture information for judgments of surface slant? Vision Res 43(24):2539–2558

Körding KP, Wolpert DM (2006) Bayesian decision theory in sensorimotor control. Trends Cognit Sci 10(7):319–326

Krubitzer L (1995) The organization of neocortex in mammals: are species differences really so different? Trends Neurosci 18(9):408–417

Krubitzer L (2007) The magnificent compromise: cortical field evolution in mammals. Neuron 56(2):201–208

Landy MS, Banks MS, Knill DC (2011) Ideal-observer models of cue integration. Sensory Cue Integration. Oxford University Press, Oxford, pp 5–29

Leichnetz GR (2001) Connections of the medial posterior parietal cortex (area 7m) in the monkey. Anat Rec 263(2):215–236

Merigan WH, Katz LM (1990) Spatial resolution across the macaque retina. Vision Res 30(7):985–991
Miller MW, Vogt BA (1984) Direct connections of rat visual cortex with sensory, motor, and association cortices. J Comp Neurol 226(2):184–202
Morris RG (1981) Spatial localization does not require the presence of local cues. Learn Motiv 12(2):239–260
Muller RU, Kubie JL (1987) The effects of changes in the environment on the spatial firing of hippocampal complex-spike cells. J Neurosci 7(7):1951–1968
Murphy RA, Mondragón E, Murphy VA (2008) Rule learning by rats. Science 319(5871):1849–1851
Naumer MJ, Kaiser J (2010) Multisensory object perception in the primate brain. Springer, New York
Nelson ME, MacIver MA (2006) Sensory acquisition in active sensing systems. J Comp Physiol A 192(6):573–586
Nikbakht N, Diamond ME (2021) Conserved visual capacity of rats under red light. ELife 10:e66429
Nikbakht N, Tafreshiha A, Zoccolan D, Diamond ME (2018) Supralinear and supramodal integration of visual and tactile signals in rats: psychophysics and neuronal mechanisms. Neuron 97(3):626–639
Northcutt RG, Kaas JH (1995) The emergence and evolution of mammalian neocortex. Trends Neurosci 18(9):373–379
O'Keefe J, Speakman A (1987) Single unit activity in the rat hippocampus during a spatial memory task. Exp Brain Res 68(1):1–27
Polley DB, Rickert JL, Frostig RD (2005) Whisker-based discrimination of object orientation determined with a rapid training paradigm. Neurobiol Learn Mem 83(2):134–142
Poo C, Agarwal G, Bonacchi N, Mainen ZF (2022) Spatial maps in piriform cortex during olfactory navigation. Nature 601(7894):595–599
Prescott TJ, Diamond ME, Wing AM (2011a) Active touch sensing. Philos Trans R Soc London B: Biol Sci 366(1581):2989–2995
Prescott TJ, Mitchinson B, Grant RA (2011b) Vibrissal behavior and function. Scholarpedia 6(10):6642
Prusky GT, Harker KT, Douglas RM, Whishaw IQ (2002) Variation in visual acuity within pigmented, and between pigmented and albino rat strains. Behav Brain Res 136(2):339–348
Prusky GT, West PW, Douglas RM (2000) Behavioral assessment of visual acuity in mice and rats. Vision Res 40(16):2201–2209
Quiroga RQ, Reddy L, Kreiman G, Koch C, Fried I (2005) Invariant visual representation by single neurons in the human brain. Nature 435(7045):1102–1107
Raposo D, Kaufman MT, Churchland AK (2014) A category-free neural population supports evolving demands during decision-making. Nat Neurosci 17:1784–1792
Raposo D, Sheppard JP, Schrater PR, Churchland AK (2012) Multisensory decision-making in rats and humans. J Neurosci 32(11):3726–3735
Reep R, Chandler H, King V, Corwin J (1994) Rat posterior parietal cortex: topography of corticocortical and thalamic connections. Exp Brain Res 100(1):67–84
Schroeder CE, Wilson DA, Radman T, Scharfman H, Lakatos P (2010) Dynamics of active sensing and perceptual selection. Curr Opin Neurobiol 20(2):172–176
Shalom S, Zaidel A (2018) Better than optimal. Neuron 97(3):484–487
Shannon CE (1948) A mathematical theory of communication. Bell Syst Tech J 22:379–423, 623–656
Shaw C, Yinon U, Auerbach E (1975) Receptive fields and response properties of neurons in the rat visual cortex. Vision Res 15(2):203–208
Stein BE, Stanford TR (2008) Multisensory integration: current issues from the perspective of the single neuron. Nat Rev Neurosci 9(4):255–266
Suzuki S, Augerinos G, Black AH (1980) Stimulus control of spatial behavior on the eight-arm maze in rats. Learn Motiv 11(1):1–18
Tafazoli S, Di Filippo A, Zoccolan D (2012) Transformation-tolerant object recognition in rats revealed by visual priming. J Neurosci 32(1):21–34
Toso A, Reinartz S, Pulecchi F, Diamond ME (2021) Time coding in rat dorsolateral striatum. Neuron 109(22):3663–3673
van Hemmen JL, van der Smagt P, Stein BE (2012) Foreword for the special issue on multimodal and sensorimotor bionics. Biol Cybern 106(11-12):615
Van Hooser SD, Heimel J. AF, Chung S, Nelson SB, Toth LJ. (2005) Orientation selectivity without orientation maps in visual cortex of a highly visual mammal. J Neurosci 25(1):19–28
Vincent SB (1912) The functions of the vibrissae in the behavior of the white rat..., vol 1. University of Chicago, Chicago
Wang Q, Sporns O, Burkhalter A (2012) Network analysis of corticocortical connections reveals ventral and dorsal processing streams in mouse visual cortex. J Neurosci 32(13):4386–4399
Whitlock JR, Sutherland RJ, Witter MP, Moser M.-B, Moser EI (2008) Navigating from hippocampus to parietal cortex. Proc Natl Acad Sci 105(39):14755–14762
Wilber AA, Clark BJ, Demecha AJ, Mesina L, Vos JM, and McNaughton BL (2014) Cortical connectivity maps reveal anatomically distinct areas in the parietal cortex of the rat. Front Neural Circuits 8. https://doi.org/10.3389/fncir.2014.00146
Woolsey TA, Welker C, Schwartz RH (1975) Comparative anatomical studies of the SmL face cortex with special reference to the occurrence of "barrels" in layer IV. J Comp Neurol 164(1):79–94
Zaidel A, Goin-Kochel RP, Angelaki DE (2015) Self-motion perception in autism is compromised by visual noise but integrated optimally across multiple senses. Proc Natl Acad Sci 112(20):6461–6466
Zoccolan D (2015) Invariant visual object recognition and shape processing in rats. Behav Brain Res 285:10–33

Zoccolan D, Oertelt N, DiCarlo JJ, Cox DD (2009) A rodent model for the study of invariant visual object recognition. Proc Natl Acad Sci 106(21):8748–8753
Zuo Y, Diamond ME (2019) Rats generate vibrissal sensory evidence until boundary crossing triggers a decision. Curr Biol 29(9):1415–1424
Zuo Y, Perkon I, Diamond ME (2011) Whisking and whisker kinematics during a texture classification task. Philos Trans R Soc London B: Biol Sci 366(1581):3058–3069

Multisensory Integration and Causal Inference in Typical and Atypical Populations

4

Samuel A. Jones and Uta Noppeney

Abstract

Multisensory perception is critical for effective interaction with the environment, but human responses to multisensory stimuli vary across the lifespan and appear changed in some atypical populations. In this review chapter, we consider multisensory integration within a normative Bayesian framework. We begin by outlining the complex computational challenges of multisensory causal inference and reliability-weighted cue integration, and discuss whether healthy young adults behave in accordance with normative Bayesian models. We then compare their behaviour with various other human populations (children, older adults, and those with neurological or neuropsychiatric disorders). In particular, we consider whether the differences seen in these groups are due only to changes in their computational parameters (such as sensory noise or perceptual priors), or whether the fundamental computational principles (such as reliability weighting) underlying multisensory perception may also be altered. We conclude by arguing that future research should aim explicitly to differentiate between these possibilities.

Keywords

Multisensory integration · Perception · Bayesian causal inference · Lifespan · Atypical

4.1 The Challenges of Multisensory Processing

Our ability to interact effectively with the world around us relies on precise, up-to-date sensory information. We therefore have an array of sensory organs that are optimised for detecting different types of information. Our eyes, for example, provide high precision when locating and identifying an object, but are of limited value for determining its temperature. The nose is an excellent tool for identifying whether food is rotten but can tell us little about how much it weighs. Usefully, most objects and events produce signals that may be detected by several sense organs simultaneously, and the brain is able to combine these to enhance our perception and reduce our uncertainty about the world around us. Sometimes this information is separate and complementary; other times, the information provided by multiple

S. A. Jones (✉)
Department of Psychology, Nottingham Trent University, Nottingham, UK
e-mail: samuel.jones@ntu.ac.uk

U. Noppeney
Donders Institute for Brain, Cognition and Behaviour, Radboud University, Nijmegen, The Netherlands
e-mail: uta.noppeney@donders.ru.nl

Y. Gu, A. Zaidel (eds.), *Advances of Multisensory Integration in the Brain*, Advances in Experimental Medicine and Biology 1437, https://doi.org/10.1007/978-981-99-7611-9_4

senses overlaps. Consider a conversation with a friend in a loud room: we are far more likely to understand what they are saying if we can both listen to their voice and watch their lips form the words. Specifically, multisensory percepts afford two major benefits. First, they are more salient, facilitating faster and more accurate detection (Diederich and Colonius 2004; Vroomen and de Gelder 2000). Second, they are more precise, improving our estimates of an object or event's properties (Ernst and Bülthoff 2004).

It is important to note, however, that our senses are constantly flooded with a mass of signals—many of them conflicting and/or unrelated—and the apparent effortlessness with which we sort and process this information belies the astounding complexity of the necessary underlying computations (and the neural processes involved in performing them). To integrate multisensory signals in an effective way, the brain needs to identify which of them belong together; only those signals that emanate from the same object or event should be processed together. This is a nontrivial process, as the observer has no access to the true causal structure of the world. Instead, sensory inputs must be matched based on *correspondence cues*—shared properties that suggest they were produced by the same source, such as appearing in the same place or at the same time (Parise and Spence 2013)—and prior beliefs and expectations about which signals belong together. This complex operation is known as *causal inference*, as it involves attempting to infer the causal structure (i.e. one source versus several) that produced the signals (Acerbi et al. 2018; Körding et al. 2007; Rohe and Noppeney 2015a). Once multiple signals have been determined to share a source, the sensory system should then integrate them into a more reliable percept that maximises the potential benefits (Alais and Burr 2004; Ernst and Banks 2002; Fetsch et al. 2012).

In this chapter, we consider these computational challenges within the framework of Bayesian probability theory (Ernst and Bülthoff 2004; Fetsch et al. 2013; Meijer and Noppeney 2020; Noppeney 2020, 2021). This is a useful approach for evaluating how humans (and other organisms) process and respond to multisensory stimuli, as it enables the specification of an ideal observer that exhibits "optimal" behaviour against which actual responses may be compared. Within this framework, the observer is said to form a generative model of how sources in the environment may have produced the signals, which is inverted during perceptual inference.

The chapter includes two parts. Part 1 discusses how observers weight and combine multiple signals that were unambiguously produced by a single source (the so-called forced-fusion scenarios). Part 2 moves beyond these forced-fusion cases and considers more complex situations in which signals can arise from common or different sources; this requires observers to infer the signals' causal structure, in order to arbitrate between sensory integration and segregation. Each part first introduces the normative Bayesian model that describes how ideal observers should combine signals during perceptual inference. We then discuss whether healthy young adults behave in accordance with these models and compare their behaviour with various other human populations (children, older adults, and those with neurological or neuropsychiatric disorders). In particular, we consider the specific ways in which these groups can diverge: populations may differ only in their computational parameters, such as sensory noise (i.e. variances) or prior expectations about environmental properties (e.g. spatial location) and structures (e.g. number of sources) that are incorporated in model priors; alternatively, some groups may vary even in the computational principles (such as reliability weighting) underlying multisensory integration (Huys et al. 2021; Jones and Noppeney 2021). At the end of each part, we briefly discuss the neural systems and mechanisms that support multisensory integration and causal inference. Though our review focuses on human work, we also briefly highlight the most relevant research in other species (for in-depth reviews, see Fetsch et al. 2013; Stein et al. 2014; Stein and Stanford 2008; Witten and Knudsen 2005).

4.2 Integrating Signals that Share a Source: Forced Fusion

All sensory inputs are corrupted to some degree by both external and internal noise, reducing their reliability or precision. The brain can attenuate this uncertainty by integrating multiple cues that contain redundant information, leading to a final multisensory percept with greater precision/reliability (i.e. less variance/noise) than any of the individual unisensory estimates. Early approaches to Bayesian modelling of sensory integration focused on situations in which all sensory cues are assumed to originate from the same source, avoiding the causal inference problem. These forced-fusion models address the question of how to weight and combine multiple cues in a way that increases the precision of the final joint percept (Alais and Burr 2004; Ernst and Banks 2002; Hillis et al. 2002; Jacobs 1999; Meijer et al. 2019).

4.2.1 Bayesian Modelling of Forced-Fusion Integration

Forced-fusion integration of multisensory signals can be illustrated using the example of locating a single object using both auditory and visual cues. In this example, the generative model specifies that a single audiovisual source (S_{AV}) simultaneously emits an auditory (x_A) and a visual signal (x_V), and that each of these unisensory signals is independently corrupted by some amount of noise. During perception, the observer computes the (posterior) probability of the source location by combining the estimates (likelihoods) of the auditory and visual signals with prior spatial expectations, according to Bayes' rule:

$$P\left(S_{AV} \mid x_A, x_V\right) \propto P\left(x_A \mid S_{AV}\right) * P\left(x_V \mid S_{AV}\right) * P\left(S_{AV}\right)$$

If we specify a uniform spatial prior $P\ (S_{AV})$, such that the source of the signals may be anywhere in space with equal probability, this computation is equivalent to maximum-likelihood estimation (MLE):

$$P\left(S_{AV} \mid x_A, x_V\right) \propto P\left(x_A \mid S_{AV}\right) * P\left(x_V \mid S_{AV}\right)$$

Assuming that the noise corrupting the unisensory signals is Gaussian, the maximum-likelihood estimate for the audiovisual source $\hat{S}_{AV}$ is the average of the unisensory estimates $\hat{S}_A$ and $\hat{S}_V$, weighted by their respective reliabilities:

$$\hat{S}_{AV} = w_A \hat{S}_A + w_V \hat{S}_V$$

where each unisensory signal's reliability is defined as the inverse of its variance, $r = 1/\sigma^2$, and the sensory weights are normalised to sum to one:

$$w_A = \frac{r_A}{r_A + r_V}$$

and

$$w_V = \frac{r_V}{r_A + r_V} = 1 - w_A$$

When a multisensory estimate is calculated in this way, the reliability of the final combined estimate is equal to the sum of the unisensory reliabilities:

$$r_{AV} = r_A + r_V = \frac{\sigma^2_A + \sigma^2_V}{\sigma^2_A * \sigma^2_V}$$

An optimal observer's integrated percept, under forced-fusion assumptions, therefore has two important properties. First, the final combined location estimate is an average of the auditory and visual estimates, weighted by their relative reliabilities. Second, the reliability of the final estimate will always be equal to, or greater than, the most reliable unisensory estimate. Though we focused here on the example of audiovisual spatial integration, the model may equally be applied to other stimulus properties, other sensory modalities, multiple cues within one modality, or situations with more than two signals (e.g. Bresciani et al. 2008; Ernst and Banks 2002; Jacobs 1999).

4.2.2 Forced-Fusion Integration in Healthy Young Adults

As discussed above, Bayesian forced-fusion models describe the way in which a theoretical "optimal" observer integrates multiple sensory cues that share a common source. They make predictions (reliability weighting and enhanced precision) against which human behaviour may be tested. To do so, we can measure the reliability of participants' responses to unisensory signals and use these measured reliabilities as known parameters in a forced-fusion model. The model predictions may then be compared with participants' true responses to multisensory stimuli.

Accumulating research has shown that humans behave in a way that is consistent with these predictions (i.e. "optimally") in many situations (Fetsch et al. 2013). For example, Ernst and Banks (2002; also Helbig and Ernst 2007) demonstrated statistically optimal integration of visual and haptic shape cues. Similar has also been shown for audiovisual localisation (Alais and Burr 2004) and rate discrimination (Raposo et al. 2012). However, it is becoming increasingly clear that this is not universal: people sometimes respond in ways that are *not* consistent with the predictions of Bayesian forced fusion models. For example, in a paradigm similar to that used by Alais and Burr, Battaglia et al. (2003) found that participants overweighted the visual information. This was later replicated by Meijer et al. (2019), and overweighting of vestibular (Butler et al. 2010) and auditory (Burr et al. 2009) cues has also been demonstrated.

4.2.3 Forced-Fusion Integration in Other Populations

A large amount of past research has compared multisensory integration behaviour between various populations, addressing how responses to multisensory stimuli change across the lifespan (Burr and Gori 2012; Jones and Noppeney 2021) and whether/how they are impacted by various pathologies (Ding et al. 2017; Feldman et al. 2018; Tseng et al. 2015). Studies often do this by comparing different groups in terms of, for example, reaction times to forced-fusion stimuli, or the strength and frequency of illusions created by conflicting cues. Though these approaches provide useful information about differences in the *outcomes* of multisensory integration, they are often unable to identify whether the integration process itself has changed. As outlined above, the outcome of forced-fusion integration is determined by both the computational process itself and by the reliabilities of the incoming unisensory signals. It is therefore possible that any observed group differences in behavioural responses to identical multisensory stimuli result from changes to unisensory reliabilities, or even response strategies, rather than from changes in the integration process itself. By applying Bayesian models, we can determine whether and how changes in computational parameters or principles can lead to group differences in responses to multisensory stimuli.

In this section, we briefly discuss research that has directly investigated population differences in forced-fusion integration by taking the approach described in the previous section on younger adults: estimating unisensory reliability from participants' responses to unisensory stimuli alone, using these reliabilities as known (i.e. fixed) parameters in a forced fusion model to predict participants' responses to multisensory stimuli, and comparing those predictions to actual behaviour.

4.2.3.1 Children

Current evidence suggests that the ability to fuse and benefit from congruent, redundant multisensory cues is not present at birth, but develops throughout childhood (Burr and Gori 2012). Nardini et al. (2008), for example, found that 4 to 8-year-olds did not integrate cues in a navigation task that relied on visual and self-motion signals, instead alternating between cues on a trial-by-trial basis. Gori et al. (2008) similarly showed that children younger than 8-year-olds were extremely sub-optimal for cue weighting in visual haptic size discrimination tasks, with one of the cues dominating entirely, while those aged 10 or older performed similarly to adults (i.e. in line with the predictions of Bayesian cue integration). Children younger than 12 have also been found to overweight the visual information in audiovisual spatial tasks (Gori et al. 2012).

These differences are not confined to integration of cues from different sensory modalities. Nardini et al. (2010) assessed integration of multiple visual cues to depth. The authors found that, while 6-year-olds responded accurately to each cue individually, they did not appear to fuse the cues to improve the precision of their responses; this only occurred in those aged 12 years and older. In agreement with these behavioural findings, evidence of cue integration in visual cortex is only apparent in children older than around 10 years old (Dekker et al. 2015).

This late development of Bayes-optimal cue integration may be explained by the fact that correspondences between senses evolve constantly throughout childhood (Burr and Gori 2012). Relevant physical properties such as interocular distance, limb length, and head size are continually changing, and the development of individual senses does not occur at the same rate. It has therefore been suggested by Burr and Gori (2012) that the brain retains flexibility—at the expense of some precision—by not fusing cues until the development of the individual senses has slowed and the correspondences between them stabilised. In later sections, we discuss how this decreased tendency to integrate may be accounted for within the parameters of a more complex Bayesian causal inference model.

4.2.3.2 Older Adults

Older adults are known to respond to multisensory stimuli differently from younger age groups in a variety of ways (see Jones and Noppeney 2021, for a review). They differ in how they perceive multisensory illusions (e.g. Bedard and Barnett-Cowan 2015; Chancel et al. 2018; DeLoss et al. 2013) and have previously been found to receive more multisensory benefit to reaction times (e.g. Laurienti et al. 2006; Peiffer et al. 2007). Despite this, research that explicitly assesses reliability weighting of multisensory cues in older adults is limited and has produced mixed results. In some cases, ageing has been found to have little impact. Braem et al. (2014), for example, found no differences between younger and older adults' cue integration performance on a visual-haptic verticality judgement task; nor did Couth et al. (2019) for visual-haptic size judgement. However, in an audiovisual rate perception paradigm, Brooks et al. (2015) found that only the younger age group received an accuracy benefit from the integration of redundant cues, despite both groups weighting them appropriately. Bates and Wolbers (2014) showed the opposite in a task that required the integration of visual and self-motion cues: while both younger and older adults' accuracies benefitted from cue integration, the latter group consistently overweighted visual information. Finally, one study of visual-haptic verticality judgement by Billino and Drewing (2018) found that older adults weighted the signals in a way that was *more* consistent with the predictions of a forced-fusion model than the younger group, who underweighted the visual signal.

4.2.3.3 Atypical Populations

Changes in multisensory perception have been reported for several neurological and neuropsychiatric disorders, but the mechanisms underlying these differences are currently unclear. As noted above, perceptual differences could arise in the absence of any changes to the computational principles (e.g. reliability weighting) as a result of, for example, reduced unisensory precision, or even due to impairments in related functions such as selective attention and response selection. Alternatively, it may be that some disorders do directly affect the computational principles governing multisensory integration. For instance, participants with neuropsychiatric disorders may resort to approximate algorithms or simple heuristics. Here we focus on autism spectrum disorder (ASD), schizophrenia (SZ), and Parkinson's disease (PD), as diverse examples that have been shown to cause some changes to multisensory perception, and consider research that has attempted to distinguish between these possibilities.

Perceptual changes are a common symptom of ASD (American Psychiatric Association

2013), and a variety of autism-related differences in multisensory perception have been reported. Collignon et al. (2013) found that, unlike non-autistic controls, autistic participants did not benefit from additional auditory cues in a visual search task. Autistic children also appear to be less able to use visual information to improve their comprehension of auditory speech stimuli (Foxe et al. 2015; Iarocci et al. 2010; Stevenson et al. 2014; Woynaroski et al. 2013), and receive less response time benefit for multisensory versus unisensory stimuli (Brandwein et al. 2013). Several differences in perception of multisensory illusions have also been documented: autistic children were found to be less susceptible to the McGurk illusion (Bebko et al. 2014; Stevenson et al. 2014), but experienced the rubber-hand (Greenfield et al. 2015) and sound-induced flash (Foss-Feig et al. 2010) illusions over a wider range of stimulus asynchronies. See Feldman et al. (2018) for a comprehensive review.

There is growing evidence that these differences in multisensory perception may not be caused by changes to sensory cue weighting. Bedford et al. (2011) compared autistic and non-autistic teenagers in their weighting of visual cues to depth and report that both performed in a way that was consistent with Bayesian forced-fusion integration. Zaidel et al. (2015) and Miller and McIntosh (2013) found similar for visual-vestibular integration. In the later section on Bayesian causal inference in atypical populations, we therefore discuss alternative explanations for multisensory processing changes in autistic individuals, particularly in terms of the role of perceptual and causal priors that define their tendency to integrate signals across the senses.

Changes in multisensory processing have also been reported in schizophrenia patients, particularly for speech cues. For example, Pearl et al. (2009) found that adolescents with schizophrenia were less susceptible to the McGurk illusion, and de Gelder et al. (2003) showed that SZ patients were impaired in an audiovisual lipreading task (despite performing similarly to controls at audiovisual localisation). Integration of multisensory emotional stimuli also appears to be changed: de Gelder et al. (2005) and de Jong et al. (2009) found that cross-sensory influence of vocal effect on emotional face categorisation (and vice-versa) was diminished in SZ patients.

The effects of schizophrenia on integration of lower level multisensory cues appear more mixed. As noted above, de Gelder et al. (2003) found no differences between SZ patients and healthy controls in an audiovisual localisation (ventriloquist) task, but schizophrenia *has* been shown to reduce multisensory reaction time benefits in classical redundant target paradigms (e.g. Williams et al. 2010). Schizophrenia patients also appear to integrate temporally incongruent audiovisual stimuli over a wider range of SOAs, as tested by sound-induced flash illusion (Haβ et al. 2017) and simultaneity judgement (Foucher et al. 2007) paradigms. See Tseng et al. (2015) for a systematic review of multisensory integration effects in SZ.

Importantly, however, there has (to our knowledge) been no research to date that has applied forced-fusion modelling to systematically test sensory cue weighting in schizophrenia.

Parkinson's disease is a neurodegenerative disease predominantly affecting the basal ganglia, which have been shown to be involved in low-level integration of multisensory signals (Nagy et al. 2006; Reig and Silberberg 2014). Ren et al. (2018) found, using a race-model analysis of reaction times, that PD patients do not benefit from multisensory stimuli in the same way as healthy controls. Parkinson's patients also seem to experience the rubber hand illusion more strongly, under a wider range of conditions (Ding et al. 2017), and to over-rely on visual information for postural and motor control (Azulay et al. 2002; Bronstein et al. 1990; Cooke et al. 1978; Halperin et al. 2021). A recent study by Yakubovich et al. (2020) found that Parkinson's patients overweighted visual cues significantly more than controls in a visual-vestibular navigation task, despite reductions in the reliability of these visual cues that correlated with the severity of disease, providing early evidence that PD may directly impact the computational processes underlying multisensory processing.

4.2.4 Neural Mechanisms of Reliability-Weighted Integration

Integration of multisensory signals has been observed across the brain. In their seminal early work, Stein and colleagues recorded from single neurons in cat superior colliculus, finding enhanced responses to audiovisual stimuli (Meredith and Stein 1983, 1985). Since then, human neuroimaging research and neurophysiological recordings in non-human primates, rodents, and other animals has shown that multisensory integration occurs pervasively throughout the cortex (Dahl et al. 2009, 2010; Foxe et al. 2002; Gau et al. 2020; Kayser et al. 2008; Lee and Noppeney 2014; Lehmann et al. 2006; Martuzzi et al. 2007; Noppeney et al. 2010; Rohe and Noppeney 2015b; Ghazanfar and Schroeder 2006; Werner and Noppeney 2010, 2011). The question of how and where reliability-weighted integration arises within this hierarchy is yet to be fully resolved; growing evidence points towards parietal and temporal association cortices, though the specific regions involved are likely to vary depending on the type of information being integrated. Neurophysiological research in non-human primates has found neurones in medial superior temporal area (MST) that respond in a way that closely approximates Bayesian integration of visuovestibular heading information (Fetsch et al. 2012; Gu et al. 2008, 2012). In humans, Helbig et al. (2012) used functional magnetic resonance imaging (fMRI) to assess neural responses during a visual-haptic shape discrimination task. The amplitude of blood-oxygen-level-dependent (BOLD) responses in areas of parietal and occipitotemporal cortices was seen to modulate in line with the weights given to the unisensory signals (derived based on behavioural responses). More recently, Rohe and Noppeney (2018) investigated the weighting of auditory and visual signals during a spatial localisation task, finding BOLD responses in intraparietal sulcus that were consistent with reliability-weighted integration.

4.3 Processing Signals from Multiple Sources: Causal Inference

In the first half of this chapter, we discussed how an optimal observer should combine sensory cues that were unambiguously generated by the same object or event and compared this with the behaviour of various human populations. In the real world, an observer does not know which signals belong together and needs to infer this from noisy sensory information: only those signals that originate from a common source should be integrated, while those from different sources must be segregated. Forced-fusion models do not provide insights into this causal inference process, as they assume that all signals are generated by one source. In this second part, we introduce the Bayesian causal inference (BCI) model, which moves beyond these approaches and accounts for the possibility that some of the incoming signals have separate causes. It does so by explicitly modelling each of the potential causal structures that could have generated the various incoming sensory signals (Körding et al. 2007; Sato et al. 2007; Shams and Beierholm 2010).

Prior to the use of BCI, research had already begun to investigate the conditions under which an observer integrates signals, finding those that are more (e.g. spatially, temporally, or semantically) congruent are more likely to be fused into a single percept or otherwise influence each other (Shams et al. 2000; Slutsky and Recanzone 2001; Thurlow and Jack 1973). BCI moves beyond these descriptive approaches to provide a principled explanation for the complex, interacting influences on observers' responses to multisensory stimuli. It models how, with increased sensory uncertainty, observers are less precise at arbitrating between integrating and segregating signals. The model also incorporates a causal prior, which quantifies the prior expectation of multiple inputs sharing a source. All else being equal, a participant with a stronger causal prior is likely to integrate stimuli under a wider range of conditions, and the presence of this parameter allows us to assess the relative contributions of past experience and incoming signal properties to the causal inference process.

4.3.1 Bayesian Modelling of Multisensory Causal Inference

If we again focus on audiovisual localisation as an example, the BCI generative model specifies that one auditory and one visual signal could have been produced either by a single object or event ($C = 1$) or by two separate objects/events ($C = 2$). These possible causal structures are sampled from a binomial distribution determined by a causal or common-source prior, P_{common}, which describes the prior probability of signals sharing a common cause. If $C = 1$ (one cause) is drawn, a single source S_{AV} is sampled from a normal distribution that quantifies the prior probability of signals originating from different areas of space. If $C = 2$ (two causes) is drawn, two separate sources S_A and S_V are sampled from the same spatial prior distribution. These sources then generate auditory and visual signals, x_A and x_V, that are independently corrupted by some amount of Gaussian noise.

This generative model is said to be inverted by the observer during perceptual inference to obtain the posterior probability over the possible causal structures and spatial locations of sensory sources. The probability of the received signals sharing a common cause is estimated by applying Bayes' rule to combine the available sensory information with the causal prior.

$$P(C=1|x_A,x_V) = \frac{P(x_A,x_V|C=1) * P_{\text{common}}}{P(x_A,x_V)}$$

In general, the possibility that the signals share a source is greater when they are closer together in space, and when the prior probability of signals sharing a source is higher.

If the observer is required to indicate whether the signals shared a source (an *explicit* causal inference judgement), a response may then be calculated based on a decision rule:

$$\hat{C} = \begin{cases} 1 \text{ if } P(C=1|x_A,x_V) > \alpha \\ 2 \text{ if } P(C=1|x_A,x_V) \le \alpha \end{cases}$$

where α is a criterion value between 0 and 1 (often set as 0.5), above which the observer responds that the signals shared a source.

As there is always some level of uncertainty about the true causal structure of the signals, auditory and visual location estimates are calculated for both possible causal structures. For the case that the signals share a common cause, reliability-weighted integration occurs in the same way as it does under forced fusion. For the case that signals have different sources, separate (segregated) spatial estimates are calculated for each modality. These two sets of estimates are then combined according to some decision strategy (Wozny et al. 2010) to provide final estimates of the auditory ($\hat{S}_A$) and visual ($\hat{S}_V$) stimulus locations.

One possible decision strategy, known as model averaging, combines the integrated and segregated estimates by weighting their influence according to estimated posterior probabilities of the possible causal structures (i.e. if it is more likely that the signals had a shared source, the integrated estimate is given more weight, and vice versa).

$$\hat{S}_A = P(C=1|x_A,x_V) * \hat{S}_{AV,C=1} + (1 - P(C=2|x_A,x_V)) * \hat{S}_{A,C=2}$$

$$\hat{S}_V = P(C=1|x_A,x_V) * \hat{S}_{AV,C=1} + (1 - P(C=2|x_A,x_V)) * \hat{S}_{V,C=2}$$

Another possible decision strategy is probability matching. This approach assumes that the observer reports either the integrated or the segregated percepts probabilistically, depending on the posterior probability of a common cause.

$$\hat{S}_A = \begin{cases} \hat{S}_{AV,C=1} \text{ if } P(C=1|x_A,x_V) > \xi \\ \hat{S}_{A,C=2} \text{ if } P(C=1|x_A,x_V) \le \xi \end{cases}$$

$$\hat{S}_V = \begin{cases} \hat{S}_{AV,C=1} \text{ if } P(C=1|x_A,x_V) > \xi \\ \hat{S}_{A,C=2} \text{ if } P(C=1|x_A,x_V) \le \xi \end{cases}$$

The Bayesian causal inference model thus builds upon forced fusion models by modelling each of the possible causal structures that may have produced a set of signals and generates responses by combining these estimates based on their relative probabilities.

4.3.2 Bayesian Causal Inference in Healthy Young Adults

A substantial body of literature shows that young, healthy adults generally perform multisensory integration in a way that is consistent with the principles of Bayesian causal inference (Acerbi et al. 2018; Körding et al. 2007; Rohe and Noppeney 2015a; Shams and Beierholm 2010; Wozny et al. 2010). To evaluate this, participants are usually presented with multisensory signals that have varying degrees of conflict, and their behavioural responses compared to those predicted by a suitable BCI model. The model makes predictions for both *explicit* causal inference tasks, in which observers decide whether two signals come from a common source, as well as for tasks in which observers need to estimate a particular environmental property (such as the location or size of an object). We refer to the latter as *implicit* causal inference because the perceptual estimates are implicitly informed by causal inference. For instance, if an observer infers that an auditory and visual signal share a common source, the perceived location of the auditory stimulus will be informed by the estimate of the visual signal's location.

In practice, such tests often involve the use of multisensory perceptual illusions that arise when two or more conflicting signals are integrated. For example, the ventriloquist illusion occurs when the perceived location of a sound is altered by a simultaneously presented visual stimulus, as with a ventriloquist and their dummy: the movement of the dummy's mouth creates the perception that the voice is emanating from the dummy itself. This illusion exploits the fact that, in humans, visual information is generally far more spatially reliable than auditory information (our eyes are much better than our ears at determining *where* something happened). It can occur even for simple stimuli, such as a simultaneous beep and flash of light. In the laboratory, stimulus properties such as the spatial reliability of the unisensory signals, and the distance between them, may be randomised on every trial. Participants are asked to make explicit ("Were the sound and the flash caused by the same source?") and/or implicit ("Where did the sound/flash come from?") causal inference judgements about each set of stimuli. BCI predicts that multisensory interactions become weaker, and are less likely to occur, as the amount of conflict between signals increases: a beep and flash that appear to originate from different locations are unlikely to share a source and should therefore have little or no influence on each other. Behavioural testing (e.g. Bertelson and Radeau 1981; Körding et al. 2007; Lewald and Guski 2003; Mohl et al. 2020; Rohe and Noppeney 2015a; Wallace et al. 2004) has repeatedly shown that this accurately describes human localisation responses to spatially discrepant audiovisual stimuli: as audiovisual spatial disparity increases, the influence of the visual stimulus on the perceived sound location declines nonlinearly (implicit causal inference) and observers are progressively less likely to perceive the two signals as sharing a common source (explicit causal inference).

The sound-induced flash illusion (SIFI; Shams et al. 2000) instead relies on *temporal* binding of signals. It involves presenting, for example, one brief flash sandwiched between two short beeps. If the onsets of these stimuli are sufficiently close together in time, participants will incorrectly report perceiving two flashes. This occurs because humans' temporal reliability is greater for auditory than visual stimuli (we are more precise at hearing, than at seeing, *when* something happened), so the auditory signals are weighted more heavily in the final percept. The illusion has also been demonstrated for more than two beeps, and for multiple flashes and one beep (resulting in fusion, rather than fission, of the visual stimuli; Andersen et al. 2004). Again, the probability of the effect occurring diminishes as the temporal distance (stimulus onset asynchrony, or SOA) between the stimuli increases, in a way that closely approximates the predictions of BCI (Shams et al. 2005).

Bayesian causal inference modelling has been successfully applied to several other paradigms and stimulus combinations, including visuovestibular heading and verticality judgement (Acerbi et al. 2018; de Winkel et al. 2018), and audiovisual speech perception (Magnotti and Beauchamp 2017; Magnotti et al. 2013).

4.3.3 Bayesian Causal Inference in Other Populations

As with forced-fusion models, Bayesian causal inference allows us to differentiate between several possible causes of inter-individual differences in multisensory integration. With BCI, we are no longer constrained to situations in which multiple signals unambiguously share a source but can also assess the causes of atypical arbitration between sensory integration and segregation. We can, for example, test whether an increase in the tendency to bind conflicting signals is due to greater sensory noise, or to a stronger causal prior. In each case, we can then consider the underlying cognitive and neural mechanisms. Reduced unisensory reliability can, for instance, be caused by changes to both peripheral and central sensory processing (see Jones and Noppeney (2021), for a discussion of how this can apply to older adults). Similarly, a stronger binding tendency could potentially be the result of attentional abnormalities that impair the ability to selectively attend to stimuli from a specific modality (for evidence of attentional effects on multisensory integration see, for example, Alsius et al. 2005, 2007).

Altered behavioural responses to multisensory stimuli may also/instead be due to differing response strategies and cost functions. Using sequential sampling models, we recently showed that older adults place a greater emphasis on accuracy when responding to multisensory stimuli in a speeded context (Jones et al. 2019). They accumulated evidence to a higher threshold before committing to a response, leading to a different speed-accuracy trade off.

Finally, individuals from atypical populations may deviate entirely from normative Bayesian principles, and instead resort to simple heuristics or approximate algorithms.

4.3.3.1 Children

In the earlier section on forced-fusion integration, we noted that children often do not benefit from congruent multisensory signals to the same degree as adults, instead relying primarily on a single sensory modality (Gori et al. 2008, 2012; Nardini et al. 2008). Within the framework of Bayesian causal inference, this decreased tendency to integrate signals would likely manifest as a weaker causal prior, which may be explained by limited multisensory experience during early neurodevelopment. Yet, children have also been reported to be more distracted by an irrelevant visual stimulus when attempting to locate a sound, which may suggest a *stronger* causal prior (Petrini et al. 2015). These considerations suggest that complex interactions between brain maturation and sensory experience must be taken into account when interpreting computational parameters of the BCI model in children. For instance, effective sensory integration and attentional mechanisms may rely on maturation of white matter tracts and frontoparietal cortices effectively. At the same time, the formation of causal priors (and other key sensory priors, such as the light-from-above prior; Stone 2011) may require sufficient exposure to the sensory statistics of the real world.

4.3.3.2 Older Adults

Two studies have recently investigated the influence of ageing on audiovisual integration for spatial localisation. In classical spatial ventriloquist paradigms, Jones et al. (2019) and Park et al. (2021) directly applied the Bayesian causal inference model to older adults' responses to multisensory stimuli. In each case, older adults were found to have lower unisensory reliabilities, but did not differ in their spatial or common-source priors and, crucially, still performed in a way that was consistent with the predictions of Bayesian causal inference.

As discussed earlier, there is also a substantial body of research assessing the effects of ageing on multisensory perception from other (i.e. non-Bayesian) perspectives, and older adults have been found to respond differently to multisensory stimuli under a variety of circumstances. In a recent review that considered this evidence alongside the limited computational research (Jones and Noppeney 2021), we tentatively concluded that ageing can affect unisensory reliability, attentional control, and response strategies, but does not seem to impact the fundamental computational principles (including causal inference) that govern the integration of multisensory stimuli. However, further research that addresses this question directly is clearly required.

4.3.3.3 Atypical Populations

In the earlier section on forced-fusion integration in atypical populations, we discussed evidence that the multisensory processing differences associated with ASD do not appear to be caused by abnormalities in sensory weighting. There is, however, increasing evidence from related areas that autism is associated with changes to perceptual (and other) priors. In an influential paper, Pellicano and Burr (2012) argue that weaker priors may be the primary cause of many perceptual symptoms of autism (see also responses by Friston et al. (2013), Van de Cruys et al. (2013)). The evidence for this view comes from studies demonstrating that autistic individuals perceive sensory stimuli in a way that is consistent with greater weight being placed on current sensory signals (the likelihood) than on past experience (the prior). For example, Skewes and Gebauer (2016) found that autistic participants were less likely to use relevant prior information when localising sounds. Similarly, when Skewes et al. (2015) presented observers with a perceptual learning task, those who scored higher for autistic traits placed more weight on the likelihood than the prior.

In the context of multisensory causal inference, this may mean that observers have a causal prior that favours neither integration nor segregation (i.e. close to 0.5). We might therefore expect autistic individuals to rely more directly on the properties of the incoming signals (conflict size, sensory noise) when arbitrating between integration and segregation. This is somewhat supported by evidence. For example, Stevenson et al. (2014; also Bebko et al. 2014) found that autistic children and adolescents with ASD are less susceptible to the McGurk illusion, which relies on the fusion of conflicting auditory and visual signals (though see Woynaroski et al. 2013, who did not find the same effect). Adolescents with ASD also appear less likely to integrate conflicting visual depth cues (Bedford et al. 2016).

Conversely, autistic individuals appear *more* likely to integrate multisensory stimuli that are spread out over time: Foss-Feig et al. (2010) report that children with ASD experienced the sound-induced flash illusion over a wider range of stimulus onset asynchronies (SOAs) than their non-autistic peers. Autism has, however, been associated with decreased performance on both auditory (Kwakye et al. 2011) and audiovisual (de Boer-Schellekens et al. 2013) temporal-order judgement tasks, suggesting this may be due to impaired temporal precision, rather than a stronger causal prior. This usefully highlights the important point that changes in different parameters can have similar effects: a participant may integrate signals with a small intersensory conflict because they have a stronger causal prior (i.e. greater binding tendency), or instead because sensory noise means they are less able to estimate the signals' true causal structure. (See Brock (2012), for a discussion of this prior/likelihood ratio problem as it relates more generally to sensory perception in autism, and Noel et al. (2020), for an example of similar priors but impaired likelihoods in autistic participants performing a visual navigation task.)

One study that does directly address this question, by applying Bayesian causal inference modelling to a multisensory perception task, also provides a specific example of how changes in different parameters can produce very similar behaviour. Noel et al. (2018) compared the ability of three age-matched groups of adolescents—some with ASD, some with schizophrenia, and a group of healthy controls—to judge the simultaneity of concurrently presented auditory and visual stimuli. The authors found that both the ASD and SZ groups integrated stimuli (i.e. judged them to be simultaneous) over a wider range of onset asynchronies than controls. However, BCI modelling revealed that this was due primarily to changed priors in the autistic group, and to a combination of both changed priors and likelihoods in participants with schizophrenia.

As with ASD, there is growing evidence that differences in priors may be responsible for some of the perceptual symptoms of schizophrenia, though in this case priors appear often to be *stronger* in SZ. A study of audiovisual speech perception showed that individuals with psychosis relied more heavily on cognitive and perceptual priors than healthy or at-risk controls (Haarsma et al. 2020). Similarly, Powers et al. (2017) conditioned participants who suffer from psychosis, and a group of healthy controls, to perceive an illusory sound when presented with a visual checkerboard stimulus. The psychosis group perceived significantly more illusory stimuli, and a hierarchical Gaussian filter model revealed that this was due to this group being more influenced by prior stimuli. This effect of stronger priors in SZ/psychosis may be limited to specific (perhaps higher-level) stimulus types, though: Kaliuzhna et al. (2019) and Valton et al. (2019) found no differences in the role of priors when perceiving low-level visual stimuli.

There is some suggestion that Parkinson's patients also differ in their use of priors for perceptual tasks. Perugini et al. (2016) applied a drift-diffusion model to assess the relative contribution of past experience and sensory information in response to a visual orientation judgement task. It was found that participants from the PD group were less able to use prior information to inform their responses. However, to our knowledge, no study has yet specifically investigated whether causal priors for multisensory integration are affected by Parkinson's disease.

4.3.4 Neural Mechanisms of Bayesian Causal Inference

Given the growing behavioural evidence that human observers integrate sensory signals in a way that is consistent with the principles of Bayesian causal inference, research has recently turned towards characterising the underlying neural mechanisms. Rohe and Noppeney (2015b, 2016) collected whole-brain functional magnetic resonance imaging (fMRI) images while presenting participants with auditory and visual stimuli at various degrees of spatial conflict and reliability (i.e. a ventriloquist paradigm). Multivariate decoding was combined with Bayesian modelling to demonstrate that the brain formed spatial representations according to different computational principles across the auditory and visual processing hierarchies. While primary sensory areas independently encoded the locations of the auditory and visual stimuli, the posterior intraparietal sulcus integrated auditory and visual signals into an integrated (i.e. forced-fusion) spatial representation. Finally, anterior intraparietal sulcus combined these into a final location estimate consistent with the predictions of Bayesian causal inference. Later work (Mihalik and Noppeney 2020) has suggested that this final estimate was informed by causal inference estimates in dorsolateral prefrontal cortex.

An electroencephalography (EEG) study, using a similar ventriloquist behavioural paradigm, aimed to establish how these various representations arise dynamically over time (Aller and Noppeney 2019). The results were highly consistent with the fMRI findings. In the early stages of processing (< 100 ms post stimulus), stimulus representations were most consistent with (segregated) unisensory estimates. These estimates then (100–200 ms) combined into a fully-integrated forced-fusion estimate, before resolving into a final estimate consistent with behavioural responses and the predictions of BCI (350-450 ms). Interestingly, the neural processes underlying Bayesian causal inference in temporal signals have been found to evolve along a similar time course (Cao et al. 2019; Rohe et al. 2019).

Collectively, these studies suggest that the brain integrates sensory signals into representations that take into account the signals' causal structure (i.e. common vs. independent sources) by dynamically encoding multiple perceptual estimates.

4.4 Conclusion

Growing evidence suggests that the human brain combines sensory signals in a way that is consistent with the principles of Bayesian causal inference. When signals come from common sources they are integrated, weighted by their relative reliabilities, into a more precise representation of the world. When they come from different sources, they are processed separately. This arbitration between sensory integration and segregation is determined by observers' causal inference, based on noisy correspondence cues such as spatial disparity or temporal asynchrony. These processes evolve throughout childhood, as children develop physiologically and build their sensory experience, and change again in later life as part of normal, healthy ageing.

Normative Bayesian models move beyond descriptive approaches, allowing us to characterise interindividual differences at the computational level. This brings the possibility of determining whether, for example, atypical populations (such as those with ASD, Parkinson's, or schizophrenia) experience multisensory illusions more (or less) frequently because of changes in sensory noise, causal priors, cost functions, or because they deviate from normative principles. Research to date suggests that individuals from some atypical populations may differ in their sensory noise and use of priors, but there is insufficient evidence to suggest that they deviate from the computational principles of Bayesian causal inference. Future research should thus aim to directly address this question, and to determine the neural and cognitive mechanisms (perhaps involving changes in selective attention and cognitive control) underlying any differences found.

References

Acerbi L, Dokka K, Angelaki DE, Ma WJ (2018) Bayesian comparison of explicit and implicit causal inference strategies in multisensory heading perception. PLoS Comput Biol 14(7):e1006110. https://doi.org/10.1371/journal.pcbi.1006110

Alais D, Burr D (2004) The ventriloquist effect results from near-optimal bimodal integration. Curr Biol 14(3):257–262. https://doi.org/10.1016/j.cub.2004.01.029

Aller M, Noppeney U (2019) To integrate or not to integrate: temporal dynamics of hierarchical Bayesian causal inference. PLoS Biol 17(4):e3000210. https://doi.org/10.1371/journal.pbio.3000210

Alsius A, Navarra J, Campbell R, Soto-Faraco S (2005) Audiovisual integration of speech falters under high attention demands. Curr Biol 15(9):839–843. https://doi.org/10.1016/j.cub.2005.03.046

Alsius A, Navarra J, Soto-Faraco S (2007) Attention to touch weakens audiovisual speech integration. Exp Brain Res 183(3):399–404. https://doi.org/10.1007/s00221-007-1110-1

American Psychiatric Association (2013) Diagnostic and statistical manual of mental disorders, 5th edn. American Psychiatric Association, Arlington, VA. https://doi.org/10.1176/appi.books.9780890425596

Andersen TS, Tiippana K, Sams M (2004) Factors influencing audiovisual fission and fusion illusions. Cogn Brain Res 21(3):301–308. https://doi.org/10.1016/j.cogbrainres.2004.06.004

Azulay JP, Mesure S, Amblard B, Pouget J (2002) Increased visual dependence in Parkinson's disease. Percept Mot Skills 95(3 Pt 2):1106–1114. https://doi.org/10.2466/pms.2002.95.3f.1106

Bates SL, Wolbers T (2014) How cognitive aging affects multisensory integration of navigational cues. Neurobiol Aging 35(12):2761–2769. https://doi.org/10.1016/j.neurobiolaging.2014.04.003

Battaglia PW, Jacobs RA, Aslin RN (2003) Bayesian integration of visual and auditory signals for spatial localization. J Opt Soc Am A 20(7):1391. https://doi.org/10.1364/JOSAA.20.001391

Bebko JM, Schroeder JH, Weiss JA (2014) The McGurk effect in children with autism and Asperger syndrome. Autism Res 7(1):50–59. https://doi.org/10.1002/aur.1343

Bedard G, Barnett-Cowan M (2015) Impaired timing of audiovisual events in the elderly. Exp Brain Res 234(1):331–340. https://doi.org/10.1007/s00221-015-4466-7

Bedford R, Pellicano E, Begus K, Mareschal D, Nardini M (2011) Integration of disparity and texture cues to slant in adolescents with an autism spectrum disorder. J Vis 11(11):440–440. https://doi.org/10.1167/11.11.440

Bedford R, Pellicano E, Mareschal D, Nardini M (2016) Flexible integration of visual cues in adolescents with autism spectrum disorder. Autism Res 9(2):272–281. https://doi.org/10.1002/aur.1509

Bertelson P, Radeau M (1981) Cross-modal bias and perceptual fusion with auditory-visual spatial discordance. Percept Psychophys 29(6):578–584. https://doi.org/10.3758/bf03207374
Billino J, Drewing K (2018) Age effects on Visuo-haptic length discrimination: evidence for optimal integration of senses in senior adults. Multisens Res 31(3–4):273–300. https://doi.org/10.1163/22134808-00002601
Braem B, Honoré J, Rousseaux M, Saj A, Coello Y (2014) Integration of visual and haptic informations in the perception of the vertical in young and old healthy adults and right brain-damaged patients. Neurophysiol Clin/Clin Neurophysiol 44(1):41–48. https://doi.org/10.1016/j.neucli.2013.10.137
Brandwein AB, Foxe JJ, Butler JS, Russo NN, Altschuler TS, Gomes H, Molholm S (2013) The development of multisensory integration in high-functioning autism: high-density electrical mapping and psychophysical measures reveal impairments in the processing of audiovisual inputs. Cereb Cortex 23(6):1329–1341. https://doi.org/10.1093/cercor/bhs109
Bresciani J-P, Dammeier F, Ernst MO (2008) Tri-modal integration of visual, tactile and auditory signals for the perception of sequences of events. Brain Res Bull 75(6):753–760. https://doi.org/10.1016/j.brainresbull.2008.01.009
Brock J (2012) Alternative Bayesian accounts of autistic perception: comment on Pellicano and Burr. Trends Cogn Sci 16(12):573–574. https://doi.org/10.1016/j.tics.2012.10.005
Bronstein AM, Hood JD, Gresty MA, Panagi C (1990) Visual control of balance in cerebellar and parkinsonian syndromes. Brain 113(3):767–779. https://doi.org/10.1093/brain/113.3.767
Brooks CJ, Anderson AJ, Roach NW, McGraw PV, McKendrick AM (2015) Age-related changes in auditory and visual interactions in temporal rate perception. J Vis 15(16):2. https://doi.org/10.1167/15.16.2
Burr D, Gori M (2012) Multisensory integration develops late in humans. In: Murray MM, Wallace MT (eds) The neural bases of multisensory processes. CRC Press/Taylor & Francis, Boca Raton, FL. http://www.ncbi.nlm.nih.gov/books/NBK92864/
Burr D, Banks MS, Morrone MC (2009) Auditory dominance over vision in the perception of interval duration. Exp Brain Res 198(1):49. https://doi.org/10.1007/s00221-009-1933-z
Butler JS, Smith ST, Campos JL, Bülthoff HH (2010) Bayesian integration of visual and vestibular signals for heading. J Vis 10(11):23–23. https://doi.org/10.1167/10.11.23
Cao Y, Summerfield C, Park H, Giordano BL, Kayser C (2019) Causal inference in the multisensory brain. Neuron 102(5):1076–1087. e8. https://doi.org/10.1016/j.neuron.2019.03.043
Chancel M, Landelle C, Blanchard C, Felician O, Guerraz M, Kavounoudias A (2018) Hand movement illusions show changes in sensory reliance and preservation of multisensory integration with age for kinaesthesia. Neuropsychologia 119:45–58. https://doi.org/10.1016/j.neuropsychologia.2018.07.027
Collignon O, Charbonneau G, Peters F, Nassim M, Lassonde M, Lepore F, Mottron L, Bertone A (2013) Reduced multisensory facilitation in persons with autism. Cortex 49(6):1704–1710. https://doi.org/10.1016/j.cortex.2012.06.001
Cooke JD, Brown JD, Brooks VB (1978) Increased dependence on visual information for movement control in patients with Parkinson's disease. Can J Neurol Sci 5(4):413–415. https://doi.org/10.1017/s0317167100024197
Couth S, Poole D, Gowen E, Champion RA, Warren PA, Poliakoff E (2019) The effect of ageing on optimal integration of conflicting and non-conflicting visual–haptic stimuli. Multisens Res 32(8):771–796. https://doi.org/10.1163/22134808-20191409
Dahl CD, Logothetis NK, Kayser C (2009) Spatial organization of multisensory responses in temporal association cortex. J Neurosci Off J Soc Neurosci 29(38):11924–11932. https://doi.org/10.1523/JNEUROSCI.3437-09.2009
Dahl CD, Logothetis NK, Kayser C (2010) Modulation of visual responses in the superior temporal sulcus by audio-visual congruency. Front Integr Neurosci 4:10. https://doi.org/10.3389/fnint.2010.00010
de Boer-Schellekens L, Eussen M, Vroomen J (2013) Diminished sensitivity of audiovisual temporal order in autism spectrum disorder. Front Integr Neurosci 7:8. https://doi.org/10.3389/fnint.2013.00008
de Gelder B, Vroomen J, Annen L, Masthof E, Hodiamont P (2003) Audio-visual integration in schizophrenia. Schizophr Res 59(2–3):211–218. https://doi.org/10.1016/s0920-9964(01)00344-9
de Gelder B, Vroomen J, de Jong SJ, Masthoff ED, Trompenaars FJ, Hodiamont P (2005) Multisensory integration of emotional faces and voices in schizophrenics. Schizophr Res 72(2):195–203. https://doi.org/10.1016/j.schres.2004.02.013
de Jong JJ, Hodiamont PPG, Van den Stock J, de Gelder B (2009) Audiovisual emotion recognition in schizophrenia: reduced integration of facial and vocal affect. Schizophr Res 107(2):286–293. https://doi.org/10.1016/j.schres.2008.10.001
de Winkel KN, Katliar M, Diers D, Bülthoff HH (2018) Causal inference in the perception of verticality. Sci Rep 8(1):5483. https://doi.org/10.1038/s41598-018-23838-w
Dekker TM, Ban H, van der Velde B, Sereno MI, Welchman AE, Nardini M (2015) Late development of Cue integration is linked to sensory fusion in cortex. Curr Biol 25(21):2856–2861. https://doi.org/10.1016/j.cub.2015.09.043
DeLoss DJ, Pierce RS, Andersen GJ (2013) Multisensory integration, aging, and the sound-induced flash illusion. Psychol Aging 28(3):802–812. https://doi.org/10.1037/a0033289
Diederich A, Colonius H (2004) Bimodal and trimodal multisensory enhancement: effects of stimulus onset and intensity on reaction time. Percept Psychophys

66(8):1388–1404. https://doi.org/10.3758/BF03195006
Ding C, Palmer CJ, Hohwy J, Youssef GJ, Paton B, Tsuchiya N, Stout JC, Thyagarajan D (2017) Parkinson's disease alters multisensory perception: insights from the rubber hand illusion. Neuropsychologia 97:38–45. https://doi.org/10.1016/j.neuropsychologia.2017.01.031
Ernst MO, Banks MS (2002) Humans integrate visual and haptic information in a statistically optimal fashion. Nature 415(6870):429–433. https://doi.org/10.1038/415429a
Ernst MO, Bülthoff HH (2004) Merging the senses into a robust percept. Trends Cogn Sci 8(4):162–169. https://doi.org/10.1016/j.tics.2004.02.002
Feldman JI, Dunham K, Cassidy M, Wallace MT, Liu Y, Woynaroski TG (2018) Audiovisual multisensory integration in individuals with autism spectrum disorder: a systematic review and meta-analysis. Neurosci Biobehav Rev 95:220–234. https://doi.org/10.1016/j.neubiorev.2018.09.020
Fetsch CR, Pouget A, DeAngelis GC, Angelaki DE (2012) Neural correlates of reliability-based cue weighting during multisensory integration. Nat Neurosci 15(1):146–154. https://doi.org/10.1038/nn.2983
Fetsch CR, DeAngelis GC, Angelaki DE (2013) Bridging the gap between theories of sensory cue integration and the physiology of multisensory neurons. Nat Rev Neurosci 14(6):429–442. https://doi.org/10.1038/nrn3503
Foss-Feig JH, Kwakye LD, Cascio CJ, Burnette CP, Kadivar H, Stone WL, Wallace MT (2010) An extended multisensory temporal binding window in autism spectrum disorders. Exp Brain Res 203(2):381–389. https://doi.org/10.1007/s00221-010-2240-4
Foucher JR, Lacambre M, Pham B-T, Giersch A, Elliott MA (2007) Low time resolution in schizophrenia: lengthened windows of simultaneity for visual, auditory and bimodal stimuli. Schizophr Res 97(1):118–127. https://doi.org/10.1016/j.schres.2007.08.013
Foxe JJ, Wylie GR, Martinez A, Schroeder CE, Javitt DC, Guilfoyle D, Ritter W, Murray MM (2002) Auditory-somatosensory multisensory processing in auditory association cortex: an fMRI study. J Neurophysiol 88(1):540–543. https://doi.org/10.1152/jn.2002.88.1.540
Foxe JJ, Molholm S, Del Bene VA, Frey H-P, Russo NN, Blanco D, Saint-Amour D, Ross LA (2015) Severe multisensory speech integration deficits in high-functioning school-aged children with autism spectrum disorder (ASD) and their resolution during early adolescence. Cereb Cortex 25(2):298–312. https://doi.org/10.1093/cercor/bht213
Friston KJ, Lawson R, Frith CD (2013) On hyperpriors and hypopriors: comment on Pellicano and Burr. Trends Cogn Sci 17(1):1. https://doi.org/10.1016/j.tics.2012.11.003
Gau R, Bazin P-L, Trampel R, Turner R, Noppeney U (2020) Resolving multisensory and attentional influences across cortical depth in sensory cortices. elife 9:e46856. https://doi.org/10.7554/eLife.46856
Ghazanfar A, Schroeder C (2006) Is neocortex essentially multisensory? Trends Cogn Sci 10(6):278–285. https://doi.org/10.1016/j.tics.2006.04.008
Gori M, Del Viva M, Sandini G, Burr DC (2008) Young children do not integrate visual and haptic form information. Curr Biol 18(9):694–698. https://doi.org/10.1016/j.cub.2008.04.036
Gori M, Sandini G, Burr D (2012) Development of visuo-auditory integration in space and time. Front Integr Neurosci 6. https://doi.org/10.3389/fnint.2012.00077
Greenfield K, Ropar D, Smith AD, Carey M, Newport R (2015) Visuo-tactile integration in autism: atypical temporal binding may underlie greater reliance on proprioceptive information. Mol Autism 6(1):51. https://doi.org/10.1186/s13229-015-0045-9
Gu Y, Angelaki DE, DeAngelis GC (2008) Neural correlates of multisensory cue integration in macaque MSTd. Nat Neurosci 11(10):1201–1210. https://doi.org/10.1038/nn.2191
Gu Y, DeAngelis GC, Angelaki DE (2012) Causal links between dorsal medial superior temporal area neurons and multisensory heading perception. J Neurosci 32(7):2299–2313. https://doi.org/10.1523/JNEUROSCI.5154-11.2012
Haarsma J, Knolle F, Griffin JD, Taverne H, Mada M, Goodyer IM, Fletcher PC, Murray GK (2020) Influence of prior beliefs on perception in early psychosis: effects of illness stage and hierarchical level of belief. J Abnorm Psychol 129(6):581–598. https://doi.org/10.1037/abn0000494
Halperin O, Israeli-Korn S, Yakubovich S, Hassin-Baer S, Zaidel A (2021) Self-motion perception in Parkinson's disease. Eur J Neurosci 53(7):2376–2387. https://doi.org/10.1111/ejn.14716
Haß K, Sinke C, Reese T, Roy M, Wiswede D, Dillo W, Oranje B, Szycik GR (2017) Enlarged temporal integration window in schizophrenia indicated by the double-flash illusion. Cogn Neuropsychiatry 22(2):145–158. https://doi.org/10.1080/13546805.2017.1287693
Helbig HB, Ernst MO (2007) Optimal integration of shape information from vision and touch. Exp Brain Res 179(4):595–606. https://doi.org/10.1007/s00221-006-0814-y
Helbig HB, Ernst MO, Ricciardi E, Pietrini P, Thielscher A, Mayer KM, Schultz J, Noppeney U (2012) The neural mechanisms of reliability weighted integration of shape information from vision and touch. NeuroImage 60(2):1063–1072. https://doi.org/10.1016/j.neuroimage.2011.09.072
Hillis JM, Ernst MO, Banks MS, Landy MS (2002) Combining sensory information: mandatory fusion within, but not between. Senses Sci 298(5598):1627–1630. https://doi.org/10.1126/science.1075396
Huys QJM, Browning M, Paulus MP, Frank MJ (2021) Advances in the computational understanding of mental illness. Neuropsychopharmacology 46(1):3–19. https://doi.org/10.1038/s41386-020-0746-4

Iarocci G, Rombough A, Yager J, Weeks DJ, Chua R (2010) Visual influences on speech perception in children with autism. Autism 14(4):305–320. https://doi.org/10.1177/1362361309353615

Jacobs RA (1999) Optimal integration of texture and motion cues to depth. Vis Res 39(21):3621–3629. https://doi.org/10.1016/S0042-6989(99)00088-7

Jones SA, Noppeney U (2021) Ageing and multisensory integration: a review of the evidence, and a computational perspective. Cortex 138:1–23. https://doi.org/10.1016/j.cortex.2021.02.001

Jones SA, Beierholm U, Meijer D, Noppeney U (2019) Older adults sacrifice response speed to preserve multisensory integration performance. Neurobiol Aging 84:148–157. https://doi.org/10.1016/j.neurobiolaging.2019.08.017

Kaliuzhna M, Stein T, Rusch T, Sekutowicz M, Sterzer P, Seymour KJ (2019) No evidence for abnormal priors in early vision in schizophrenia. Schizophr Res 210:245–254. https://doi.org/10.1016/j.schres.2018.12.027

Kayser C, Petkov CI, Logothetis NK (2008) Visual modulation of neurons in auditory cortex. Cerebral Cortex (New York, N.Y.: 1991) 18(7):1560–1574. https://doi.org/10.1093/cercor/bhm187

Körding KP, Beierholm U, Ma WJ, Quartz S, Tenenbaum JB, Shams L (2007) Causal inference in multisensory perception. PLoS One 2(9):e943. https://doi.org/10.1371/journal.pone.0000943

Kwakye LD, Foss-Feig JH, Cascio CJ, Stone WL, Wallace MT (2011) Altered auditory and multisensory temporal processing in autism spectrum disorders. Front Integr Neurosci 4. https://doi.org/10.3389/fnint.2010.00129

Laurienti PJ, Burdette JH, Maldjian JA, Wallace MT (2006) Enhanced multisensory integration in older adults. Neurobiol Aging 27(8):1155–1163. https://doi.org/10.1016/j.neurobiolaging.2005.05.024

Lee H, Noppeney U (2014) Temporal prediction errors in visual and auditory cortices. Curr Biol 24(8):R309–R310. https://doi.org/10.1016/j.cub.2014.02.007

Lehmann C, Herdener M, Esposito F, Hubl D, di Salle F, Scheffler K, Bach DR, Federspiel A, Kretz R, Dierks T, Seifritz E (2006) Differential patterns of multisensory interactions in core and belt areas of human auditory cortex. NeuroImage 31(1):294–300. https://doi.org/10.1016/j.neuroimage.2005.12.038

Lewald J, Guski R (2003) Cross-modal perceptual integration of spatially and temporally disparate auditory and visual stimuli. Cogn Brain Res 16(3):468–478. https://doi.org/10.1016/S0926-6410(03)00074-0

Magnotti JF, Beauchamp MS (2017) A causal inference model explains perception of the McGurk effect and other incongruent audiovisual speech. PLoS Comput Biol 13(2):e1005229. https://doi.org/10.1371/journal.pcbi.1005229

Magnotti JF, Ma WJ, Beauchamp MS (2013) Causal inference of asynchronous audiovisual speech. Front Psychol 4. https://doi.org/10.3389/fpsyg.2013.00798

Martuzzi R, Murray MM, Michel CM, Thiran J-P, Maeder PP, Clarke S, Meuli RA (2007) Multisensory interactions within human primary cortices revealed by BOLD dynamics. Cereb Cortex 17(7):1672–1679. https://doi.org/10.1093/cercor/bhl077

Meijer D, Noppeney U (2020) Computational models of multisensory integration. In: Multisensory perception: from laboratory to clinic. Academic Press, pp 113–133

Meijer D, Veselič S, Calafiore C, Noppeney U (2019) Integration of audiovisual spatial signals is not consistent with maximum likelihood estimation. Cortex 119:74–88. https://doi.org/10.1016/j.cortex.2019.03.026

Meredith M, Stein B (1983) Interactions among converging sensory inputs in the superior colliculus. Science 221(4608):389–391. https://doi.org/10.1126/science.6867718

Meredith MA, Stein BE (1985) Spatial factors determine the activity of multisensory neurons in cat superior colliculus. Brain Res 365:5

Mihalik A, Noppeney U (2020) Causal inference in audiovisual perception. J Neurosci 40(34):6600–6612. https://doi.org/10.1523/JNEUROSCI.0051-20.2020

Miller L, McIntosh RD (2013) Visual and proprioceptive Cue weighting in children with developmental coordination disorder, autism Spectrum disorder and typical development. I-Perception 4(7):486. https://doi.org/10.1068/ig10

Mohl JT, Pearson JM, Groh JM (2020) Monkeys and humans implement causal inference to simultaneously localize auditory and visual stimuli. J Neurophysiol 124(3):715–727. https://doi.org/10.1152/jn.00046.2020

Nagy A, Eördegh G, Paróczy Z, Márkus Z, Benedek G (2006) Multisensory integration in the basal ganglia. Eur J Neurosci 24(3):917–924. https://doi.org/10.1111/j.1460-9568.2006.04942.x

Nardini M, Jones P, Bedford R, Braddick O (2008) Development of Cue integration in human navigation. Curr Biol 18(9):689–693. https://doi.org/10.1016/j.cub.2008.04.021

Nardini M, Bedford R, Mareschal D (2010) Fusion of visual cues is not mandatory in children. Proc Natl Acad Sci 107(39):17041–17046. https://doi.org/10.1073/pnas.1001699107

Noel J-P, Stevenson RA, Wallace MT (2018) Atypical audiovisual temporal function in autism and schizophrenia: similar phenotype, different cause. Eur J Neurosci 47(10):1230–1241. https://doi.org/10.1111/ejn.13911

Noel J-P, Lakshminarasimhan KJ, Park H, Angelaki DE (2020) Increased variability but intact integration during visual navigation in autism Spectrum disorder. Proc Natl Acad Sci 117(20):11158–11166. https://doi.org/10.1073/pnas.2000216117

Noppeney U (2020) Multisensory perception: behavior, computations and neural mechanisms. In: Poeppel D, Mangun GR, Gazzaniga MS (eds) The cognitive neurosciences, 6th edn. The MIT Press

Noppeney U (2021) Perceptual inference, learning, and attention in a multisensory world. Annu Rev

Neurosci 44:449–473. https://doi.org/10.1146/annurev-neuro-100120-085519
Noppeney U, Ostwald D, Werner S (2010) Perceptual decisions formed by accumulation of audiovisual evidence in prefrontal cortex. J Neurosci 30(21):7434–7446. https://doi.org/10.1523/JNEUROSCI.0455-10.2010
Parise C, Spence C (2013) Audiovisual cross-modal correspondences in the general population. In: The Oxford handbook of synesthesia. Oxford University Press, pp 790–815. https://doi.org/10.1093/oxfordhb/9780199603329.001.0001
Park H, Nannt J, Kayser C (2021) Sensory- and memory-related drivers for altered ventriloquism effects and aftereffects in older adults. Cortex 135:298–310. https://doi.org/10.1016/j.cortex.2020.12.001
Pearl D, Yodashkin-Porat D, Katz N, Valevski A, Aizenberg D, Sigler M, Weizman A, Kikinzon L (2009) Differences in audiovisual integration, as measured by McGurk phenomenon, among adult and adolescent patients with schizophrenia and age-matched healthy control groups. Compr Psychiatry 50(2):186–192. https://doi.org/10.1016/j.comppsych.2008.06.004
Peiffer AM, Mozolic JL, Hugenschmidt CE, Laurienti PJ (2007) Age-related multisensory enhancement in a simple audiovisual detection task. Neuroreport 18(10):1077–1081
Pellicano E, Burr D (2012) When the world becomes 'too real': a Bayesian explanation of autistic perception. Trends Cogn Sci 16(10):504–510. https://doi.org/10.1016/j.tics.2012.08.009
Perugini A, Ditterich J, Basso MA (2016) Patients with Parkinson's disease show impaired use of priors in conditions of sensory uncertainty. Curr Biol 26(14):1902–1910. https://doi.org/10.1016/j.cub.2016.05.039
Petrini K, Jones PR, Smith L, Nardini M (2015) Hearing where the eyes see: children use an irrelevant visual Cue when localizing sounds. Child Dev 86(5):1449–1457. https://doi.org/10.1111/cdev.12397
Powers AR, Mathys C, Corlett PR (2017) Pavlovian conditioning–induced hallucinations result from overweighting of perceptual priors. Science 357(6351):596–600. https://doi.org/10.1126/science.aan3458
Raposo D, Sheppard JP, Schrater PR, Churchland AK (2012) Multisensory decision-making in rats and humans. J Neurosci 32(11):3726–3735. https://doi.org/10.1523/JNEUROSCI.4998-11.2012
Reig R, Silberberg G (2014) Multisensory integration in the mouse striatum. Neuron 83(5):1200–1212. https://doi.org/10.1016/j.neuron.2014.07.033
Ren Y, Suzuki K, Yang W, Ren Y, Wu F, Yang J, Takahashi S, Ejima Y, Wu J, Hirata K (2018) Absent audiovisual integration elicited by peripheral stimuli in Parkinson's disease. Parkinson's Dis 2018:e1648017. https://doi.org/10.1155/2018/1648017
Rohe T, Noppeney U (2015a) Sensory reliability shapes perceptual inference via two mechanisms. J Vis 15(5):22
Rohe T, Noppeney U (2015b) Cortical hierarchies perform Bayesian causal inference in multisensory perception. PLoS Biol 13(2):e1002073. https://doi.org/10.1371/journal.pbio.1002073
Rohe T, Noppeney U (2016) Distinct computational principles govern multisensory integration in primary sensory and association cortices. Curr Biol 26(4):509–514. https://doi.org/10.1016/j.cub.2015.12.056
Rohe T, Noppeney U (2018) Reliability-weighted integration of audiovisual signals can be modulated by top-down control. ENeuro 5(1). https://doi.org/10.1523/ENEURO.0315-17.2018
Rohe T, Ehlis A-C, Noppeney U (2019) The neural dynamics of hierarchical Bayesian causal inference in multisensory perception. Nat Commun 10(1):1907. https://doi.org/10.1038/s41467-019-09664-2
Sato Y, Toyoizumi T, Aihara K (2007) Bayesian inference explains perception of Unity and Ventriloquism aftereffect: identification of common sources of audiovisual stimuli. Neural Comput 19(12):3335–3355. https://doi.org/10.1162/neco.2007.19.12.3335
Shams L, Beierholm UR (2010) Causal inference in perception. Trends Cogn Sci 14(9):425–432. https://doi.org/10.1016/j.tics.2010.07.001
Shams L, Kamitani Y, Shimojo S (2000) Illusions: what you see is what you hear. Nature 408(6814):788. https://doi.org/10.1038/35048669
Shams L, Ma WJ, Beierholm U (2005) Sound-induced flash illusion as an optimal percept. Neuroreport 16(17):1923–1927
Skewes JC, Gebauer L (2016) Brief report: suboptimal auditory localization in autism spectrum disorder: support for the Bayesian account of sensory symptoms. J Autism Dev Disord 46(7):2539–2547. https://doi.org/10.1007/s10803-016-2774-9
Skewes JC, Jegindø E-M, Gebauer L (2015) Perceptual inference and autistic traits. Autism 19(3):301–307. https://doi.org/10.1177/1362361313519872
Slutsky DA, Recanzone GH (2001) Temporal and spatial dependency of the ventriloquism effect. Neuroreport 12(1):7–10
Stein BE, Stanford TR (2008) Multisensory integration: current issues from the perspective of the single neuron. Nat Rev Neurosci 9(4):255–266. https://doi.org/10.1038/nrn2331
Stein BE, Stanford TR, Rowland BA (2014) Development of multisensory integration from the perspective of the individual neuron. Nat Rev Neurosci 15(8):520–535. https://doi.org/10.1038/nrn3742
Stevenson RA, Siemann JK, Schneider BC, Eberly HE, Woynaroski TG, Camarata SM, Wallace MT (2014) Multisensory temporal integration in autism spectrum disorders. J Neurosci 34(3):691–697. https://doi.org/10.1523/JNEUROSCI.3615-13.2014
Stone JV (2011) Footprints sticking out of the sand. Part 2: Children's Bayesian priors for shape and lighting direction. Perception 40(2):175–190. https://doi.org/10.1068/p6776
Thurlow WR, Jack CE (1973) Certain determinants of the "ventriloquism effect". Percept Mot Skills 36(Suppl. 3):1171–1184. https://doi.org/10.2466/pms.1973.36.3c.1171

Tseng H-H, Bossong MG, Modinos G, Chen K-M, McGuire P, Allen P (2015) A systematic review of multisensory cognitive–affective integration in schizophrenia. Neurosci Biobehav Rev 55:444–452. https://doi.org/10.1016/j.neubiorev.2015.04.019

Valton V, Karvelis P, Richards KL, Seitz AR, Lawrie SM, Seriès P (2019) Acquisition of visual priors and induced hallucinations in chronic schizophrenia. Brain 142(8):2523–2537. https://doi.org/10.1093/brain/awz171

Van de Cruys S, de Wit L, Evers K, Boets B, Wagemans J (2013) Weak priors versus overfitting of predictions in autism: reply to Pellicano and Burr (TICS, 2012). I-Perception 4(2):95–97. https://doi.org/10.1068/i0580ic

Vroomen J, de Gelder B (2000) Sound enhances visual perception: cross-modal effects of auditory organization on vision. J Exp Psychol Hum Percept Perform 26(5):1583–1590. https://doi.org/10.1037//0096-1523.26.5.1583

Wallace MT, Roberson GE, Hairston WD, Stein BE, Vaughan JW, Schirillo JA (2004) Unifying multisensory signals across time and space. Exp Brain Res 158(2). https://doi.org/10.1007/s00221-004-1899-9

Werner S, Noppeney U (2010) Distinct functional contributions of primary sensory and association areas to audiovisual integration in object categorization. J Neurosci 30(7):2662–2675. https://doi.org/10.1523/JNEUROSCI.5091-09.2010

Werner S, Noppeney U (2011) The contributions of transient and sustained response codes to audiovisual integration. Cereb Cortex 21(4):920–931. https://doi.org/10.1093/cercor/bhq161

Williams LE, Light GA, Braff DL, Ramachandran VS (2010) Reduced multisensory integration in patients with schizophrenia on a target detection task. Neuropsychologia 48(10):3128–3136. https://doi.org/10.1016/j.neuropsychologia.2010.06.028

Witten IB, Knudsen EI (2005) Why seeing is believing: merging auditory and visual worlds. Neuron 48(3):489–496. https://doi.org/10.1016/j.neuron.2005.10.020

Woynaroski TG, Kwakye LD, Foss-Feig JH, Stevenson RA, Stone WL, Wallace MT (2013) Multisensory speech perception in children with autism spectrum disorders. J Autism Dev Disord 43(12):2891–2902. https://doi.org/10.1007/s10803-013-1836-5

Wozny DR, Beierholm UR, Shams L (2010) Probability matching as a computational strategy used in perception. PLoS Comput Biol 6(8):e1000871. https://doi.org/10.1371/journal.pcbi.1000871

Yakubovich S, Israeli-Korn S, Halperin O, Yahalom G, Hassin-Baer S, Zaidel A (2020) Visual self-motion cues are impaired yet overweighted during visual-vestib ular integration in Parkinson's disease. Brain Commun 2(1). https://doi.org/10.1093/braincomms/fcaa035

Zaidel A, Goin-Kochel RP, Angelaki DE (2015) Self-motion perception in autism is compromised by visual noise but integrated optimally across multiple senses. Proc Natl Acad Sci 112(20):6461–6466. https://doi.org/10.1073/pnas.1506582112

Multisensory Integration in Body Representation

5

Wen Fang, Yuqi Liu, and Liping Wang

Abstract

To be aware of and to move one's body, the brain must maintain a coherent representation of the body. While the body and the brain are connected by dense ascending and descending sensory and motor pathways, representation of the body is not hardwired. This is demonstrated by the well-known rubber hand illusion in which a visible fake hand is erroneously felt as one's own hand when it is stroked in synchrony with the viewer's unseen actual hand. Thus, body representation in the brain is not mere maps of tactile and proprioceptive inputs, but a construct resulting from the interpretation and integration of inputs across sensory modalities.

Keywords

Causal inference · Macaques · Body representation · Self-consciousness

W. Fang (✉) · Y. Liu · L. Wang
Institute of Neuroscience, Key Laboratory of Primate Neurobiology, CAS Center for Excellence in Brain Science and Intelligence Technology, Chinese Academy of Sciences, Shanghai, China
e-mail: wenfang@ion.ac.cn; liuyq@ion.ac.cn; liping.wang@ion.ac.cn

Accurate integration of multisensory information serves essential purposes in daily life. For example, we recognize ourselves in the mirror and distinguish our own shadow from others by matching movements we intend to generate and movements seen on the visual image. Successful motor control, such as grabbing a coffee mug, critically relies on the integration of visual and proprioceptive information on one's arm and hand position, and of visual and tactile information on where the fingers are on the handle. Therefore, accurate integration of multisensory information is crucial for distinguishing between oneself and external world as well as for interacting with the environment. As the brain is constantly flooded with sensory information both from one's own body and the environment, how it properly integrates and segregates sensory information becomes an important question for understanding the underlying mechanism of body representation. In this chapter, we first discuss behavioral work that investigates the constraints and principles underlying multisensory integration regarding the body. We then introduce a Bayesian framework that theorizes multisensory integration as inferring the source of the sensory inputs by an optimal observer. Finally, we review evidence from neuroimaging and neurophysiological studies on the neural correlates and computational principles of body representation.

Y. Gu, A. Zaidel (eds.), *Advances of Multisensory Integration in the Brain*, Advances in Experimental Medicine and Biology 1437, https://doi.org/10.1007/978-981-99-7611-9_5

5.1 Temporal and Spatial Constraints on Multisensory Integration

Multisensory integration is typically studied by creating a mismatch between inputs from different modalities. Early studies on the integration of visual and auditory information have identified spatial and temporal rules of multisensory integration, such that inputs are more likely to be combined if spatially and temporally closer (Meredith 1987; Meredith and Stein 1986). The same rules also apply to multisensory integration regarding the body. Studies have shown that the processing of bodily signals is influenced by external stimuli occurring within a limited space around the body part, also known as the peripersonal space (PPS) (Làdavas et al. 1998; Spence et al. 2004). For example, the perceived location of tactile stimuli is more strongly biased toward a concurrent visual stimulus when the visual stimulus occurs close to the body versus far (Spence et al. 2004). These findings demonstrate the multisensory nature of body representation such that tactile perception is automatically biased by task-irrelevant visual information. Moreover, there is a spatial limit within which sensory inputs are considered bodily-related and integrated together.

Multisensory integration not only underlies the perception of bodily sensory inputs, but also how the brain represents the body itself. Because it is not always possible to separate information from one's own body, bodily illusions are often used. In these paradigms, participants' actual hand is hidden from view while seeing a fake hand. In this way, proprioceptive information and visual information are dissociated. Integration between visual and proprioceptive information is indexed by the illusory embodiment of the viewed hand, i.e., the viewed hand feels like one's own body—referred to as body ownership, and the hand is perceived closer to where the viewer sees it (Botvinick and Cohen 1998; Holmes et al. 2004). By examining the effect of spatial and temporal congruence between the viewed hand and unseen actual hand on the strength of illusion, researchers can identify factors influencing multisensory integration.

One important factor influencing multisensory integration is the spatial location of the viewed fake hand. As the distance between the viewed hand and unseen actual hand increases, hence more difficult to reconcile the discrepancy between visual and proprioceptive information, participants are less likely to feel the viewed hand as their own hand (Medina et al. 2015) or is located where the viewed hand is (Holmes et al. 2004, 2006; Holmes and Spence 2005). By gradually displacing the rubber hand further from participants' unseen actual hand, Lloyd (2007) found decreasing illusion with increasing distance. Importantly, there was an exponential non-linear decrease as the rubber hand was positioned outside of the participants' reachable space, marking the boundary of the spatial range within which an object could be embodied.

In addition to spatial location, a congruent posture of the viewed and unseen actual hand is critical. For example, with a fixed distance between the rubber hand and the unseen actual hand, subjective ownership of the rubber hand decreased with mismatch in the hand angle between the two hands (e.g., fingers pointing forward vs. pointing 30° leftward) despite synchronous visuotactile stimulation (Costantini and Haggard 2007; Ide 2013).

It has been well-established that the temporal synchrony between the viewed hand and the unseen actual hand plays an important role. In the rubber hand illusion, synchronous strokes on the unseen hidden hand and the rubber hand, such that strokes seen on the viewed hand match tactile sensation on the actual hand, elicit strong illusion. Asynchronous strokes, however, abolish the illusion (Botvinick and Cohen 1998; Tsakiris and Haggard 2005). Using a different and more powerful paradigm, researchers can manipulate whether the unseen hand and the viewed hand are performing congruent movements (mirror box illusion; Ramachandran and Rogers-Ramachandran 1996; Medina et al. 2015). Whereas participants experience strong ownership of the viewed hand when the seen movements on the viewed hand is congruent with movements performed by the unseen actual hand, the illusion is much weaker when the movements are out of phase (Holmes and Spence 2005; Liu and Medina 2018; Medina et al. 2015). These findings provide evidence for the importance of temporal synchrony in multisensory integration.

The evidence discussed above suggests a key role of cross-modal congruence in multisensory integration and body representation. However, matching between bottom-up sensory information is not sufficient for the brain to embody an object. It makes intuitive sense that one would not perceive a dog's paw as their own hand regardless of sensory information, implying additional constraints from prior knowledge of our body independent of incoming sensory inputs. The following section discusses the influence of prior information on multisensory integration.

5.2 Prior Knowledge of the Body Influences Multisensory Integration

At the level of visual features, the human body has specific anatomical and structural properties that differ from other objects (e.g., "a hand has five fingers sticking out"). At the semantic level, we use labels and descriptive language to distinguish objects of different categories (e.g., a human body versus a tree). These forms of prior knowledge are learned in life experience and exist independent of online sensory information. Using the rubber hand illusion paradigm, studies found weaker illusory embodiment when individuals view an object (e.g., a wood stick) versus a realistic rubber hand (Holmes and Spence 2005; Tsakiris et al. 2008; Tsakiris and Haggard 2005). The illusion was also weaker for neutral objects whose shape deviates from a standard hand (Haans et al. 2008; Tsakiris et al. 2010). These findings indicate that individuals have prior expectations of what features constitute a hand based on stored visual body representation (Tsakiris 2010). Interestingly, visual features that are more specific to one's own body, such as size and color, do not have a dramatic effect on the embodiment of the viewed hand or body (Austen et al. 2004; Farmer et al. 2012, 2014; Peck et al. 2013; but see Pavani and Zampini 2007). Based on these findings, it was proposed that the stored visual body representation encodes general shape information of body parts instead of self-specific features (Kilteni et al. 2015; Tsakiris 2010).

Another source of prior knowledge comes from the body schema that represents online body position in space as the body moves (Head and Holmes 1911; Schwoebel and Coslett 2005). The movement of the body is limited by biomechanical constraints—resistance caused by joint and muscle structures that determines the difficulty in and the possible range of body movement (Parsons 1987, 1994). As such, biomechanical constraints are also encoded in the body schema. Importantly, biomechanical constraints not only affect physical movements but manifests in mental body representations. For example, when individuals are asked to judge the chirality of a hand image in a selected orientation, the reaction time is longer if the rotation from the individuals' own hand posture to the hand image is more biomechanically constrained, even if the rotation angle is the same (Cooper and Shepard 1975; Parsons 1987, 1994; Zapparoli et al. 2014). These findings provide evidence that the participants performed the task by mentally simulating body movements, in which biomechanical constraints are encoded.

There is evidence that biomechanical constraints between the unseen actual hand and viewed hand influences multisensory integration. With the angular difference between the viewed hand and unseen actual hand fixed, biomechanical constraints can be manipulated such that the rotation from the actual to the viewed hand is less biomechanically constrained in one condition but more constrained in another (Ide 2013; Liu and Medina 2017). Participants reported weaker illusions in the more-biomechanically-constrained condition despite matched angular differences, indicating that the amount of biomechanical constraint is also computed into the overall discrepancy between visual and proprioceptive information. These findings prove that multisensory information is influenced not only by bottom-up sensory information but also prior information stored in the body schema.

Another type of constraint regards anatomical plausibility of body configuration, i.e., whether a body posture is possible to occur. It was found that the rubber hand illusion was abolished when the viewed hand is in anatomically implausible postures (e.g., fingers pointing straight toward one's body along the sagittal axis) despite synchronous visuotactile stimulation (Ehrsson et al. 2004; Ide 2013; Tsakiris and Haggard 2005). These findings can be accounted for by the encoding of biomechanical constraints discussed above, such that the viewed hand is in an infinitely biomechanically constrained position that proprioceptive information cannot be biased toward. Alternatively, the brain may refer to a body structural description, the stored knowledge of the relative position of body parts on the body surface, for example, the arm is attached to the trunk (Buxbaum and Coslett 2010). Anatomically implausible hand postures often imply breaking of joints, hence violating the body structural description, leading to lower degrees of embodiment (Kilteni et al. 2015).

In summary, the brain uses both incoming sensory information and prior knowledge of the body in multisensory information. The interplay of both bottom-up sensory inputs and top-down knowledge is summarized in a neurocognitive model of body ownership (Tsakiris 2010) that conceptualizes body representation as a set of information comparators. What factors determine the relative importance of each source of information, and how do these factors lead to the final percept that a hand belongs to one's own? In the next section, we discuss a Bayesian framework that addresses the computational principles of multisensory integration.

5.3 A Bayesian Framework of Multisensory Integration

Forming a coherent body representation is a process of inferring the state of the body based on incoming sensory information. The Bayesian framework posits that upon receiving sensory inputs from multiple modalities, the brain generates hypotheses on the cause of these sensory inputs. For example, inferring the hand position gives rise to the perceived visual and proprioceptive information. Given the noise in both external inputs and internal sensory systems, each hypothesis is only correct at a certain probability. These probabilities are called *posterior probability* as they are conditional on a particular set of sensory inputs, written as $P(S|x_v, x_p)$ (the probability of state S, for example, hand position, given visual (x_v), and proprioceptive (x_p) information). The goal of an optimal observer is to find the state S with the maximum posterior probability. Following Bayes' rule, posterior probability depends on the product of two components: the likelihood of obtaining a particular set of sensory inputs given state S ($P(x_v, x_a|S)$), and the prior probability of this hypothesis based on prior knowledge ($P(S)$) (see Eq. 5.1).

$$P\left(S|x_v,\, x_p\right) = \frac{P\left(x_v, x_p|S\right)P\left(S\right)}{P\left(x_v, x_p\right)} \quad (5.1)$$

As the denominator $P(x_v, x_p)$ is independent of the state S, Eq. (5.1) is simplified as:

$$P\left(S|x_v, x_p\right) \propto P\left(x_v, x_p|S\right)P\left(S\right) \quad (5.2)$$

Assuming a uniform distribution $P(S)$, denoting equal prior probability that the hand appears anywhere in space, maximizing the posterior probability equals to maximizing the likelihood. Further assume that each sensory modality is corrupted by independent Gaussian noise, Eq. (5.2) is simplified as:

$$P\left(S|x_v, x_p\right) \propto P\left(x_v|S\right)P\left(x_p|S\right) \quad (5.3)$$

Maximum-likelihood estimation of this equation then becomes a weighted sum of visual and proprioceptive input based on their relative reliability, with reliability the inverse of the standard deviation of each input:

$$\hat{S} = \frac{\frac{1}{\sigma_v}}{\frac{1}{\sigma_v} + \frac{1}{\sigma_p}} x_v + \frac{\frac{1}{\sigma_p}}{\frac{1}{\sigma_v} + \frac{1}{\sigma_p}} x_p \quad (5.4)$$

As a result, the final percept is biased toward the more reliable unimodal estimate, and the reliability of the final percept is maximized (Ernst and Bülthoff 2004). This model therefore accounts for the findings that participants perceive their hand toward where the visual hand is (i.e., visual capture), presumably because visual information is typically less noisy than proprioceptive information. As an optimal weighting principle, maximum-likelihood estimation has been supported by multiple studies on various sensory modalities (Alais and Burr 2004; Ernst and Banks 2002; Van Beers et al. 1999; Witten and Knudsen 2005). For example, in one such study, participants were asked to estimate the height of a block based on haptic and visual information (Ernst and Banks 2002). Small discrepancies between visual and haptic inputs were introduced by manipulating the visual image viewed from a pair of goggles. Consistent with the model, participants' estimate was biased toward visual information, with the weight of the visual information decreasing with visual noise.

An important consequence of this model is that unimodal estimates are always integrated to a single final percept, hence the model is referred to as the forced-fusion model (Körding et al. 2007; Shams and Beierholm 2010). Although it well explains human behavior when the discrepancy between unimodal inputs is small and easy to resolve, it does not apply to all real-world situations. Organized representations require not only accurate integration of inputs belonging to the same object, but segregation of inputs that come from different sources. For example, when playing a duet, it is equally important to integrate information on one's own hand and to segregate information from the partner's hand. In experiments on bodily illusions, the viewed hand is not embodied if it is placed too far from the participant's actual hand (Lloyd 2007; Medina et al. 2015), indicating segregation of visual and proprioceptive information. The underlying problem is to infer whether inputs are emitted by the same object, or in other words, have a common cause, a process referred to as "causal inference."

Researchers have proposed a Bayesian causal inference model to account for how the brain infers the causal structure of sensory inputs (Fig. 5.1; Körding et al. 2007; Shams and Beierholm 2010). In this model, the underlying causal structure of whether multisensory inputs come from a common cause is denoted by the posterior probability ($P(C|x_v, x_p)$). Two causal hypotheses are tested: that the inputs are caused by a common source ($C = 1$), which leads to complete integration of inputs from both modalities, or that the inputs are caused by independent sources ($C = 2$), which leads to complete segregation of information from different modalities. In this way, the Bayesian causal inference model has a hierarchical structure in which the brain first infers the causal structure of sensory inputs, and then makes an estimate of the object state (e.g., location) under the causal structure. Following Bayes' rule, the posterior probability of each causal hypothesis depends on the likelihood of receiving the current sensory inputs given this causal hypothesis ($P(x_v, x_a|C)$, and the prior probability of the causal hypothesis ($P(C)$). An important cue that informs the brain about the causal structure of inputs from multiple modalities is the "similarity" between the inputs: inputs that are closer in time, space, or other dimensions tend to have a higher likelihood under the common cause hypothesis, making them more likely to be integrated.

How is the estimate of object location contingent on the inferred causal structure? One strategy is to follow the most likely causal hypothesis (model selection, Wozny et al. 2008). If the probability of the common cause hypothesis is higher than the hypothesis of the independent cause, inputs are fully integrated by maximum-likelihood estimation. Otherwise, the brain estimates each modality independently without combing them. Another strategy is to weight the estimate under each causal hypothesis in proportion to each hypothesis's posterior probability (model averaging, see Eqs. (5.5a) and (5.5b), Wozny et al. 2010; Körding et al. 2007). By considering the causal structure of inputs from multiple modalities, the Bayesian causal inference model can account for the full range of multisensory integration from complete integration to complete segregation.

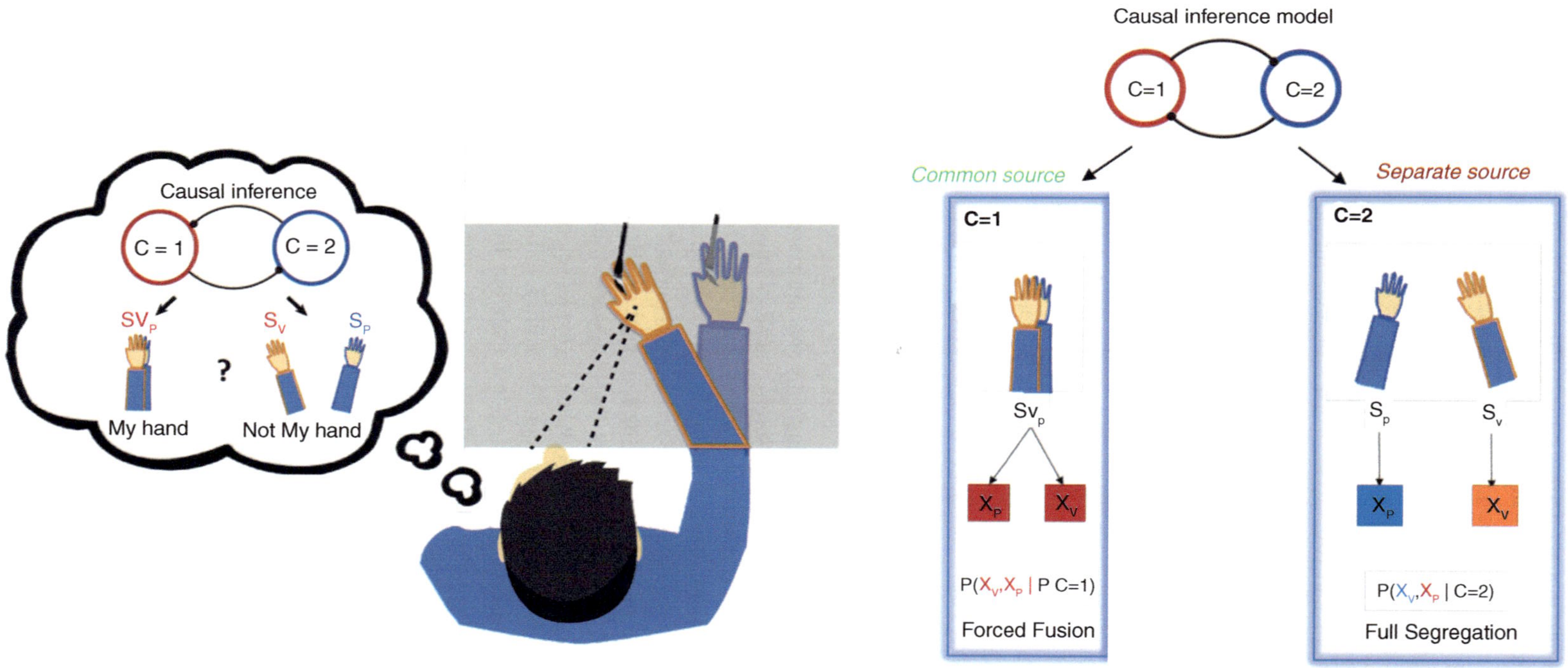

Fig. 5.1 Causal inference problem in body representation. Upon receiving visual and proprioceptive information about the hand, the brain infers the probability that the two sources of information come from a common cause

$$\hat{S}_p = P\left(C = 1|x_p, x_v\right)_{vp,c=1} + \hat{S}P\left(C = 2|x_p, x_v\right)\hat{S}_{p,c=2} \tag{5.5a}$$

$$\hat{S}_v = P\left(C = 1|x_p, x_v\right)\hat{S}_{vp,c=1} + P\left(C = 2|x_p, x_v\right)\hat{S}_{v,c=2} \tag{5.5b}$$

The Bayesian causal inference model can account for the effect of various factors on multisensory integration regarding the body (Fig. 5.2; Fang et al. 2019; Kilteni et al. 2015; Samad et al. 2015). Provided with multisensory information about the hand including vision, proprioceptive, touch, etc., the brain is faced with a causal inference problem of whether all sources of information come from the same hand, i.e., "my hand." The posterior probability of a common hand depends on the likelihood of receiving the current sensory information if they belong to the same hand and the prior probability of there being one hand. The closer the visual and proprioceptive hand positions are, the more likely they define the same hand. Similarly, other spatial and temporal congruence factors discussed above contribute to the likelihood of a common cause. On the other hand, prior body knowledge constrains how strongly a viewed object can be embodied. For example, the model-fitted prior probability of a common cause was lower when participants viewed a woodblock versus a verisimilar hand (Fang et al. 2019). The higher the posterior probability of a common cause, the more strongly the viewed hand is perceived as

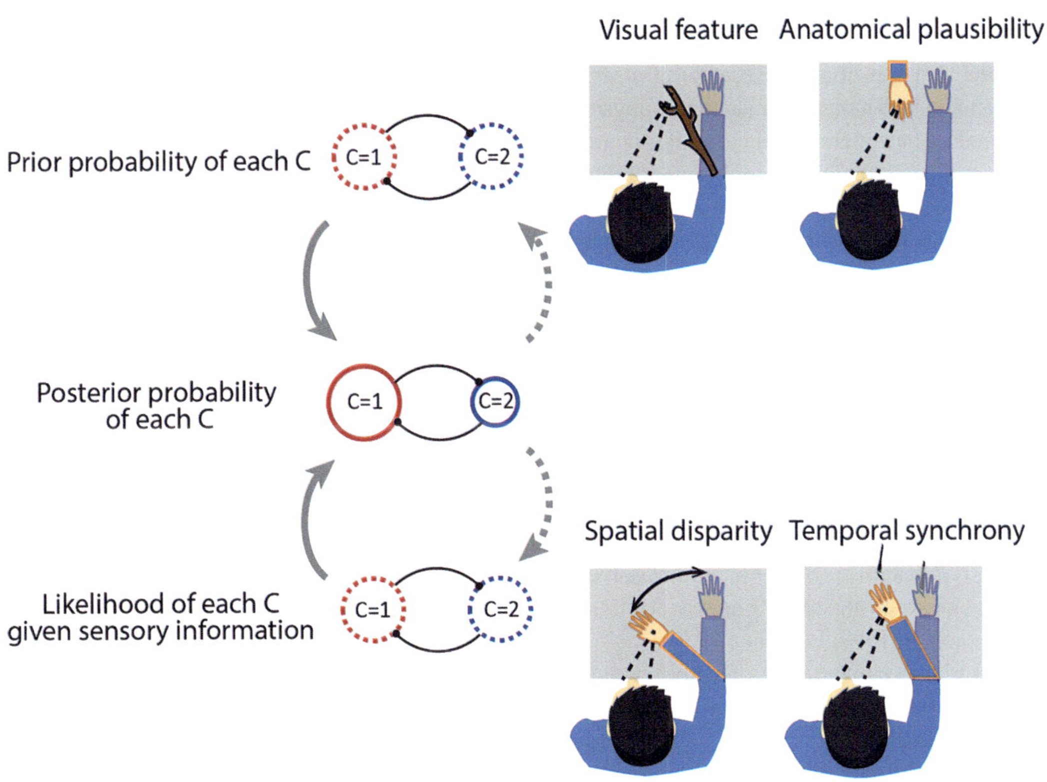

Fig. 5.2 Bayesian causal inference in body representation. The posterior probability of a common cause depends on the prior probability and likelihood. Prior information refers to existent knowledge including the visual feature and anatomical plausibility of the body. Likelihood depends on the congruence between information across modalities, such as spatial disparity and visual-tactile synchrony

one's own (Fang et al. 2019), suggesting that causal inference may be a mechanism of how the brain distinguishes oneself from the environment.

5.4 Neurophysiological Evidence of Body Representation

5.4.1 Neuronal Basis of Multisensory Integration of Body-Related Signals

The last section focuses on the neurophysiological and neuroimaging evidence related to multisensory body representation. Since numerous behavioral findings have demonstrated that the body representation depends on the integration of body-related signals from multiple modalities, most studies exploring the neural basis of body representation have focused on the multisensory neurons (Noel et al. 2018; Blanke et al. 2015). Human and non-human primate neurophysiological studies have demonstrated a high degree of overlap of brain regions in multisensory integration and body representation (Blanke et al. 2015). Using single-unit recordings in animals, numerous kinds of multisensory neurons have been identified in various brain regions, including the parietal association cortex, premotor cortex, insula, and superior colliculus (Mountcastle et al. 1975; Avillac et al. 2007; Stein and Stanford 2008). In particular, body-related multisensory neurons are predominantly located in the posterior parietal cortex, including the ventral intraparietal (VIP) area, area 5 and area 7, and premotor cortex (Graziano et al. 1994; Graziano et al. 1997; Avillac et al. 2005; Fogassi et al. 1996; Graziano et al. 1999; Leinonen 1980).

Multisensory integration of body-related signals at the single neuron level has been well studied on the bimodal neurons responding to somatosensory stimuli and visual (auditory) stimuli near the body. In most of these neurons, the visual (auditory) receptive field is localized on a given body part (Graziano et al. 1994, 1999; Avillac et al. 2005). One important multisensory property of these neurons is that the neural response to the tactile stimulus is modulated by visual (auditory) stimuli presented within their receptive field. Similar to other multisensory neurons (e.g., visual-vestibular multisensory neurons), such multisensory modulation on neural response can be super-additive (increased firing rate) or sub-additive (decreased firing rate) compared to the arithmetic sum of the responses in unimodal conditions (Avillac et al. 2007). In analogy to peripersonal space in human psychological studies, the size of the visual (auditory) receptive field is proportional to the tactile receptive field on a given body part. The visual receptive field of monkey premotor neurons typically extends 40 cm from the upper limb (Graziano et al. 1994, 1999). It ranges from 5 cm to 1 m depending on the size of other body parts (Avillac et al. 2005; Schlack et al. 2005; Jiang et al. 2013). Furthermore, the visual (auditory) receptive field of these multisensory neurons is anchored on the corresponding body part despite the movement of the body (Fogassi et al. 1996; Graziano et al. 1997; Graziano 1999, 2000). For example, Graziano and colleagues found that the visual receptive field of monkey premotor neurons shifts to the new spatial location where the limb is placed (Graziano 1999). Such binding receptive fields can also be observed in other body-centered multisensory neurons, such as face-centered neurons in VIP (Avillac et al. 2005) and trunk-centered in area 7 (Iriki et al. 1996). Thus, these multisensory neurons with an anchored visual (auditory) receptive field encode body-related signals in body-part centered reference frames. Taken together, the visual (auditory) receptive field of these multisensory neurons is thus conceived as the neural basis of peripersonal space in human psychological studies, which constitute an interface for the body–environment interaction (Noel et al. 2018).

Human neuroimaging studies have consistently highlighted the premotor and posterior parietal cortex in integrating body-related signals (Makin et al. 2008). Similar to the single neuron responses in non-human primates, super-additive and sub-additive responses are also observed in the intra-parietal sulcus (IPS) and ventral premotor cortex when the visual stimulus is integrated within the peripersonal space (Gentile et al. 2011). For example, human fMRI studies found enhanced BOLD response in IPS when participants were presented with a visual object near the body (Makin et al. 2007; Sereno and Huang 2006). Using the BOLD adaptation paradigm, Brozzoli and colleagues examined the neural activations when consecutive visual stimuli were presented in peripersonal space around the participants' hand. Since the adaptation paradigm (reduced neural activity in response to repeatedly presented stimuli) has been well established in electrophysiology and fMRI studies to reflect selectivity of neural responses to a specific stimulus, it is a useful method to examine the neural populations that respond to the visual stimuli within PPS. The adaptation effect was observed in IPS, inferior parietal lobe, and premotor cortex when the visual object was presented near the right hand, but not far from the hand (Brozzoli et al. 2011), indicating that these brain regions selectively respond to inputs near the body. Taken together, both human and animal neurophysiological findings suggest two key regions, the posterior parietal cortex, and premotor cortex, in multisensory processing of bodily signals. Single neuron and population activities in these areas offer a common neural basis in humans and non-human primates for multisensory integration in peripersonal space.

5.4.2 Neuronal Representation of One's Own Body

Despite the rich neurophysiology studies about multisensory processing within peripersonal space in body representation, direct evidence about the neural basis of subjective recognition of one's body is obtained from body illusion studies on humans. Using synchronous visual-tactile stimulation on a viewed hand and the participant's actual hidden hand within fMRI scanner, Ehrsson and colleagues examined brain activation when the participants experienced the rubber hand illusion. They found activation in IPS, ventral premotor cortex, cerebellum (Ehrsson et al. 2004, 2005), and right posterior insula (Tsakiris et al. 2007) was closely correlated to the change of limb ownership in RHI. These neural correlates can be further confirmed by conducting the threatening test, which is commonly used in the behavioral test in body ownership. Lloyd and colleagues found increased activity in the posterior parietal cortex and supplementary motor cortex when the threatening visual object approached the fake limb in RHI (Lloyd et al. 2006), as if the participant's own body was being threatened.

Furthermore, several studies examined the shift of peripersonal space toward the viewed hand after the rubber hand illusion has been induced. As expected, peripersonal space around the limb is remapped onto the fake limb under the RHI, and the adaptation effect in IPS and premotor cortex was observed when visual stimuli were repeatedly presented near the fake limb, but the adaptation effect is not observed if the contralateral fake limb is presented (Brozzoli et al. 2012).

The most direct evidence about the relationship between the body-related multisensory integration and body representation on a single neuron level has been examined by Graziano and colleagues in monkeys. Using the single-unit recording, the authors examined the neuronal response in monkey area 5 when the animal experienced the visual-somatosensory multisensory condition similar to the rubber hand illusion in humans (Graziano 2000). The multisensory neurons showed spatial selectivity to proprioceptive (monkey's veridical limb) information and visual (fake limb) information of limb positions. For instance, if a neuron has a higher firing rate when the monkey's veridical limb (proprioceptive input) is positioned on the left side versus the right side, its firing rate also increased when the fake visual limb is presented on the left side and decreased when it was shown on the right side. Intriguingly, such tuning modulation by visual information depended on the physiological feature of the visual limb. Non-body objects (e.g., a trunk of wood) and physically impossible limb position did not modulate the neural response. Furthermore, electrical stimulation of these body-related multisensory neurons in non-human primates results in defensive-like actions (Cooke and Graziano 2003; Graziano and Cooke 2006). These results are comparable to the RHI results in human behavioral and imaging studies.

5.4.3 Electrophysiological Evidence of Causal Inference in Body Representation

In the last session, we take the Bayesian theory of multisensory integration into account to explain the neural implementation of the subjective experience of body representation and self-other discrimination. In a recent study, the authors established a moving rubber hand illusion paradigm based on reaching movements under a virtual reality system. They recoded single neuron responses in the premotor cortex in awake behaving monkeys. By introducing various disparities between a monkey's real limb and a visual fake limb, authors can examine the proprioceptive drift, which is the probe of the illusion strength under dynamic multisensory conditions. The behavioral result showed that the limb's integration of visual and proprioceptive information could be well explained by Bayesian causal inference theory, consistent with the human behavioral results (Fang et al. 2019).

More importantly, under the Bayesian causal inference framework, the ownership of the visual fake limb is determined by the posterior probability that the sensory signals come from a common source. That is, the central neural system should integrate the signals when the visual-proprioceptive limbs are aligned and segregate the signals when the disparity is too large. To examine the neural correlates of the posterior probability, the authors conducted two control conditions to establish the ideal neuronal responses of integration and segregation. In the integration control condition, the visual and proprioceptive limb position was always perfectly aligned, while the visual limb was not presented in the segregation control condition. Thus, when the disparity was systematically changed across trials, the representation of the common source probability on a single neuron level can be approached according to how similar the neural response is to these ideal neural responses. The single neuron analysis revealed a considerable population in the premotor cortex associated with the posterior probability of common source predicted by the Bayesian causal inference model. The dynamics of the posterior probability of common source across trials can also be decoded by the population neuronal activities. More importantly, the probability of integration at both the behavioral and neural levels was decreased when the visual feedback was replaced by a piece of wood (Fang et al. 2019).

The neural mechanism of causal inference was further extended by a recent artificial neural network study. The authors trained a neural network model to solve causal inference for motion estimation (Rideaux et al. 2021). It was suggested that the multisensory neurons with congruent Gaussian tuning may account for multisensory integration, whereas those with incongruent multisensory tuning have been considered to account for segregating (French and DeAngelis 2020). In line with this prediction, the neural network develops multisensory neurons with congruent and opposite tunings and demonstrated both congruent and opposite neurons contribute to the multisensory behavior. This simulation thus showed that to determine whether the signals should be integrated or segregated, the causal inference problem can be solved by balancing between the activities of these two populations.

References

Alais D, Burr D (2004) Ventriloquist effect results from near-optimal bimodal integration. Curr Biol 14(3):257–262. https://doi.org/10.1016/S0960-9822(04)00043-0

Antal, Haans Wijnand A., IJsselsteijn Yvonne A.W., de Kort (2008) The effect of similarities in skin texture and hand shape on perceived ownership of a fake limb Body Image 5(4) 389–394 https://doi.org/10.1016/j.bodyim.2008.04.003

Austen EL, Soto-Faraco S, Enns JT, Kingstone A (2004) Mislocalizations of touch to a fake hand. Cogn Affect Behav Neurosci 4(2):170–181. https://doi.org/10.3758/CABN.4.2.170

Avillac M, Deneve S, Olivier E, Pouget A, Duhamel JR (2005) Reference frames for representing visual and tactile locations in parietal cortex. Nat Neurosci 8(7):941–949. https://doi.org/10.1038/nn1480

Avillac M, Ben Hamed S, Duhamel JR (2007) Multisensory integration in the ventral intraparietal area of the macaque monkey. J Neuroscience 27(8):1922–1932. https://doi.org/10.1523/jneurosci.2646-06.2007

Blanke O, Slater M, Serino A (2015) Behavioral, neural, and computational principles of bodily self-consciousness. Neuron 88(1):145–166. https://doi.org/10.1016/j.neuron.2015.09.029

Botvinick M, Cohen J (1998) Rubber hands 'feel' touch that eyes see. Nature 1998 391:6669 391(6669):756–756. https://doi.org/10.1038/35784

Brozzoli C, Gentile G, Petkova VI, Ehrsson HH (2011) FMRI adaptation reveals a cortical mechanism for the coding of space near the hand. J Neurosci 31(24):9023–9031. https://doi.org/10.1523/JNEUROSCI.1172-11.2011

Brozzoli C, Gentile G, Ehrsson HH (2012) That's near my hand! Parietal and premotor coding of hand-centered space contributes to localization and self-attribution of the hand. J Neurosci 32(42):14573–14582. https://doi.org/10.1523/jneurosci.2660-12.2012

Buxbaum LJ, Coslett HB (2010) Specialised structural descriptions for human body parts: evidence from autotopagnosia. Cogn Neuropsychol 18(4):289–306. https://doi.org/10.1080/02643290126172

Cooke DF, Graziano MS (2003) Defensive movements evoked by air puff in monkeys. J Neurophysiol 90(5):3317–3329. https://doi.org/10.1152/jn.00513.2003

Cooper LA, Shepard RN (1975) Mental transformation in the identification of left and right hands. J Exp Psychol Hum Percept Perform 1(1):48–56. https://doi.org/10.1037/0096-1523.1.1.48

Costantini M, Haggard P (2007) The rubber hand illusion: sensitivity and reference frame for body ownership. Conscious Cogn 16(2):229–240. https://doi.org/10.1016/j.concog.2007.01.001

David R., Wozny Ulrik R., Beierholm Ladan, Shams (2010) Probability Matching as a Computational Strategy Used in Perception PLoS Computational Biology 6(8):e1000871. https://doi.org/10.1371/journal.pcbi.1000871

Ehrsson HH, Spence C, Passingham RE (2004) That's my hand! Activity in premotor cortex reflects feeling of ownership of a limb. Science 305(5685):875–877. https://doi.org/10.1126/SCIENCE.1097011

Ehrsson HH, Holmes NP, Passingham RE (2005) Touching a rubber hand: feeling of body ownership is associated with activity in multisensory brain areas. J Neurosci 25(45):10564–10573. https://doi.org/10.1523/JNEUROSCI.0800-05.2005

Ernst MO, Banks MS (2002) Humans integrate visual and haptic information in a statistically optimal fashion. Nature 415(6870):429–433. https://doi.org/10.1038/415429a

Ernst MO, Bülthoff HH (2004) Merging the senses into a robust percept. Trends Cogn Sci 8(4):162–169. https://doi.org/10.1016/j.tics.2004.02.002

Fang W, Li J, Qi G, Li S, Sigman M, Wang L (2019) Statistical inference of body representation in the macaque brain. Proc Natl Acad Sci U S A 116(40):20151–20157. https://doi.org/10.1073/PNAS.1902334116/-/DCSUPPLEMENTAL

Farmer H, Tajadura-Jiménez A, Tsakiris M (2012) Beyond the colour of my skin: how skin colour affects the sense of body-ownership. Conscious Cogn 21(3):1242–1256. https://doi.org/10.1016/j.concog.2012.04.011

Farmer H, Maister L, Tsakiris M (2014) Change my body, change my mind: the effects of illusory ownership of an outgroup hand on implicit attitudes toward that outgroup. Front Psychol 4(JAN):1016. https://doi.org/10.3389/FPSYG.2013.01016/BIBTEX

Fogassi L, Gallese V, Fadiga L, Luppino G, Matelli M, Rizzolatti G (1996) Coding of peripersonal space in inferior premotor cortex (area F4). J Neurophysiol 76(1):141–157. https://doi.org/10.1152/jn.1996.76.1.141

French RL, DeAngelis GC (2020) Multisensory neural processing: from cue integration to causal inference. Curr Opin Physio 16:8–13. https://doi.org/10.1016/j.cophys.2020.04.004

Gentile G, Petkova VI, Ehrsson HH (2011) Integration of visual and tactile signals from the hand in the human brain: an FMRI study. J Neurophysiol 105(2):910–922. Retrieved from http://jn.physiology.org/content/105/2/910.full.pdf

Graziano MSA (1999) Where is my arm? The relative role of vision and proprioception in the neuronal representation of limb position. Proc Natl Acad Sci U S A 96(18):10418–10421. https://doi.org/10.1073/pnas.96.18.10418

Graziano MSA (2000) Coding the location of the arm by sight. Science 290(5497):1782–1786. https://doi.org/10.1126/science.290.5497.1782

Graziano MS, Cooke DF (2006) Parieto-frontal interactions, personal space, and defensive behavior. Neuropsychologia 44(13):2621–2635. Retrieved from https://www.ncbi.nlm.nih.gov/pubmed/17128446

Graziano MS, Yap GS, Gross CG (1994) Coding of visual space by premotor neurons. Science 266(5187):1054–1057. https://doi.org/10.1126/science.7973661

Graziano MS, Hu XT, Gross CG (1997) Visuospatial properties of ventral premotor cortex. J Neurophysiol 77(5):2268–2292. https://doi.org/10.1152/jn.1997.77.5.2268

Graziano MS, Reiss LA, Gross CG (1999) A neuronal representation of the location of nearby sounds. Nature 397(6718):428–430. https://doi.org/10.1038/17115

Head and Holmes (1911) Sensory Disturbances From Cerebral Lesions Brain 34(2–3):102–254. https://doi.org/10.1093/brain/34.2-3.102

Holmes NP, Spence C (2005) Visual bias of unseen hand position with a mirror: spatial and temporal factors. Exp Brain Res 166(3–4):489–497. https://doi.org/10.1007/s00221-005-2389-4

Holmes NP, Crozier G, Spence C (2004) When mirrors lie: "visual capture" of arm position impairs reaching

performance. Cogn Affect Behav Neurosci 4(2):193–200. https://doi.org/10.3758/CABN.4.2.193
Holmes NP, Snijders HJ, Spence C (2006) Reaching with alien limbs: visual exposure to prosthetic hands in a mirror biases proprioception without accompanying illusions of ownership. Perception & Psychophysics 2006 68:4 68(4):685–701. https://doi.org/10.3758/BF03208768
Ide M (2013) The effect of "anatomical plausibility" of hand angle on the rubber-hand illusion. Perception 42:103–111. https://doi.org/10.1068/p7322
Iriki A, Tanaka M, Iwamura Y (1996) Coding of modified body schema during tool use by macaque postcentral neurones. Neuroreport 7(14):2325–2330. Retrieved from http://www.ncbi.nlm.nih.gov/pubmed/8951846
Jiang HH, Hu YZ, Wang JH, Ma YY, Hu XT (2013) Visuospatial properties of caudal area 7b in Macaca fascicularis. Dongwuxue Yanjiu 34(E2):E50–E61. https://doi.org/10.3724/SP.J.1141.2013.E02E50
Kilteni K, Maselli A, Kording KP, Slater M (2015) Over my fake body: body ownership illusions for studying the multisensory basis of own-body perception. Front Hum Neurosci 9(MAR):141. https://doi.org/10.3389/FNHUM.2015.00141/BIBTEX
Körding KP, Beierholm U, Ma WJ, Quartz S, Tenenbaum JB, Shams L (2007) Causal inference in multisensory perception. PLoS One 2(9):e943. https://doi.org/10.1371/journal.pone.0000943
Làdavas et al. (1998) Visual peripersonal space centred on the face in humans Brain 121(12):2317–2326. https://doi.org/10.1093/brain/121.12.2317
Leinonen L (1980) Functional properties of neurones in the posterior part of area 7 in awake monkey. Acta Physiol Scand 108(3):301–308. https://doi.org/10.1111/j.1748-1716.1980.tb06536.x
Liu Y, Medina J (2017) Influence of the body schema on multisensory integration: evidence from the Mirror box illusion. Scientific Reports 2017 7:1 7(1):1–11. https://doi.org/10.1038/s41598-017-04797-0
Liu Y, Medina J (2018) Integrating multisensory information across external and motor-based frames of reference. Cognition 173:75–86. https://doi.org/10.1016/j.cognition.2018.01.005
Lloyd DM (2007) Spatial limits on referred touch to an alien limb may reflect boundaries of visuo-tactile peripersonal space surrounding the hand. Brain Cogn 64(1):104–109. https://doi.org/10.1016/j.bandc.2006.09.013
Lloyd D, Morrison I, Roberts N (2006) Role for human posterior parietal cortex in visual processing of aversive objects in peripersonal space. J Neurophysiol 95(1):205–214. https://doi.org/10.1152/jn.00614.2005
Makin TR, Holmes NP, Zohary E (2007) Is that near my hand? Multisensory representation of peripersonal space in human intraparietal sulcus. J Neurosci 27(4):731–740. https://doi.org/10.1523/JNEUROSCI.3653-06.2007
Makin TR, Holmes NP, Ehrsson HH (2008) On the other hand: dummy hands and peripersonal space. Behav Brain Res 191(1):1–10. https://doi.org/10.1016/j.bbr.2008.02.041
Medina J, Khurana P, Coslett HB (2015) The influence of embodiment on multisensory integration using the mirror box illusion. Conscious Cogn 37:71–82. https://doi.org/10.1016/j.concog.2015.08.011
Meredith MA, Nemitz JW, Stein BE (1987) Determinants of multisensory integration in superior colliculus neurons. I. Temporal factors. J Neurosci 7(10):3215–3229. doi: citeulike-article-id:409430
Meredith MA, Stein BE (1986) Spatial factors determine the activity of multisensory neurons in cat superior colliculus. Brain Res 5:350–354. https://doi.org/10.1016/0006-8993(86)91648-3
Mountcastle VB, Lynch JC, Georgopoulos A, Sakata H, Acuna C (1975) Posterior parietal association cortex of the monkey: command functions for operations within extrapersonal space. J Neurophysiol 38(4):871–908. Retrieved from http://www.ncbi.nlm.nih.gov/pubmed/808592
Noel JP, Blanke O, Serino A (2018) From multisensory integration in peripersonal space to bodily self-consciousness: from statistical regularities to statistical inference. Ann N Y Acad Sci 1426(1):146–165. https://doi.org/10.1111/nyas.13867
Parsons LM (1994) Temporal and kinematic properties of motor behavior reflected in mentally simulated action. J Exp Psychol Hum Percept Perform 20(4):709–730. https://doi.org/10.1037/0096-1523.20.4.709
Parsons (1987) Imagined spatial transformation of one's body. Journal of Experimental Psychology: General 116(2):172–191. https://doi.org/10.1037/0096-3445.116.2.172
Pavani and Zampini (2007) The Role of Hand Size in the Fake-Hand Illusion Paradigm Perception 36(10):1547–1554. https://doi.org/10.1068/p5853
Peck TC, Seinfeld S, Aglioti SM, Slater M (2013) Putting yourself in the skin of a black avatar reduces implicit racial bias. Conscious Cogn 22(3):779–787. https://doi.org/10.1016/j.concog.2013.04.016
Ramachandran and Rogers-Ramachandran (1996) Synaesthesia in phantom limbs induced with mirrors Proceedings of the Royal Society of London. Series B: Biological Sciences 263(1369):377–386. https://doi.org/10.1098/rspb.1996.0058
Rideaux R, Storrs KR, Maiello G, Welchman AE (2021) How multisensory neurons solve causal inference. Proc Natl Acad Sci U S A 118(32):e2106235118. https://doi.org/10.1073/pnas.2106235118
Samad M, Chung AJ, Shams L (2015) Perception of body ownership is driven by Bayesian sensory inference. PLoS One 10(2):1–23. https://doi.org/10.1371/journal.pone.0117178
Schlack A, Sterbing-D'Angelo SJ, Hartung K, Hoffmann KP, Bremmer F (2005) Multisensory space representations in the macaque ventral intraparietal area. J Neurosci 25(18):4616–4625. https://doi.org/10.1523/JNEUROSCI.0455-05.2005
Schwoebel J, Coslett HB (2005) Evidence for multiple, distinct representations of the human body.

J Cogn Neurosci 17(4):543–553. https://doi.org/10.1162/0898929053467587
Sereno MI, Huang RS (2006) A human parietal face area contains aligned head-centered visual and tactile maps. Nat Neurosci 9(10):1337–1343. https://doi.org/10.1038/nn1777
Shams L, Beierholm UR (2010) Causal inference in perception. Trends Cogn Sci 14(9):425–432. https://doi.org/10.1016/j.tics.2010.07.001
Spence C, Pavani F, Driver J (2004) Spatial constraints on visual-tactile cross-modal distractor congruency effects. Cogn Affect Behav Neurosci 4(2):148–169. https://doi.org/10.3758/CABN.4.2.148
Stein BE, Stanford TR (2008) Multisensory integration: current issues from the perspective of the single neuron. Nat Rev Neurosci 9(4):255–266. https://doi.org/10.1038/nrn2331
Tsakiris M (2010) My body in the brain: a neurocognitive model of body-ownership. Neuropsychologia 48(3):703–712. https://doi.org/10.1016/j.neuropsychologia.2009.09.034
Tsakiris M, Haggard P (2005) The rubber hand illusion revisited: Visuotactile integration and self-attribution. J Exp Psychol Hum Percept Perform 31(1):80–91. https://doi.org/10.1037/0096-1523.31.1.80
Tsakiris M, Hesse MD, Boy C, Haggard P, Fink GR (2007) Neural signatures of body ownership: a sensory network for bodily self-consciousness. Cereb Cortex 17(10):2235–2244. https://doi.org/10.1093/cercor/bhl131
Tsakiris M, Costantini M, Haggard P (2008) The role of the right temporo-parietal junction in maintaining a coherent sense of one's body. Neuropsychologia 46(12):3014–3018. https://doi.org/10.1016/j.neuropsychologia.2008.06.004
Tsakiris M, Carpenter L, James D, Fotopoulou A (2010) Hands only illusion: multisensory integration elicits sense of ownership for body parts but not for non-corporeal objects. Exp Brain Res 204(3):343–352. https://doi.org/10.1007/s00221-009-2039-3
Van Beers RJ, Sittig AC, Denier Van Der Gon JJ (1999) Integration of proprioceptive and visual position-information: an experimentally supported model. J Neurophysiol 81(3):1355–1364. https://www.ncbi.nlm.nih.gov/pubmed/10085361
Witten IB, Knudsen EI (2005) Why seeing is believing: merging auditory and visual worlds. Neuron 48(3):489–496. https://doi.org/10.1016/j.neuron.2005.10.020
Wozny DR, Beierholm UR, Shams L (2008) Human trimodal perception follows optimal statistical inference. J Vis 8(3):24. https://doi.org/10.1167/8.3.24
Zapparoli L, Invernizzi P, Gandola M, Berlingeri M, De Santis A, Zerbi A et al (2014) Like the back of the (right) hand? A new fMRI look on the hand laterality task. Exp Brain Res 232(12):3873–3895

Crossmodal Associations and Working Memory in the Brain

6

Yixuan Ku and Yongdi Zhou

Abstract

Crossmodal associations between stimuli from different sensory modalities could emerge in non-synesthetic people and be stored in working memory to guide goal-directed behaviors. This chapter reviews a plethora of studies in this field to summarize where, when, and how crossmodal associations and working memory are processed. It has been found that in those brain regions that are traditionally considered as unimodal primary sensory areas, neural activity could be influenced by crossmodal sensory signals at temporally very early stage of information processing. This phenomenon could not be due to feedback projections from higher level associative areas. Sequentially, neural processes would then occur in associative cortical areas including the posterior parietal cortex and prefrontal cortex. Neural oscillations in multiple frequency bands may reflect brain activity in crossmodal associations, and it is likely that neural synchrony is related to potential neural mechanisms underlying these processes. Primary sensory areas and associative areas coordinate together through neural synchrony to fulfil crossmodal associations and to guide working memory performance.

Keywords

Crossmodal associations · Working memory · Neural synchrony · Primary sensory cortices · Multisensory associative areas

Chinese cuisine is well known all over the world, and decent taste always goes together with perfect color and pleasant smell. Thus *delicious* needs multisensory integration when visual and olfactory information facilitates gustatory perception in this case. Such crossmodal associations between different sensory modalities are usually processed and stored in working memory to guide goal-directed behaviors. How does the crossmodal association come from and how is it stored in working memory? And how can we allocate brain resources to deal with the crossmodal association between different sources of information and to process the associations dynamically? We will discuss different aspects of crossmodal associations and working memory in this Chapter. Where (in the brain), When (how

Y. Ku (✉)
Department of Psychology, Center for Brain and Mental Well-being, Sun Yat-sen University, Guangzhou, China

Peng Cheng Laboratory, Shenzhen, China
e-mail: kuyixuan@sysu.edu.cn

Y. Zhou
School of Psychology, Shenzhen University, Shenzhen, China

Y. Gu, A. Zaidel (eds.), *Advances of Multisensory Integration in the Brain*, Advances in Experimental Medicine and Biology 1437, https://doi.org/10.1007/978-981-99-7611-9_6

dynamically in temporal scales), and How (the underlying mechanisms) do crossmodal associations and working memory emerge, and where, when, and how are they processed?

6.1 Crossmodal Associations and Working Memory

Some people naturally acquire associations between pieces of information from different senses, which is called synesthesia. In this situation, one sense involuntarily and automatically elicits another. The brilliant pianist and composer Liszt Ferenc, who was thought to have synesthesia, when he reminded performers to achieve a certain effect of music performance, always talked about the color, e.g., that this the tone needed to be bluer (Blom 2010). There were also many artists who were seeing colors from words, or even "hearing" colors or "tasting" shapes (Ramachandran and Hubbard 2003). It has been estimated that the incidence of synesthesia, in most cases strong synesthesia, is 1/2000 among the general population, when females outnumbered males with 6 to 1 (Baron-Cohen et al. 1996). In contrast, weak synesthesia, which is analogous to the crossmodal association mentioned here, is much more prevalent(Martino and Marks 2001).

Crossmodal associations have long been examined ever since ancient Greece philosophers. Aristotle (B.C. 384–322) proposed that harmony of colors was like harmony of sounds (Marks 1975). Many philosophers discussed crossmodal associations afterwards. Among them, Associationism, as first proposed by philosopher Hume, was put forward to explain mental processes in which pairs of thoughts became associated based on past experience. Therefore, stimuli presented from different sensory modalities are more likely to be associated when they occur closer in time or space (Driver and Spence 2000; McDonald et al. 2000), which is also called crossmodal binding. Moreover, crossmodal associations could be established by semantic congruency (Laurienti et al. 2004). For instance, people associated warm with warm colors (red, orange, and yellow) and cool with cool colors (blue and green) long time ago (Sully 1879).

In addition to these "strong association" situations, associations between stimuli can also be momentarily established and implemented for goal-directed behavior. For example, when we play Wisconsin Card Sorting game, we need to associate different cards with distinctive rules: sometimes sorting cards with numbers; sometimes with shapes of elements. These unimodal associations can also be generalized to crossmodal conditions. Studies found that both synesthetes and non-synesthetes could predict color preferences based on the acoustic characteristics of vowels, though more predictive in the former group (Moos et al. 2014). These associations, especially in non-synesthetes, could be learned from experiences. Moreover, even animals could be trained to associate different stimuli (Miyashita and Chang 1988). However, lesions in certain brain regions (e.g., amygdala) only influence crossmodal associations but not unimodal associations (Murray and Mishkin 1985). This dissociation between crossmodal and unimodal conditions leads us to think that the neural mechanisms underlying crossmodal associations may be different from those underlying unimodal associations. Here a couple of immediate questions come: in the brain, where are crossmodal associations processed? and where do they emerge?

A typical paradigm to investigate crossmodal associations is the "delayed matching/non-matching to sample" task, which is one type of working memory task. Working memory is the cognitive process that retains sensory information in the short term (often for seconds) in preparation for upcoming goal-directed activities (Baddeley 2012). Unlike the massive capacity of long-term memory, working memory capacity is severely restricted (Cowan 2001; Miller 1956). Yet, working memory is highly flexible and strongly correlated with general cognitive capabilities (Engle et al. 1999). Comparisons between crossmodal and unimodal delayed matching to sample tasks reveal processes in crossmodal associations and working memory (Fig. 6.1). These task-related crossmodal associations,

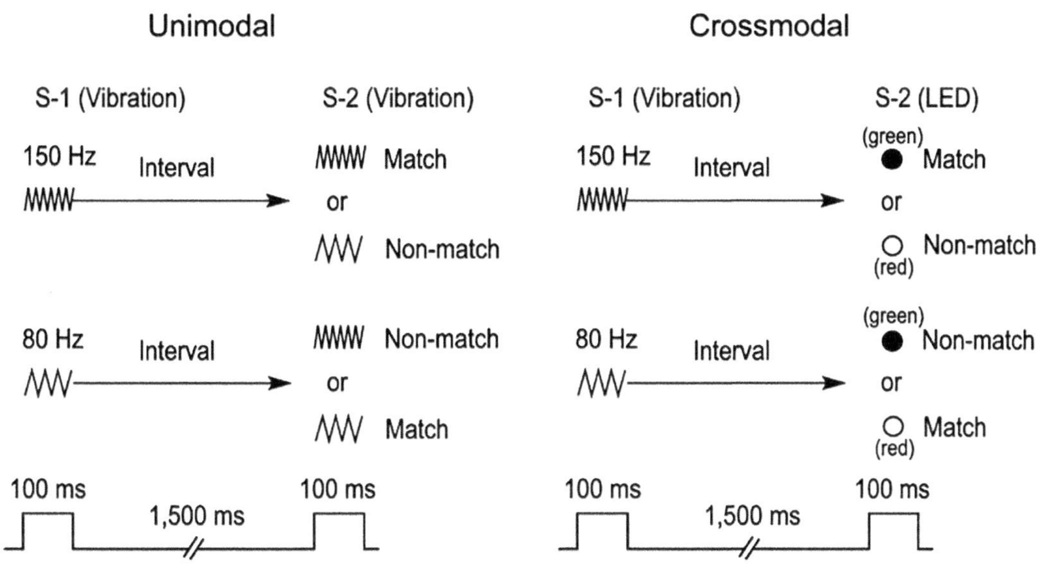

Fig. 6.1 The unimodal and crossmodal delayed matching-to-sample tasks. During the unimodal matching test, S-1 is a 80 Hz or 150 Hz tactile vibration that is applied to the participant's left index fingertip, and S-2 is another tactile vibration (high frequency or low frequency). In contrast, during the crossmodal matching activity, S-2 is an LED light (either red or green) that is presented in front of the subject at eye level. Red and green lights match high and low frequencies, respectively

which are not stored in long-term memory but transiently established and stored in working memory, bridge stimuli presented across time to guide goal-directed behavior (Fuster et al. 2000; Wang et al. 2015).

6.2 Where Are Crossmodal Associations and Working Memory in the Brain?

Now we get to know that representations of working memory are spread out in the brain (Christophel et al. 2017). But back to 1970s, the first study that found memory cells in the brain was performed with rhesus monkeys (Fuster and Alexander 1971). Neurons in the dorsolateral prefrontal cortex kept firing during the delay when the monkey was maintaining sample information and waiting for matching it with subsequent probe information. After choice was made, the sustained firing dropped back to the baseline. Sustained activity in the dorsolateral prefrontal cortex was also found to represent other features such as visuospatial information (Funahashi et al. 1989). Therefore the neural circuit and neural activity in the dorsolateral prefrontal cortex were proposed as the underlying neural mechanisms of working memory (Goldman-Rakic 1990).

Apart from these associations of features in the same sensory domain, crossmodal associations between different sensory modalities were also found in the dorsolateral prefrontal cortex (Fuster et al. 2000). After monkeys were trained to associate tones and colors, they needed to decide whether a color matched a tone that was presented as a sample stimulus 10 s earlier. The prefrontal cells selectively responded to tones or colors during the delay based on the task rule. More importantly, the selectivity was correlated with tones and colors only in correct trials but not when the monkey made an error. Thus, neural circuits in the dorsolateral prefrontal cortex were proposed to represent crossmodal associations (Fuster et al. 2000).

The prefrontal cortex lies in a vantage position to associate one sense with the other as it receives inputs from multiple sensory areas (Jones and

Powell 1970; Romanski et al. 1999). Crossmodal associations and crossmodal integration also occur in other multisensory areas such as the superior colliculus (Meredith and Stein 1983), superior temporal sulcus (Padberg et al. 2003), intraparietal sulcus (Andersen et al. 1997), and premotor cortex (Graziano et al. 1994). Human neuroimaging studies confirmed the roles of these multisensory areas in crossmodal associations in the human brain (Driver and Noesselt 2008).

It should be noted that crossmodal associations have also been found in primary sensory areas, for instances the primary visual cortex (Morrell 1972), primary auditory cortex (Calvert et al. 1997), and primary somatosensory cortex (Zhou and Fuster 1997, 2000). Given these findings, researchers even proposed that our neocortices were essentially multisensory (Ghazanfar and Schroeder 2006). However, there could be other possibilities that crossmodal associations originated from sensory-surrounding areas newly identified as crossmodal or came from feedback projections from associative areas to sensory areas (Driver and Noesselt 2008). Moreover, it has recently been proposed that subcortical areas such as the pulvinar play important roles in crossmodal associations since they are also connected with multiple sensory areas (Froesel et al. 2021).

Taken together, crossmodal associations are distributed across the brain, including both unimodal sensory and multimodal associative areas. Then, the second question comes, how do these areas interplay with each other during crossmodal associations and working memory? First, in order to allocate the origin of crossmodal associations, we need to dissociate the temporal dynamics of activity in these brain areas.

6.3 When Are Crossmodal Associations Stored in Working Memory?

It is well accepted that unimodal sensory information is first processed in areas that have been traditionally thought of as pure-sensory ones. When crossmodal associations emerge in these areas, roles of bottom-up or top-down influence on information transfer between brain areas need to be clarified (Calvert et al. 1997; Morrell 1972; Zhou and Fuster 1997, 2000). If crossmodal associations occur at the early stage of information processes, the bottom-up influence is likely more prominent; if crossmodal associations occur at the late processing stage, the top-down influence is likely more probable. Apparently, electrophysiological methods with high temporal resolution are needed to address these questions.

In the field of multimodal object recognition, event related potentials (ERPs) on humans first revealed that auditory-visual multisensory integration could influence activity in the visual areas as early as 40 ms after stimulus onset (Giard and Peronnet 1999). Later, activity in the auditory cortex has been found to be influenced by somatosensory information as early as 50 ms after stimulus onset through multisensory interaction (Murray 2005). These changes in early activity of different sensory cortices were thought to represent the feedforward sweep of processing, and thus exclude the possibility of feedback projections from higher level associative areas.

The first ERP study about the influence of crossmodal associations using working memory paradigm was carried out in the tactile domain (Ku et al. 2007). Combining independent component analysis and source localization methods, Ku et al. identified an early ERP component P45 localized in the primary somatosensory cortex, and they found that crossmodal associations could enhance this component as early as 50 ms, compared with unimodal conditions. Thus crossmodal influence from other sensory modalities could affect early sensory processing, which was not possibly due to feedback projections originated from higher level brain areas.

More accurate conclusions could be drawn from electrophysiological studies in animals with high resolutions in both temporal and spatial domains (Lakatos et al. 2007). Neocortical areas in primates have six layers and using laminar recordings we can delineate different sources of information as we know that layer 4 in primary sensory areas receives feedforward input from the thalamus, while superior layers 2/3 and inferior layers 5/6 convey cortico-cortical projections, both validated in animals (Douglas and Martin 2004; Miller et al. 2001) and humans (Yu et al. 2019). With laminar recordings in the primary auditory cortex (A1), Lakatos et al. nicely demonstrated when the somatosensory stimulus enhanced auditory responses in A1 (Lakatos et al. 2007). The enhancement was not simply 1 + 1 > 2. The somatosensory input reset the phase of auditory responses into an optimal excitation period. And most interestingly, the stimulus onset asynchrony (SOA) between auditory and somatosensory stimuli mattered. The somatosensory stimulus only facilitated auditory response at certain SOAs, and the crossmodal enhancement pattern depended on SOA with an oscillation cycle. These results implicate that timing (question "when") is crucial for crossmodal associations.

Using the crossmodal delayed-matching-to-sample task (Fig. 6.1), we could delineate sequential processes of crossmodal associations and working memory (Ohara et al. 2006). Apart from the crossmodal influence at early sensory stage (Ku et al. 2007), discrete processes representing crossmodal associations and working memory were sequentially observed during the delay (Ohara et al. 2006). The late positive component at posterior parietal areas represented crossmodal associations and the late negative component at frontal areas indicated working memory maintenance. Moreover, using paired associative learning task, we further delineated when and how these processes emerged (Gui et al. 2017). Frontal N400, which was found to be related with familiarity (Yovel and Paller 2004) but not recollection in memory, reflected semantic associations during a recognition test (Voss and Federmeier 2011). Interestingly, we observed increment in FN400 with learning in both unimodal and crossmodal conditions, indicating that modality-free associations were established in frontal areas (Gui et al. 2017). In contrast, the posterior component at around 600 ms after stimulus onset differentiated along with distinctive learning phases only in crossmodal conditions (Gui et al. 2017), indicating that visuo-tactile crossmodal associations emerged in parieto-occipital areas. Furthermore, oscillations in the alpha band (8–13 Hz) accompanied with learning only in the crossmodal task in both frontal and posterior areas (Gui et al. 2017). The oscillation emerged from around 800 ms after sample stimulus onset until the response to the probe stimulus.

Taken together, crossmodal associations and working memory include multiple brain areas from uni-sensory to multisensory associative areas. Sequential processes have also been observed during encoding and delay periods, ranging from early sensory processing to late working memory maintenance. Then the remaining question is, how are these brain areas coordinated during crossmodal associations and working memory?

6.4 How Are Crossmodal Associations and Working Memory Processed?

We have seen that crossmodal associations and working memory recruit both primary sensory areas and higher level associative areas in the above sections. How do they interplay with each other? Neuronal coherence through oscillations could be a possible substrate (Fries 2005). There exist multiple oscillations throughout a spectrum of frequencies that coordinate cognitive processes, including delta (1–4 Hz), theta (4–8 Hz), alpha (8–13 Hz), beta (13–30 Hz), and gamma (>30 Hz) bands. Oscillatory synchronization underlies communications between wide ranges of brain areas (Varela et al. 2001). For example, cortical theta rhythms represent hippocampal–cortical interactions (Klimesch 1999), whose power covariates with working memory load (Raghavachari et al. 2001). Meanwhile, cortical alpha rhythms are produced by thalamocortical and cortico-cortical loops (da Silva 2013) and also increase with the number of distractors (Bonnefond and Jensen 2012) and working memory load in multiple sensory modalities (Haegens et al. 2010; Jensen et al. 2002; Spitzer and Blankenburg 2012). Moreover, cortical beta oscillations, which are often observed in motor tasks, have also been found in working memory tasks, especially in animal neurophysiological studies. In conjunction with beta oscillations, gamma oscillations (Lundqvist et al. 2016) have been proposed to subserve perceptual binding of features in the same sensory modality (Singer and Gray 1995) and across modalities (Senkowski et al. 2007).

In addition, oscillatory synchronization between different areas predicts working memory performance. For example, theta coherence between the prefrontal cortex and V4 (the associative sensory cortex) correlates with working memory performance (Liebe et al. 2012). Interestingly, higher frequency oscillations in the gamma band are suggested to reflect feedforward information transmission from lower level brain areas to higher level brain areas; lower frequency oscillations in alpha and beta bands are suggested to represent feedback projections from higher level areas to lower level areas (Bastos et al. 2015; van Kerkoerle et al. 2014, 2017; Michalareas et al. 2016; Richter et al. 2017). Our previous study discovered that alpha band oscillations in both frontal and posterior areas were only associated with crossmodal learning in the working memory task (Gui et al. 2017), which suggested that information went from frontal sites to posterior sites in the brain in top-down processing. Indeed, frontal alpha oscillations indicate top-down regulation on perceptual gains (Misselhorn et al. 2019) and interregional alpha synchrony upholds temporal crossmodal integration (van Driel et al. 2014).

Furthermore, phase coupling between different frequencies is also predominant in the brain and linked to cognitive processes (Canolty and Knight 2010). However, there are at least two kinds of coupling between oscillations, which are phase resetting and neural entrainment (Bauer et al. 2020). As we mentioned above, Lakatos et al. demonstrated that the phase of auditory induced oscillations in A1 is reset by somatosensory inputs (Lakatos et al. 2007). The same group also explored the role of entrainment of neuronal oscillations in attentional selection (Lakatos et al. 2008). Although the two processes have the similar outcome in neural synchrony but they are different in mechanisms. Specifically, phase resetting indicates transient influence of one stimulus on neural oscillations, while neural entrainment designates a period of time during which two oscillations become resonant and one oscillation shifts from one frequency to the other oscillation. It should be noted that the entrainment in low-frequency oscillations including delta and theta has normally been observed, but not in high-frequency oscillations. Yet, it is still hard to dissociate these two processes in the above-mentioned neural synchrony results. Future studies need to incorporate more manipulations of sensory input, for example, transient without rhythm vs. sustained with rhythm, to dissociate these underlying mechanisms.

6.5 Causal Evidence in Neural Mechanisms

We have summarized where, when, and how do crossmodal associations and working memory processed. However, the above-mentioned literatures are mostly correlational evidence. It is still not clear whether neural activity in those brain regions causally influences crossmodal associations and working memory. We all know that correlations cannot lead to causality for granted. Thus we need to implement causal tools to explore answers for these questions. The most commonly used causal tools for human studies include transcranial current stimulation (tCS) and transcranial magnetic stimulation (TMS).

Repetitive TMS and transcranial direct current stimulation (tDCS) are widely used in clinical investigation as it can change neural plasticity of the brain and induce long-lasting (long-term potentiation or long-term depression) effects (Lauro et al. 2014; Lefaucheur et al. 2014). However, if we want to explore the timing of cognitive processes, single-pulse TMS (spTMS) is the best choice. Applying spTMS over different brain areas, we explore the roles played by these areas in crossmodal associations and working memory. The logic is if the areas participate in the processes and have a causal role, spTMS will disrupt the processes in the areas and lead to a decline in related behavioral performance. As we have seen that the primary somatosensory cortex (Ku et al. 2007), the posterior parietal cortex, and prefrontal cortex (Gui et al. 2017) have differential activity in the crossmodal working memory task compared with the unimodal working memory task. We are curious about whether these brain areas have causal influences on behaviors or they are just redundantly activated during the cognitive processes. Interestingly, we have observed that the primary somatosensory cortex plays a causal role at an early stage of 300 ms after the sample stimulus; the posterior parietal cortex plays a causal role at a late stage of 600 ms after the sample stimulus (Ku et al. 2015); the prefrontal cortex plays a causal role in bridging the two stimuli (sample and probe) (Zhao and Ku 2018). However, it remains unclear how these areas integrate sequential processes. Future studies need to explore these questions by combining TMS and/or electroencephalography (EEG) and/or functional magnetic resonance imaging (fMRI) techniques.

6.6 Concluding Remarks

We have reviewed a plethora of studies to demonstrate where, when, and how crossmodal associations and working memory are processed. It is clear now that both primary sensory areas and associative areas, or even subcortical areas participate in these processes. Interestingly, the crossmodal influence in primary sensory areas emerges at a temporally early stage, which possibly reflects feedforward sweeping but not feedback influences from higher level areas. Meanwhile, crossmodal associations can also be established in posterior parietal areas and stored in both frontal and parietal areas. There exist sequential processes during crossmodal working memory, and related neural synchrony may represent how different brain areas are coordinated in these processes. Almost all frequency bands have been observed in the processes of crossmodal associations and working memory, but alpha oscillations are most prominent in these processes, which possibly reflect top-down information transfer. Primary sensory and associative areas are coordinated together through neural synchrony to fulfil crossmodal associations among stimuli from different sensory modalities and to store information in working memory dynamically to guide goal-directed behavior, such as eating delicious Chinese cuisine.

References

Andersen RA, Snyder LH, Bradley DC, Xing J (1997) Multimodal representation of space in the posterior parietal cortex and its use in planning movements. Annu Rev Neurosci 20:303–330. https://doi.org/10.1146/annurev.neuro.20.1.303

Baddeley A (2012) Working memory: theories, models, and controversies. Annu Rev Psychol 63:1–29. https://doi.org/10.1146/annurev-psych-120710-100422

Baron-Cohen S, Burt L, Smith-Laittan F, Harrison J, Bolton P (1996) Synaesthesia: prevalence and familiarity. Perception 25(9):1073–1079. https://doi.org/10.1068/p251073

Bastos AM, Vezoli J, Bosman CA, Schoffelen J-M, Oostenveld R, Dowdall JR, Weerd PD, Kennedy H, Fries P (2015) Visual areas exert feedforward and feedback influences through distinct frequency channels. Neuron 85(2):390–401. https://doi.org/10.1016/j.neuron.2014.12.018

Bauer A-KR, Debener S, Nobre AC (2020) Synchronisation of neural oscillations and crossmodal influences. Trends Cogn Sci 24(6):481–495. https://doi.org/10.1016/j.tics.2020.03.003

Blom JD (2010) A dictionary of hallucinations. Springer, New York

Bonnefond M, Jensen O (2012) Alpha oscillations serve to protect working memory maintenance against anticipated distracters. Curr Biol CB 22(20):1969–1974. https://doi.org/10.1016/j.cub.2012.08.029

Calvert GA, Bullmore ET, Brammer MJ, Campbell R, Williams SCR, McGuire PK, Woodruff PWR, Iversen SD, David AS (1997) Activation of auditory cortex during silent lipreading. Science 276(5312):593–596. https://doi.org/10.1126/science.276.5312.593

Canolty RT, Knight RT (2010) The functional role of cross-frequency coupling. Trends Cogn Sci 14(11):506–515. https://doi.org/10.1016/j.tics.2010.09.001

Christophel TB, Klink PC, Spitzer B, Roelfsema PR, Haynes J-D (2017) The distributed nature of working memory. Trends Cogn Sci 21(2):111–124. https://doi.org/10.1016/j.tics.2016.12.007

Cowan N (2001) The magical number 4 in short-term memory: a reconsideration of mental storage capacity. Behav Brain Sci 24(1):87–114; discussion 114–85. http://eutils.ncbi.nlm.nih.gov/entrez/eutils/elink.fcgi?dbfrom=pubmed&id=11515286&retmode=ref&cmd=prlinks

da Silva FL (2013) EEG and MEG: relevance to neuroscience. Neuron 80(5):1112–1128. https://doi.org/10.1016/j.neuron.2013.10.017

Douglas RJ, Martin KAC (2004) Neuronal circuits of the neocortex. Neuroscience 27(1):419–451. https://doi.org/10.1146/annurev.neuro.27.070203.144152

Driver J, Noesselt T (2008) Multisensory interplay reveals crossmodal influences on "sensory-specific" brain regions, neural responses, and judgments. Neuron 57(1):11–23. https://doi.org/10.1016/j.neuron.2007.12.013

Driver J, Spence C (2000) Multisensory perception: beyond modularity and convergence. Curr Biol 10(20):R731–R735. https://doi.org/10.1016/S0960-9822(00)00740-5

Engle RW, Tuholski SW, Laughlin JE, Conway ARA (1999) Working memory, short-term memory, and general fluid intelligence: a latent-variable approach. J Exp Psychol Gen 128(3):309–331. https://doi.org/10.1037/0096-3445.128.3.309

Fries P (2005) A mechanism for cognitive dynamics: neuronal communication through neuronal coherence. Trends Cogn Sci 9(10):474–480. https://doi.org/10.1016/j.tics.2005.08.011

Froesel M, Cappe C, Hamed SB (2021) A multisensory perspective onto primate pulvinar functions. Neurosci Biobehav Rev 125:231–243. https://doi.org/10.1016/j.neubiorev.2021.02.043

Funahashi S, Bruce CJ, Goldman-Rakic PS (1989) Mnemonic coding of visual space in the monkey's dorsolateral prefrontal cortex. J Neurophysiol 61(2):331–349. https://doi.org/10.1152/jn.1989.61.2.331

Fuster JM, Alexander GE (1971) Neuron activity related to short-term memory. Science 173(3997):652–654. http://eutils.ncbi.nlm.nih.gov/entrez/eutils/elink.fcgi?dbfrom=pubmed&id=4998337&retmode=ref&cmd=prlinks

Fuster JM, Bodner M, Kroger JK (2000) Cross-modal and cross-temporal association in neurons of frontal cortex. Nature 405(6784):347–351. https://doi.org/10.1038/35012613

Ghazanfar AA, Schroeder CE (2006) Is neocortex essentially multisensory? Trends Cogn Sci 10(6):278–285. https://doi.org/10.1016/j.tics.2006.04.008

Giard MH, Peronnet F (1999) Auditory-visual integration during multimodal object recognition in humans: a behavioral and electrophysiological study. J Cogn Neurosci 11(5):473–490. https://doi.org/10.1162/089892999563544

Goldman-Rakic PS (1990) Cellular and circuit basis of working memory in prefrontal cortex of nonhuman primates. Prog Brain Res 85:325–335; discussion 335–6. http://eutils.ncbi.nlm.nih.gov/entrez/eutils/elink.fcgi?dbfrom=pubmed&id=2094903&retmode=ref&cmd=prlinks

Graziano MSA, Yap GS, Gross CG (1994) Coding of visual space by premotor neurons. Science 266(5187):1054–1057. https://doi.org/10.1126/science.7973661

Gui P, Ku Y, Li L, Li X, Bodner M, Lenz FA, Wang L, Zhou Y-D (2017) Neural correlates of visuo-tactile crossmodal paired-associate learning and memory in humans. Neuroscience 362(24):181–195. https://doi.org/10.1016/j.neuroscience.2017.08.035

Haegens S, Osipova D, Oostenveld R, Jensen O (2010) Somatosensory working memory performance in humans depends on both engagement and disengagement of regions in a distributed network. Hum Brain Mapp 31(1):26–35. https://doi.org/10.1002/hbm.20842

Jensen O, Gelfand J, Kounios J, Lisman JE (2002) Oscillations in the alpha band (9-12 Hz) increase with memory load during retention in a short-term memory task. Cerebral cortex (New York, N.Y. : 1991) 12(8):877–882. http://eutils.ncbi.nlm.nih.gov/entrez/eutils/elink.fcgi?dbfrom=pubmed&id=12122036&retmode=ref&cmd=prlinks

Jones EG, Powell TPS (1970) An anatomical study of Converging sensory pathways within the cerebral cortex of the monkey. Brain 93(4):793–820. https://doi.org/10.1093/brain/93.4.793

Klimesch W (1999) EEG alpha and theta oscillations reflect cognitive and memory performance: a review and analysis. Brain Res Brain Res Rev 29(2–3):169–195. http://eutils.ncbi.nlm.nih.gov/entrez/eutils/elink.fcgi?dbfrom=pubmed&id=10209231&retmode=ref&cmd=prlinks

Ku Y, Ohara S, Wang L, Lenz FA, Hsiao SS, Bodner M, Hong B, Zhou Y-D (2007) Prefrontal cortex and somatosensory cortex in tactile crossmodal association: an independent component analysis of ERP recordings. PLoS One 2(8):e771. https://doi.org/10.1371/journal.pone.0000771

Ku Y, Zhao D, Hao N, Hu Y, Bodner M, Zhou Y-D (2015) Sequential roles of primary somatosensory cortex and posterior parietal cortex in tactile-visual cross-modal working memory: a single-pulse transcranial magnetic stimulation (spTMS) study. Brain Stimul 8(1):88–91. https://doi.org/10.1016/j.brs.2014.08.009

Lakatos P, Chen C-M, O'Connell MN, Mills A, Schroeder CE (2007) Neuronal oscillations and multisensory interaction in primary auditory cortex. Neuron 53(2):279–292. https://doi.org/10.1016/j.neuron.2006.12.011

Lakatos P, Karmos G, Mehta AD, Ulbert I, Schroeder CE (2008) Entrainment of neuronal oscillations as a mechanism of attentional selection. Science 320(5872):110–113. https://doi.org/10.1126/science.1154735

Laurienti PJ, Kraft RA, Maldjian JA, Burdette JH, Wallace MT (2004) Semantic congruence is a critical factor in multisensory behavioral performance. Exp Brain Res 158(4):405–414. https://doi.org/10.1007/s00221-004-1913-2

Lauro LJR, Rosanova M, Mattavelli G, Convento S, Pisoni A, Opitz A, Bolognini N, Vallar G (2014) TDCS increases cortical excitability: direct evidence from TMS–EEG. Cortex 58:99–111. https://doi.org/10.1016/j.cortex.2014.05.003

Lefaucheur JP, Andre-Obadia N, Antal A, Ayache SS, Baeken C, Benninger DH, Cantello RM, Cincotta M, de Carvalho M, Ridder DD, Devanne H, Lazzaro

VD, Filipovic SR, Hummel FC, Jääskeläinen SK, Kimiskidis VK, Koch G, Langguth B, Nyffeler T et al (2014) Evidence-based guidelines on the therapeutic use of repetitive transcranial magnetic stimulation (rTMS). Clin Neurophysiol 125(11):2150–2206. https://doi.org/10.1016/j.clinph.2014.05.021

Liebe S, Hoerzer GM, Logothetis NK, Rainer G (2012) Theta coupling between V4 and prefrontal cortex predicts visual short-term memory performance. Nat Neurosci 15(3):456–462, S1-2. https://doi.org/10.1038/nn.3038

Lundqvist M, Rose J, Herman P, Brincat SL, Buschman TJ, Miller EK (2016) Gamma and Beta bursts underlie working memory. Neuron 90(1):152–164. https://doi.org/10.1016/j.neuron.2016.02.028

Marks LE (1975) On colored-hearing synesthesia: cross-modal translations of sensory dimensions. Psychol Bull 82(3):303–331. https://doi.org/10.1037/0033-2909.82.3.303

Martino G, Marks LE (2001) Synesthesia: strong and weak. Curr Dir Psychol Sci 10(2):61–65. https://doi.org/10.1111/1467-8721.00116

McDonald JJ, Teder-Sälejärvi WA, Hillyard SA (2000) Involuntary orienting to sound improves visual perception. Nature 407(6806):906–908. https://doi.org/10.1038/35038085

Meredith MA, Stein BE (1983) Interactions among converging sensory inputs in the superior colliculus. Science 221(4608):389–391. https://doi.org/10.1126/science.6867718

Michalareas G, Vezoli J, Van Pelt S, Schoffelen J-M, Kennedy H, Fries P (2016) Alpha-beta and gamma rhythms subserve feedback and feedforward influences among human visual cortical areas. Neuron 89(2):384–397. https://doi.org/10.1016/j.neuron.2015.12.018

Miller GA (1956) The magical number seven, plus or minus two: some limits on our capacity for processing information. Psychol Rev 63(2):81–97. https://doi.org/10.1037/h0043158

Miller KD, Pinto DJ, Simons DJ (2001) Processing in layer 4 of the neocortical circuit: new insights from visual and somatosensory cortex. Curr Opin Neurobiol 11(4):488–497. https://doi.org/10.1016/s0959-4388(00)00239-7

Misselhorn J, Friese U, Engel AK (2019) Frontal and parietal alpha oscillations reflect attentional modulation of cross-modal matching. Sci Rep 9(1):5030. https://doi.org/10.1038/s41598-019-41636-w

Miyashita Y, Chang HS (1988) Neuronal correlate of pictorial short-term memory in the primate temporal cortex. Nature 331(6151):68–70. https://doi.org/10.1038/331068a0

Moos A, Smith R, Miller SR, Simmons DR (2014) Cross-modal associations in synaesthesia: vowel colours in the ear of the beholder. I-Perception 5(2):132–142. https://doi.org/10.1068/i0626

Morrell F (1972) Visual systems view of acoustic space. Nature 238(5358):44. http://eutils.ncbi.nlm.nih.gov/entrez/eutils/elink.fcgi?dbfrom=pubmed&id=12635274&retmode=ref&cmd=prlinks

Murray MM (2005) Grabbing your ear: rapid auditory-somatosensory multisensory interactions in low-level sensory cortices are not constrained by stimulus alignment. Cereb Cortex (New York, NY: 1991) 15(7):963–974. https://doi.org/10.1093/cercor/bhh197

Murray EA, Mishkin M (1985) Amygdalectomy impairs crossmodal association in monkeys. Science 228(4699):604–606. https://doi.org/10.1126/science.3983648

Ohara S, Lenz F, Zhou Y-D (2006) Sequential neural processes of tactile-visual crossmodal working memory. Neuroscience 139(1):299–309. https://doi.org/10.1016/j.neuroscience.2005.05.058

Padberg J, Seltzer B, Cusick CG (2003) Architectonics and cortical connections of the upper bank of the superior temporal sulcus in the rhesus monkey: an analysis in the tangential plane. J Comp Neurol 467(3):418–434. https://doi.org/10.1002/cne.10932

Raghavachari S, Kahana MJ, Rizzuto DS, Caplan JB, Kirschen MP, Bourgeois B, Madsen JR, Lisman JE (2001) Gating of human theta oscillations by a working memory task. J Neurosci 21(9):3175–3183. http://eutils.ncbi.nlm.nih.gov/entrez/eutils/elink.fcgi?dbfrom=pubmed&id=11312302&retmode=ref&cmd=prlinks

Ramachandran VS, Hubbard EM (2003) Hearing colors, tasting shapes. Sci Am 288(5):52–59. https://doi.org/10.1038/scientificamerican0503-52

Richter CG, Thompson WH, Bosman CA, Fries P (2017) Top-down beta enhances bottom-up gamma. J Neurosci 37(28):6698–6711. https://doi.org/10.1523/jneurosci.3771-16.2017

Romanski LM, Tian B, Fritz J, Mishkin M, Goldman-Rakic PS, Rauschecker JP (1999) Dual streams of auditory afferents target multiple domains in the primate prefrontal cortex. Nat Neurosci 2(12):1131–1136. https://doi.org/10.1038/16056

Senkowski D, Talsma D, Grigutsch M, Herrmann CS, Woldorff MG (2007) Good times for multisensory integration: effects of the precision of temporal synchrony as revealed by gamma-band oscillations. Neuropsychologia 45(3):561–571. https://doi.org/10.1016/j.neuropsychologia.2006.01.013

Singer W, Gray CM (1995) Visual feature integration and the temporal correlation hypothesis. Annu Rev Neurosci 18:555–586. https://doi.org/10.1146/annurev.ne.18.030195.003011

Spitzer B, Blankenburg F (2012) Supramodal parametric working memory processing in humans. J Neurosci 32(10):3287–3295. https://doi.org/10.1523/jneurosci.5280-11.2012

Sully J (1879) Harmony of colors. Mind os-4(14):172–191. https://doi.org/10.1093/mind/os-4.14.172

van Driel J, Knapen T, van Es DM, Cohen MX (2014) Interregional alpha-band synchrony supports temporal cross-modal integration. NeuroImage 101:404–415. https://doi.org/10.1016/j.neuroimage.2014.07.022

van Kerkoerle T, Self MW, Dagnino B, Gariel-Mathis M-A, Poort J, van der Togt C, Roelfsema PR (2014) Alpha and gamma oscillations characterize feedback and feedforward processing in monkey visual cortex. Proc Natl Acad Sci 111(40):14332–14341. https://doi.org/10.1073/pnas.1402773111

van Kerkoerle T, Self MW, Roelfsema PR (2017) Layer-specificity in the effects of attention and working memory on activity in primary visual cortex. Nat Commun 8:13804. https://doi.org/10.1038/ncomms13804

Varela F, Lachaux JP, Rodriguez E, Martinerie J (2001) The brainweb: phase synchronization and large-scale integration. Nat Rev Neurosci 2(4):229–239. https://doi.org/10.1038/35067550

Voss JL, Federmeier KD (2011) FN400 potentials are functionally identical to N400 potentials and reflect semantic processing during recognition testing. Psychophysiology 48(4):532–546. https://doi.org/10.1111/j.1469-8986.2010.01085.x

Wang L, Li X, Hsiao SS, Lenz FA, Bodner M, Zhou Y-D, Fuster JM (2015) Differential roles of delay-period neural activity in the monkey dorsolateral prefrontal cortex in visual-haptic crossmodal working memory. Proc Natl Acad Sci 112(2):E214–E219. https://doi.org/10.1073/pnas.1410130112

Yovel G, Paller KA (2004) The neural basis of the butcher-on-the-bus phenomenon: when a face seems familiar but is not remembered. NeuroImage 21(2):789–800. https://doi.org/10.1016/j.neuroimage.2003.09.034

Yu Y, Huber L, Yang J, Jangraw DC, Handwerker DA, Molfese PJ, Chen G, Ejima Y, Wu J, Bandettini PA (2019) Layer-specific activation of sensory input and predictive feedback in the human primary somatosensory cortex. Science. Advances 5(5):eaav9053. https://doi.org/10.1126/sciadv.aav9053

Zhao D, Ku Y (2018) Dorsolateral prefrontal cortex bridges bilateral primary somatosensory cortices during cross-modal working memory. Behav Brain Res 350:116–121. https://doi.org/10.1016/j.bbr.2018.04.053

Zhou Y-D, Fuster JM (1997) Neuronal activity of somatosensory cortex in a cross-modal (visuo-haptic) memory task. Exp Brain Res 116(3):551–555. http://eutils.ncbi.nlm.nih.gov/entrez/eutils/elink.fcgi?dbfrom=pubmed&id=9372304&retmode=ref&cmd=prlinks

Zhou Y-D, Fuster JM (2000) Visuo-tactile cross-modal associations in cortical somatosensory cells. Proc Natl Acad Sci U S A 97(17):9777–9782. http://www.ncbi.nlm.nih.gov/pmc/articles/PMC16941/

Synesthetic Correspondence: An Overview

7

Lihan Chen

Abstract

Intramodal and cross-modal perceptual grouping based on the spatial proximity and temporal closeness between multiple sensory stimuli, as an operational principle has built a coherent and meaningful representation of the multisensory event/object. To implement and investigate the cross-modal perceptual grouping, researchers have employed excellent paradigms of spatial/temporal ventriloquism and cross-modal dynamic capture and have revealed the conditional constraints as well as the functional facilitations among various correspondence of sensory properties, with featured behavioral evidence, computational framework as well as brain oscillation patterns. Typically, synesthetic correspondence as a special type of cross-modal correspondence can shape the efficiency and effect-size of cross-modal interaction. For example, factors such as pitch/loudness in the auditory dimension with size/brightness in the visual dimension could modulate the strength of the cross-modal temporal capture. The empirical behavioral findings, as well as psychophysical and neurophysiological evidence to address the cross-modal perceptual grouping and synesthetic correspondence, were summarized in this review. Finally, the potential applications (such as artificial synesthesia device) and how synesthetic correspondence interface with semantics (sensory linguistics), as well as the promising research questions in this field have been discussed.

Keywords

Cross-modal · Perceptual grouping · Synesthetic correspondence · Individual differences · Sensory linguistics

7.1 Introduction

Cross-modal integration entails the correspondence between different sensory properties to reach a coherent and meaningful representation of the environment as well as the target events/objects (Ernst and Bülthoff 2004; Stein and Meredith 1993). Cross-modal correspondences

L. Chen (✉)
School of Psychological and Cognitive Sciences, Peking University, Beijing, China

Beijing Key Laboratory of Behavior and Mental Health, Peking University, Beijing, China

Key Laboratory of Machine Perception (Ministry of Education), Peking University, Beijing, China

National Key Laboratory of General Artificial Intelligence, Peking University, Beijing, China

National Engineering Laboratory for Big Data Analysis and Applications, Peking University, Beijing, China
e-mail: CLH@pku.edu.cn

Y. Gu, A. Zaidel (eds.), *Advances of Multisensory Integration in the Brain*, Advances in Experimental Medicine and Biology 1437, https://doi.org/10.1007/978-981-99-7611-9_7

(CMC) are defined as pairs of associations between two sensory or cognitive processes that are generally agreed upon by most individuals within a population (Brang et al. 2013). In a seminal tutorial by Charles Spence (Spence 2011), CMC has been defined to contain three forms—statistical correspondences, structural correspondence, and semantically mediated correspondences. The three forms have been examined and materialized in a number of studies (Bien et al. 2012; Bremner et al. 2013; Guo et al. 2017; Ngo et al. 2013; Parise and Spence 2012; Piqueras-Fiszman and Spence 2011; Spence and Deroy 2013; Wan et al. 2014; Zeljko et al. 2019; Zhang and Chen 2016). Although the seminal categorization of CMC has been proposed almost a decade ago, the empirical research evidence as well as the concepts developed over the years, could still be nicely fit into this framework. Accumulated evidence has shown the perceptual grouping and intersensory binding of sensory properties indeed affect the ultimate outcome of multisensory integration. The behavioral manifestation and perceptual benefits (as well as interferences with incongruent CMC) are solid from empirical experiments though the underlying mechanism of CMC is far from clear.

Among the kaleidoscope of CMC, the synesthetic correspondence is the most intriguing type and modulates the efficiency and effect-size of cross-modal interaction (Bien et al. 2012; Chen et al. 2016; Grossenbacher and Lovelace 2001; Hidaka and Yaguchi 2018; Parise and Spence 2009; Saenz and Koch 2008). Below I address respectively, the manifestations of synesthetic correspondence, the underlying neurocognitive mechanisms, sound symbolism as an example and the individual differences research approach, as well as the potential applications in this field.

7.2 Synesthetic Correspondence and Cross-modal Integration

The correspondence of properties among different sensory modalities raises two prominent questions as well as challenges. On the one hand, how to compare and measure the qualia (properties) associated with different sensory modalities. For example, as we learn, the properties for different sensory stimuli (e.g., sound pitch vs. visual size) are not comparable apparently, so that how to standardize them to reach "correspondence"? On the other hand, it pits the challenge to describe this correspondence with appropriate words—both with overt/precise words and with covert/implicit descriptions. Linguistic sensory words are usually deficient to describe and capture the large volumes of sensory properties. "The word we use for things like the redness of a rose, beauty, leadership, and charisma. We usually know exactly what we mean, but we have no words to describe let alone define it" (Ross 2018).

The concept of "correspondence" has originally arisen and henceforth extensively exploited in unimodal—visual domain. In the dynamic visual scene, during the continuous eye movements for visual perception, an outstanding question is that how the visual system could establish and maintain the identity of an object under above synergistic "interaction" between environment and observers (say, juggling), i.e., *correspondence problem* (Ullman 1979). Studies using motion correspondence have explored the modulating factors including featural aspect. In Hein and Cavanagh (2012) study, they used Ternus display (a form of ambiguous visual apparent motion) and found contrast polarity, spatial frequency, and size modulate the perceived motion categorization (such as bias in element motion), and hence demonstrated a spatiotopic-based feature bias (Hein and Cavanagh 2012). The Ternus display has been exploited to investigate the perceptual grouping in auditory modality (Wang et al. 2014), tactile modality (Chen et al. 2010), and multisensory processing (Fig. 7.1) (Chen et al. 2018).

Cross-modal correspondences (CMCs) have been defined as "a tendency for a feature, attribute, dimension, or stimulus in one sensory modality, either physically present or merely imagined, to be matched (or associated) with a feature, attribute, dimension, or stimulus in another sensory modality" (Parise 2016). Usually, audiovisual correspondences have been adopted to investigate CMC. Below, synesthetic corre-

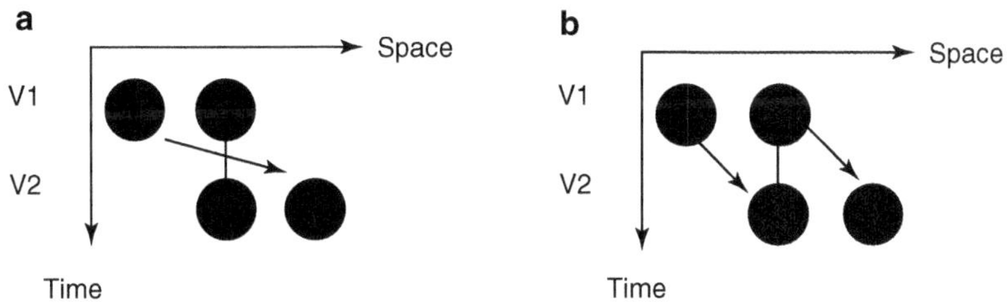

Fig. 7.1 Ternus display and stimulus configurations. Visual Ternus display contained two types of exclusive percepts: (**a, b**). "Group" motion in which the two frames move from one side to the other as a whole. (From Fig. 1, Chen et al. (2018), Journal of Experimental Psychology: General, with permission)

spondence was illustrated with appropriate examples following the categorization proposed by Spence (2011).

The "structural" correspondence, has been classically materialized in spatial/temporal ventriloquism paradigm, in which the perceived temporal onset has been shifted due to the presence of concurrent task-irrelevant distractors (temporal ventriloquism) or been mislocated due to the proximate distractors (spatial ventriloquism) (Bertelson and Aschersleben 2003; Bertelson et al. 2000; Chen and Vroomen 2013; Morein-Zamir et al. 2003; Ogulmus et al. 2018; Orchard-Mills et al. 2016; Slutsky and Recanzone 2001; Tian et al. 2020; Vroomen et al. 2004; Vroomen and Keetels 2006). According to this structural account, with the ventriloquism paradigm, perceptual grouping across space and time is operating in an automatic fashion to arrange stimuli according to their magnitudes/intensities. Or put in another way, the correspondence modulates the spatiotemporal factors in intra- and cross-modal groupings. For instance, Parise and Spence (2009) once reported that spatial and temporal ventriloquism effects are enhanced for cross-modally congruent pairs of auditory and visual stimuli as compared with pairs of incongruent stimuli (Parise and Spence 2009), prominently, pitch-size cross-modal correspondence modulated the spatial ventriloquism effect as well (Parise and Spence 2008). As a counterpart for spatial ventriloquism, Orchard-Mills et al. (2016) investigated whether cross-modal correspondence between auditory pitch and visual elevation modulates temporal ventriloquism. Using temporal order judgment, they asked participants to judge the order of two visual stimuli (above and below fixation) across a range of stimulus onset asynchronies (SOAs). The results show that incongruent pairings between pitch and elevation abolish temporal ventriloquism. In contrast, the potential "congruent" ("facilitation") effect was dependent on the saliency of the cross-modal mapping (Orchard-Mills et al. 2016).

In statistical account, cross-modal correspondences can be thought of as the internalization of the statistical regularities of the stimuli associations (combinations) and the environment. Human observers use the similarity in the temporal structure of multisensory signals to solve the correspondence problem—inferring causation from correlation (Parise et al. 2012). With an elegant design, Parise et al. (2012) uncovered the role of correlation between the fine temporal structure of auditory and visual signals in causal inference (i.e., correspondence problem). They found that in a localization task with visual, auditory, and combined audiovisual targets, the improvement in precision for combined (auditory and visual stimuli) relative to unimodal targets (only with auditory stimuli) was statistically optimal only when audiovisual signals were correlated. In a following study, Parise and Ernst further generalized the research scope and proposed that "correlation" detection as a general mechanism for multisensory integration. They termed the multisensory correlation detector which integrates related multisensory signals through a set of temporal filters followed by linear combination. The "correlation" detection mechanism successfully explains a range of phenomena with physiologically plausible model

(Parise and Ernst 2016). The above two representative studies suggested that the "statistical" relations between signals from two different modalities, have essentially built "correspondence" in temporal relations to facilitate perceptual decision-making, such as spatial localization for auditory target.

In cross-modal interaction scenario, Zhang and Chen (2016) applied the "correspondence" principle and manipulated the statistical relationship between the temporal onsets of two beeps and two visual Ternus frames, in which the visual frames with different colors ("red" or "black") were associated with certain audiovisual temporal structures (Fig. 7.2). They found that by binding the relevant temporal information and stimuli properties—through manipulating the probabilities of the occurrences of audiovisual events, the perception of visual Ternus motion could be quickly recalibrated. Observers could implicitly employ the temporal (interval) relations between the target events as a prior and demonstrate the

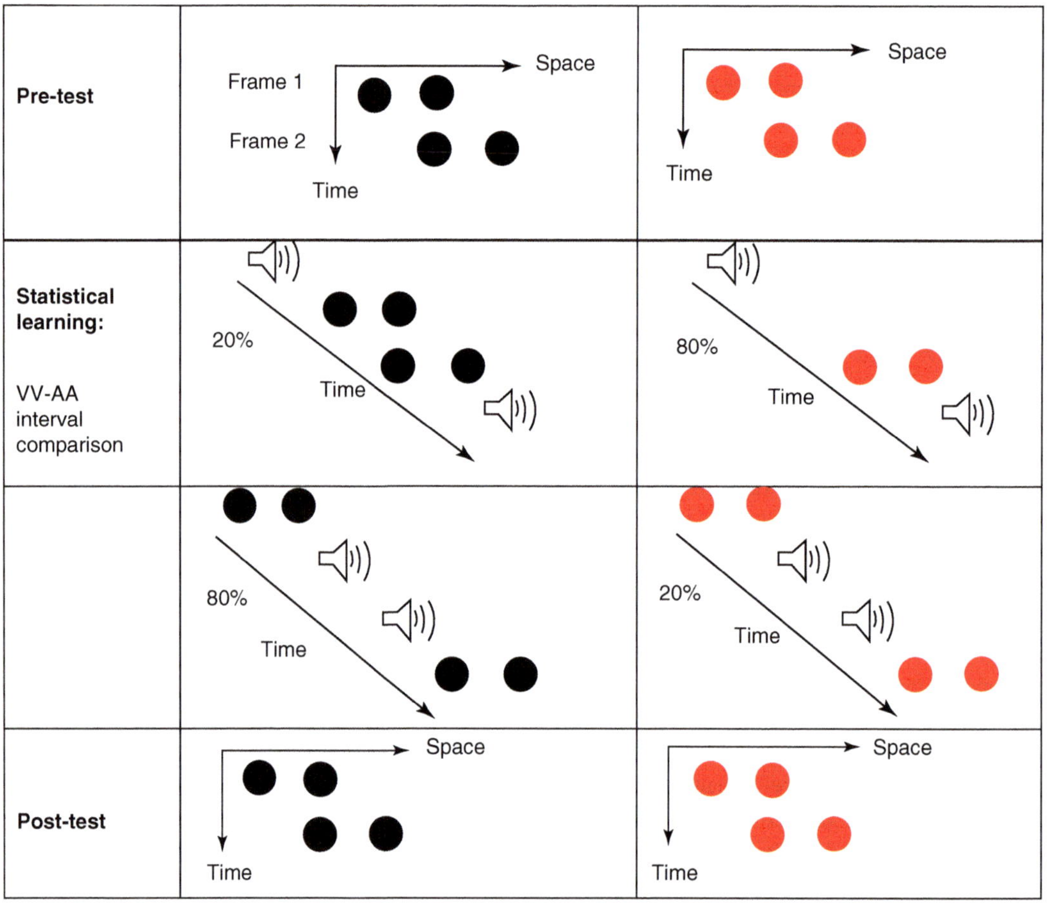

Fig. 7.2 Ternus displays (for pretest and posttest) and illustrations of the stimuli configurations for Experiment 1 in Zhang and Chen (2016). Two kinds of Ternus display were adopted (with black frames or red frames). In the training session, for black-black (BB) configuration, in 80% trials, first sound preceded the first red visual frame and the second sound trailed the second red visual frame by 80 ms (i.e., "AVVA" condition). In red-red (RR) configuration, in 80% trials, the first visual frame preceded the first beep and the second visual frame trailed the second beep by 80 ms (hereafter "VAAV" condition). The inter-stimulus-interval (ISI) between two Ternus frames were randomly set from 50 to 230 ms. For the rest 20% trials, the RR and BB configurations were associated with temporal structures of AVVA and VAAV, respectively. (From Fig. 2 in Zhang and Chen (2016), Frontiers in Psychology, with permission)

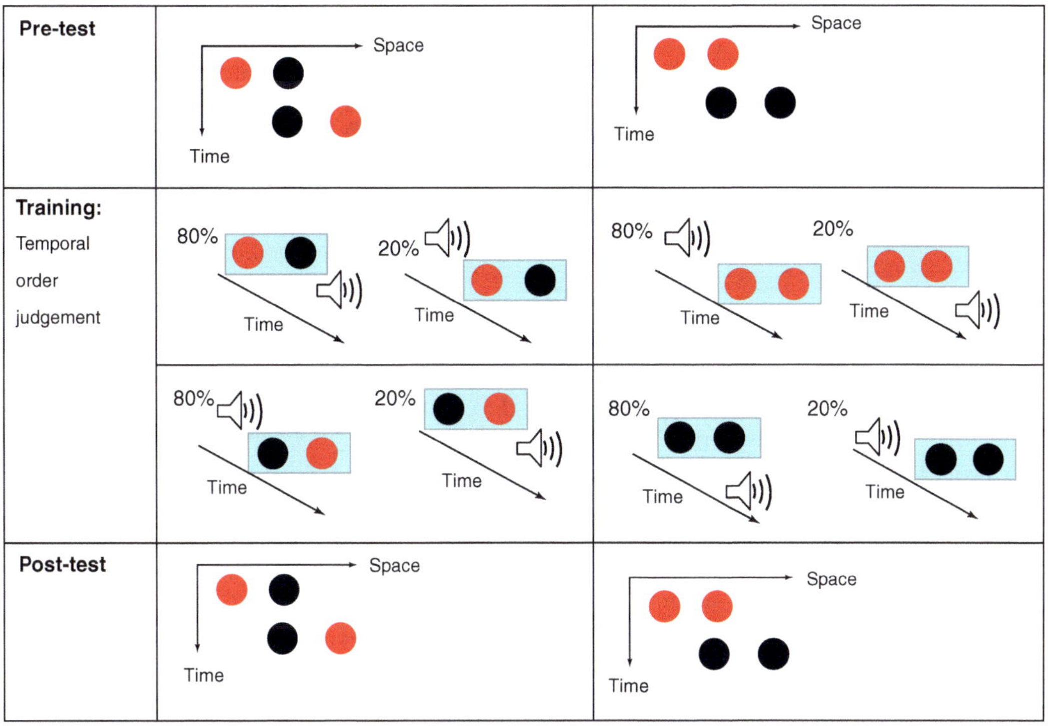

Fig. 7.3 Ternus displays (for pretest and posttest) and illustrations of the stimuli configurations for Experiment 2 in Zhang and Chen (2016). Two kinds of Ternus display were adopted. In the training session, for RB-BR configuration, in 80% trials, the RB frames preceded the sound beep and BR frames trailed the beep. In contrast, 20% of RB frames trailed the sound beep and 20% BR frames preceded the sound beep. For RR-BB configuration, 80% of the RR frames trailed the sound beep and BB frames preceded the beep. In contrast, 20% of RR frames preceded the sound beep and BB frames trailed the sound beep (From Fig. 5 in Zhang and Chen (2016), Frontiers in Psychology, with permission)

observable temporal aftereffects which give rise to the featured proportional changes of reporting element motion vs. group motion of Ternus display (Zhang and Chen 2016) (Fig. 7.3).

Synesthetic correspondence represents a natural tendency of associating sensory properties between two typical modalities and contains semantic meanings (values). For example, elevation serves as a fundamental organizational dimension for many cross-modal correspondences. Visual elevation corresponds to the high pitch and while the low visual location (height) naturally corresponds to the low auditory pitch, according to our gained world knowledge. People often show a systematic tendency to associate moving objects with changing pitches.

Saenz and Koch reported that auditory synesthesia does indeed exist with evidence from four healthy adults for whom seeing visual flashes or visual motion automatically causes the perception of sound (Saenz and Koch 2008).

By employing the synesthetic audiovisual correspondences between the visual Ternus movement directions (upward and downward) and the changes of pitches of concurrent glides, Guo et al. (2017) examined how human observers' cognitive abilities, typically, the cognitive styles (field-dependent vs. field-independent) modulate/make distinction the relations between the properties of pitch glides and the perception of visual apparent motion (Fig. 7.4). The results indicated that for pitch ascending (decreasing) glides, when they were paired with moving upward (downward) visual Ternus, observers would perceive dominant percept of "element motion." Importantly, field-independent observ-

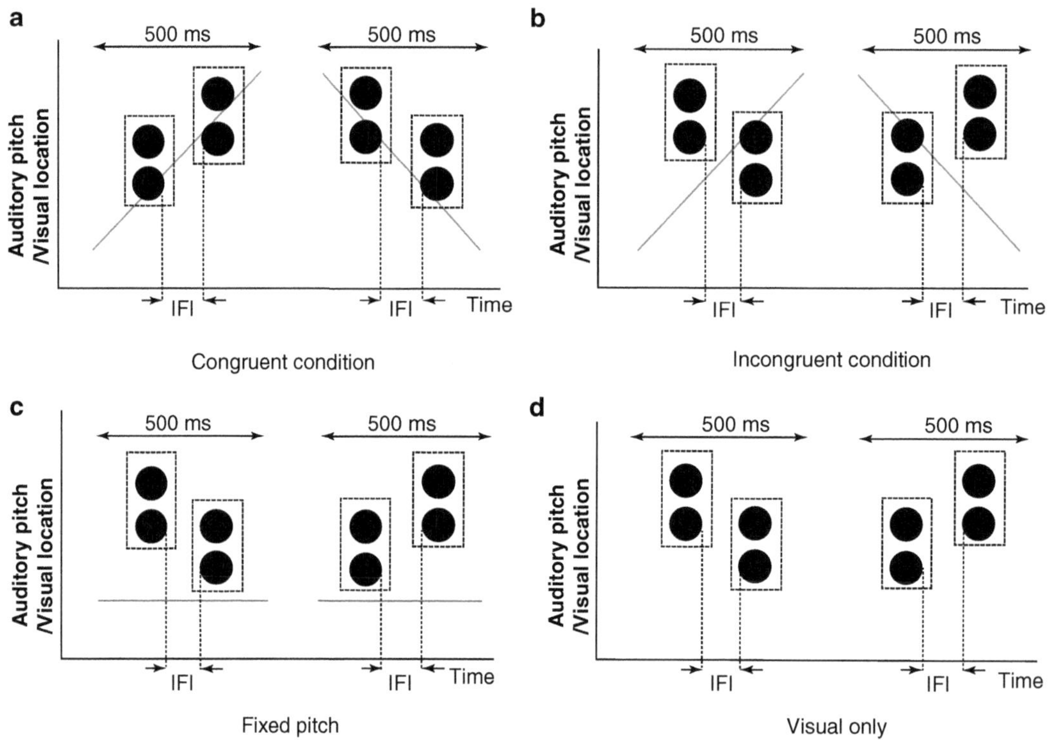

Fig. 7.4 An illustration of the stimuli configuration for Guo et al. (2017). There were four configurations: "Congruent," "Incongruent," "Fixed-pitched," and "Visual only." The former three conditions contained audiovisual stimuli. With the differential directions of the solid lines, the glide was either one of the three types: ascending, descending, or invariant. "Visual only" referred to visual Ternus display without accompanying sounds. The durations of audiovisual stimuli were constant at 500 ms. For "congruent" configuration, Ternus displays with upward/downward motion were given together with ascending-pitched/descending-pitched glides. For "incongruent" condition, upward (downward) Ternus motion was synchronized with descending (ascending) pitch glide. For "fixed-pitched" condition, a glide with pure tone (500 Hz) was given. IFI = inter-frame-interval. (From Fig. 2 in Guo et al. (2017), Multisensory Research, with permission)

ers tended to readily identify the audiovisual events and use the principle of cross-modal correspondence to demonstrate the above "biased" perception (Guo et al. 2017).

Synesthetic correspondence with semantic (linguistic) relations is a higher level and abstract process. Often stimuli may be related because of the similar terms we sometimes use to describe stimuli presented in different senses, such as the "high"/"low" could be used as common words for both pitch and spatial elevation in English (also in Chinese) (Martino and Marks 1999; Sadaghiani et al. 2009). One typical example of semantic correspondence is sound symbolism (addressed in detail in the following session), referring to the non-acoustic information (literal meaning) of auditory property. The "higher-level" nature is supported by the evidence that the sensitivities to sound symbolism increased with acquired (greater) language experience.

7.3 Sound Symbolism as a Form of Synesthetic Correspondence

According to the Cassierer's framework, language provides two very different ways of expressing interrelationships among the senses. One is analogical and the other is symbolic. Language serves as a formal representation to communicate cross-modal equivalence of (sensory) properties (Cassirer 1953). In sensory linguistic, sound symbolism is not referring to onomatopoeic expressions like words—"buzz," but reflects some nonacoustical property of nature. The relationship between sound and meaning is indeed intrinsic. The most significant and popular exemplar of sound symbolism is the "bouba-kiki" effect, which refers to the *shape-sound* correspondence (Barton and Halberstadt 2018; Cuskley et al. 2017; De Carolis et al. 2018; Fort et al. 2015; Fryer et al. 2014; Gold and Segal 2017; Karthikeyan et al. 2016; Köhler 1929; Maurer et al. 2006; Peiffer-Smadja and Cohen 2019; Sakamoto and Watanabe 2018). People usually associate "Bouba" to an image of a rounded object and "Kiki" to an image of a spiky object. Indeed, features, phonemes, and letters may all contribute to sound symbolism. Sound symbolism can be found in different literature forms and across different levels.

In his seminal book "The Unity of the Senses", Lawrence E. Marks introduced a very interesting topic for Sound Symbolism in Poetry, in which he addressed the expressions of sensory analogy and correspondence in literature, and mainly with examples from the poetry of the nineteenth century in the western world (Marks 2014). Poetry is a container that showcases the sound symbolism in speech. Sounds suggest meanings through suprasensory perceptual attributes and one way that sounds symbolize meaning is by means of affect. As reported by earlier studies, /a/ and /o/ were judged more pleasant than /u/ (Roblee and Washburn 1912). Speech sounds could represent more general and delicate sensory properties such as the gradations in brightness and changes in spatial size. Macdermott (1940) summarized how low-pitched and high-pitched vowels correspond to the sensory attributes by the method of semantic differentiation. On the other hand, synesthetic metaphor conveys intrinsic cross-modal relationship and suggests a trend of the evolution for semantic (language) adjectives that applied to typical sensations, such as "sharp" and "dull" for descriptions of tactile and visual sensations (Macdermott 1940). Westbury et al. (2018) constructed graphic representation of correlations between hybrid model estimates (adjectives) of the categories, revealing the overlapping of estimations to different degrees. Therefore, sound symbolism operates at broad categorical levels (Westbury et al. 2018).

7.4 Neurocognitive Mechanism of Synesthetic Correspondence

Researchers have developed experimental paradigms of explicit (cross-modal matching, including forced-choice) (Kohler 1947) and implicit associations (with reaction times) (Hung et al. 2017; Parise and Spence 2012; Westbury 2005) as well as Transcranial Magnetic Stimulation (TMS) (Bien et al. 2012), to address the neural substrates which are responsible for the (synesthetic) correspondence.

Bien et al. (2012) had carried out a representative study to uncover the cross-modal binding in pitch-size synesthesia by adopting a combined TMS, EEG, and psychophysics approach. Using an auditory spatial localization task (Ventriloquist effect), they found that the performance of localization was impaired with the congruent mapping of pitch-size. Synesthetic congruency results in multisensory binding, and consequently auditory sources are misallocated to the visual source. To pin down the neural substrates, they further purposely disturbed the right intraparietal sulcus (via TMS) and observed the ventriloquist effect was abolished and largely identified right intraparietal sulcus being the critical cerebral area to account for the pitch-size mapping. Further correlation analysis (together with ERP) indicated the origin of synesthetic pitch-size mappings to a right intraparietal involvement around 250 ms (Bien et al. 2012).

Synesthetic correspondence is linked with the adjacency of the neural anatomies between two brain areas, which are responsible for encoding/representing the individual (two) sensory properties, respectively. An interesting question is whether there is functional dominant of a sensory modality over the other as in typical multisensory interaction. Evidence has shown that pitch-height associations are vision-dependent, suggesting the visual experience plays an important role in the synesthetic correspondence (Deroy et al. 2016a, b; Fryer et al. 2014). The dependency on dominant sensory modality (visual modality) in synesthetic correspondence has been studied with special group who have deficiency in one sensory modality (such as vision). For example, Bottini et al. (2019) revealed an intriguing possibility that people with early blindness may be more sensitive to (some) iconic features of language compared to sighted (Bottini et al. 2019). In the absence of a direct visual stimulus, visual imagery plays a role in cross-modal integration. As an important replication of "Bouba-Kiki" effect that has been demonstrated in visual-auditory modalities over 80 years ago, Fryer et al. (2014) implemented "Bouba-Kiki" effect in auditory-haptic modalities in which participants received "felt" shapes but did not see them. The sighted participants showed a robust "Bouba-Kiki" effect while people with visual impairments had less pronounced effect.

The neural mechanisms for synesthesia provide valuable information for the corresponding neural mechanism of synesthetic correspondence since synesthesia has been recognized as a specialized/extreme form of "synesthetic correspondence" in its wide spectrum (Marks and Mulvenna 2013). The above cross-talk between two brain areas suggests one representative hypothesis regarding synesthesia: direct cross-activation. The cross-activation model suggests that synesthesia has been attributed to specific anatomical pathways, which are weak or absent for non-synesthetes. Those pathways provide neural underpinnings that a direct link between one sensory modality to another becomes viable (Hubbard et al. 2005, 2011; Ramachandran and Hubbard 2001a, b, 2003).

The hyperbinding account, however, assumes common brain areas for synesthetes and non-synesthetes. However, the two groups differed in the function as well as efficiency of the cross-talk between two senses. Especially, the synesthetes generally have weaker inhibition of the feedback from the inducers (Grossenbacher and Lovelace 2001).

7.5 Cognitive Benefits of Synesthetic Associations and the Role of Training (Experience)

Synesthetic associations bring forth cognitive benefits or differentiations in cognitive performance, including executive control functions, number-space perceptions, and memory functions. Synesthesia broadly impacts perception but has differential impacts (McCarthy and Caplovitz 2014). Rouw et al. (2013) investigated the relationship between synesthesia and executive control functions. Using traditional executive control paradigms and Stroop tasks, they suggested no clear relationship between executive control functions and synesthetic behavioral effects (Rouw et al. 2013).

Gertner et al. (2013) found that synesthetes' automaticity of processing given semantic meaning of numerals were affected by their number-space perceptions. Specifically, both synesthetes and their control peer exhibit the similar, classic size congruency effect (SiCE) for numerical block. However, for the case of physical block, synesthetes demonstrated weak automaticity in processing numerical magnitude when the numbers for comparison were placed incompatibly concerning their relative position in the format of "synesthetic" number space (Gertner et al. 2013).

Ramachandra (2016) investigated the influence of lexical-gustatory synesthesia on memory, using a paired-associate learning task. A single subject (synesthete) predicted that her learning would be better in the "taste" condition when compared to the "no taste" condition, showing a "foresight bias" in the vein of metamemory task (judgement-of-learning) although no significant differences emerged between the "no taste" and "taste" conditions (Ramachandra 2016).

Synesthetic associations are experience dependent. Using sound symbolic novel words and foreign words meaning round and pointy, and conducting a cross-modal matching task (i.e., word-referent mapping, picking round/pointy picture to correspond given word), Tzeng et al. found that while 3-year-old participants showed chance performance, 5- and 7-year-olds showed reliably above-chance performance. In this means, sound symbolism sensitivity is a manifestation of the pre-linguistically available cross-modal mappings (Tzeng et al. 2017). Intersensory correspondences are based on neural connections present very early in life. However, the correspondences are dependent on the specific combinations of cross-modal properties pair.

Haryu and Kajikawa (2012) found 10-month-olds paired a higher frequency tone with an object of a brighter color and associated a lower frequency tone with darker color. However, for the size dimension, the same group did not always pair a higher frequency tone with a smaller object, nor did they pair a lower frequency tone with a larger object. Those findings indicated infants have an initial tendency to match pitch with brightness, while the pitch-size correspondences were loosely connected but could be formed tightly through statistical learning of the simultaneously presented stimuli pair ("pitch-size") (Haryu and Kajikawa 2012).

Extensive cognitive training has been shown to generate synesthesia-like phenomenology. To address the question that whether these experiences are accompanied by neurophysiological changes characteristic of synesthesia will provide a unique opportunity to elucidate the neural basis of perceptual plasticity relevant to conscious experiences. Rothen et al. (2018) confirmed that overtraining synesthetic associations results in synesthetic phenomenology, by using Stroop tasks to reveal synesthesia-like performance following training (Rothen et al. 2018). Goodhew et al. (2015) found that synesthetes produce substantially greater semantic priming magnitudes (using a lexical decision task and a semantic categorization task), unrelated to their specific synesthetic experience (Goodhew et al. 2015). van Petersen et al. (2020) found individuals with sequence-space synesthesia (SSS) who perceive sequences like months, days, and numbers in certain spatial arrangements, have enhanced spatial navigation skills, in addition to the established cognitive benefits such as enhanced mental rotation, more vivid visual imagery and an advantage in spatial processing (van Petersen et al. 2020).

Synesthetic correspondence entails the cognitive ability that appropriately binds the elementary properties of visual and auditory events, the weak correspondence might exist in atypical developing groups (including dyslexics and autistic individuals).

For a starter in reading, they should initially form a basic correspondence between orthographic tokens and phonemic utterances. The associations between letters and sounds remained intact for dyslexics (Blomert 2011). —Synesthetic correspondence, as a type of cross-modal association, has been revealed to remain robustly for dyslexic children—match larger visual shapes with lower auditory pitch, or smaller visual forms with higher auditory pitches (Chen et al. 2016). With the visual temporal order judgment (TOJ) task (two consecutively presented discs were enclosed by two beeps), Chen et al. (2016) also found that the congruent audiovisual pair boosted the TOJ performance by reducing the just noticeable difference (JND), whereas the incongruent audiovisual pair degraded the performance by increasing the JND. The modulation magnitudes were larger in dyslexic children than in normal peer. Importantly, this very basic perceptual performance of TOJ, has been revealed to be correlated with the higher order cognitive expertise such as reading skills (measured by scores of character recognition test and reading fluency) (Chen et al. 2016).

Occelli et al. (2012) reported that when children with Autism Spectrum Disorders (ASDs, 6–15 years) were asked to indicate which of two bouncing visual patterns was making a centrally located sound. In this parametric design, the visual size, surface brightness as well as visual shape were manipulated; the sounds were modified in pitches. The performance of ASDs were at chance levels, indicating that the general poor capabilities to integrate auditory and visual inputs, might constrain the synesthetic associations.

7.6 Synesthetic Correspondence in Sensory Branding

Assigning the right name to a given product is extremely important, as it can help enhance the overall product experience, the product satisfaction as well as ensure the subsequent, repeated consumption (Spence 2012). Sound symbolism has been exploited extensively in sensory branding field for the last decade. In a range of commercial fruit pulps/juices, Ngo et al. (2013) examined shape and sound symbolism. British and Colombian participants associate sweet juices with lower sourness to rounder shapes, while they consistently map sour ones to angular shapes (Ngo et al. 2013). Those findings suggested that in packaging and labeling, it is important to obey these correspondences, especially when the customers are unfamiliar with those products. A "crossmodal-cognitive" perception model of uniqueness has been developed to account for the appropriate food names in terms of sound symbolism (Favalli et al. 2013). Within the evaluation procedure, the investigation of the product and its definition was composed of the combination of a cross-modal and cognitive approach. The moderating influence of the sound symbolism has been recently discovered in willingness-to-pay (WTP) behavior, in which back vowels and voiced consonants (linked with a larger size of retail store) would lead to higher WTP (Ketron and Spears 2019).

As suggested in the former section, the synesthetic correspondence is experience-dependent. However, the evidence for the cultural variation in synesthetic correspondence is rare. Interestingly, the "Bouba-Kiki" effect has also been observed in the Himba of Northern Namibia, a remote population with little exposure to Western cultural influences, though the effect has been robustly demonstrated in Western participants. However, it is not consistent with the mapping of shape to taste (Himba-associated carbonation to angular shape rather than rounded shape). This finding indicates both general and special (individualized) pattern for cross-modal correspondences under different cultural contexts (Bremner et al. 2013).

Investigations have been made as regards with the role of personality in synesthetic correspondences (or synesthesia). Generally, the synesthetes showed increased intelligence as compared with matched non-synesthetes. This was a general effect rather than bounded to a specific cognitive domain or to a specific (synesthesia-type to stimulus material) relationship. By tapping on the "big five" personality (the NEO-PI-R personality inventory), the expected effect of increased "Openness" in synesthesia was obtained, as well as two unexpected effects in personality traits (increased "Neuroticism" and decreased "Conscientiousness"). The personality and cognitive characteristics were found related to having synesthesia (in general) rather than to particular synesthesia subtypes. Therefore, this piece of finding supports the existence of a general synesthetic "trait," over the notion of relatively independent "types" of synesthesia (Rouw and Scholte 2016).

7.7 Sensory Linguistics Perspective: Plurality or Ambiguous Meanings

As seen from its literally meaning, "synesthetic correspondence" implies a common basis for sensory comparison. This idea was supported by ATOM (A Theory of Magnitude) model (Bueti and Walsh 2009; Walsh 2003). ATOM assumes a unified, cross-modal magnitude system for all kinds of sensory magnitudes (including numerical quantities) (Winter et al. 2015). Therefore, "magnitude" could be a common currency as well as cognitive quality among different sensory modalities.

However, there is plurality or ambiguous meanings in both lexicon and forms of correspondences. For example, the word of "hot" have at least two differential meanings. Its basic meaning is referring to higher "temperature." The other one could be "spicy" for foods.

Interestingly, Rakova (2003) investigated the neural basis for the two perceptual meanings and found they a had been sub-served by the same underlying neural system (Rakova 2003). The plurality of words has found its root in "shared" sensations among different sensory channels. For instance, "Words denoting taste cannot always be separated clearly from those for feel and smell" (Staniewski 2017). This common sharing could not be based on perceptual dimension but also on emotional one. For example, the associations between music and color are based more on emotional correspondences (e.g., between major mode and happiness) rather than on perceptual correspondences (Palmer et al. 2013).

In a sense of "embodied cognition," different senses diversify in their relatedness to one's own body. The perceptual experiences through vision and hearing have been mainly projected onto objects (see also Shen 2008, p. 302).

Touch, taste, and smell, on the other hand, are argued to be relatively more subjective and more related to one's bodily consciousness. Zhao et al. (2019) examined the mapping directionality tendencies of linguistic synesthesia in Mandarin. They adopted a corpus-based approach and categorized three types of directional tendencies for Mandarin synesthesia: unidirectional biased-directional and bidirectional (Zhao et al. 2019). The directional flow is determined by the sensory dominance of one modality over the other. For instance, mapping from TASTE to HEARING rather than the opposite is typical for "unidirectional" tendency.

This directional flow of sensory mapping indeed constrains the synesthetic correspondence among different cohorts of sensory properties.

On the other hand, efficient codability is a psycholinguistic measure of the relative ease of expressing certain percepts. It is generally thought that sight is the most codable sensory modality. Levinson and Majid (2014) say that "in English at least, it seems generally easier to linguistically code colors than (nonmusical) sounds, sounds than tastes, tastes than smells" (p. 415) (Levinson and Majid 2014). A persistent challenge in mapping odors to names has been named as a "muted sense" and hence it is generally believed that smell is the most ineffable sensory modality. Sight is overall more codable in English, which actually mean that sight has more distinct perceptual qualities that are codable. With that said, it seems an "unbalanced" correspondence between literal terms and the sensory properties. Humans can distinguish between millions of different colors (up to ten million distinct colors (Judd and Wyszecki 1975), but languages generally have only very basic color terms (Andrea 2007; Brent and Paul 1969; Siegfried 2007). Likewise, humans can similarly distinguish thousands of different smells (Agapakis and Tolaas 2012; Yeshurun and Sobel 2010), but there are very few smell words at least in English. This mismatch of words to sensory properties indicates the potential ineffability of fine perceptual details and linguistically impoverished in sensory descriptions (naming) (Agapakis and Tolaas 2012; Yeshurun and Sobel 2010).

7.8 Applications: Synesthesia Device

Human senses could be simulated, digitalized, and even visualized. The emergence of artificial cognitive system and digital technology has made the interactive visualization of human senses viable and benefit the applications of synesthesia device as well. Kim et al. (2019) presented flexible artificial synesthesia electronics that visualize continuous and complicated sounds (Kim et al. 2019). It is by far an excellent example to illustrate the synesthetic correspondence by designing an elegant device. They made an electronic device (FASSEL) which contains a thin composite film of a piezoelectric polymer for sound generation and inorganic electroluminescence (EL) microparticles for direct visualization of input sound signals (Fig. 7.5a). Field-induced EL responded to the source sound wave. The main

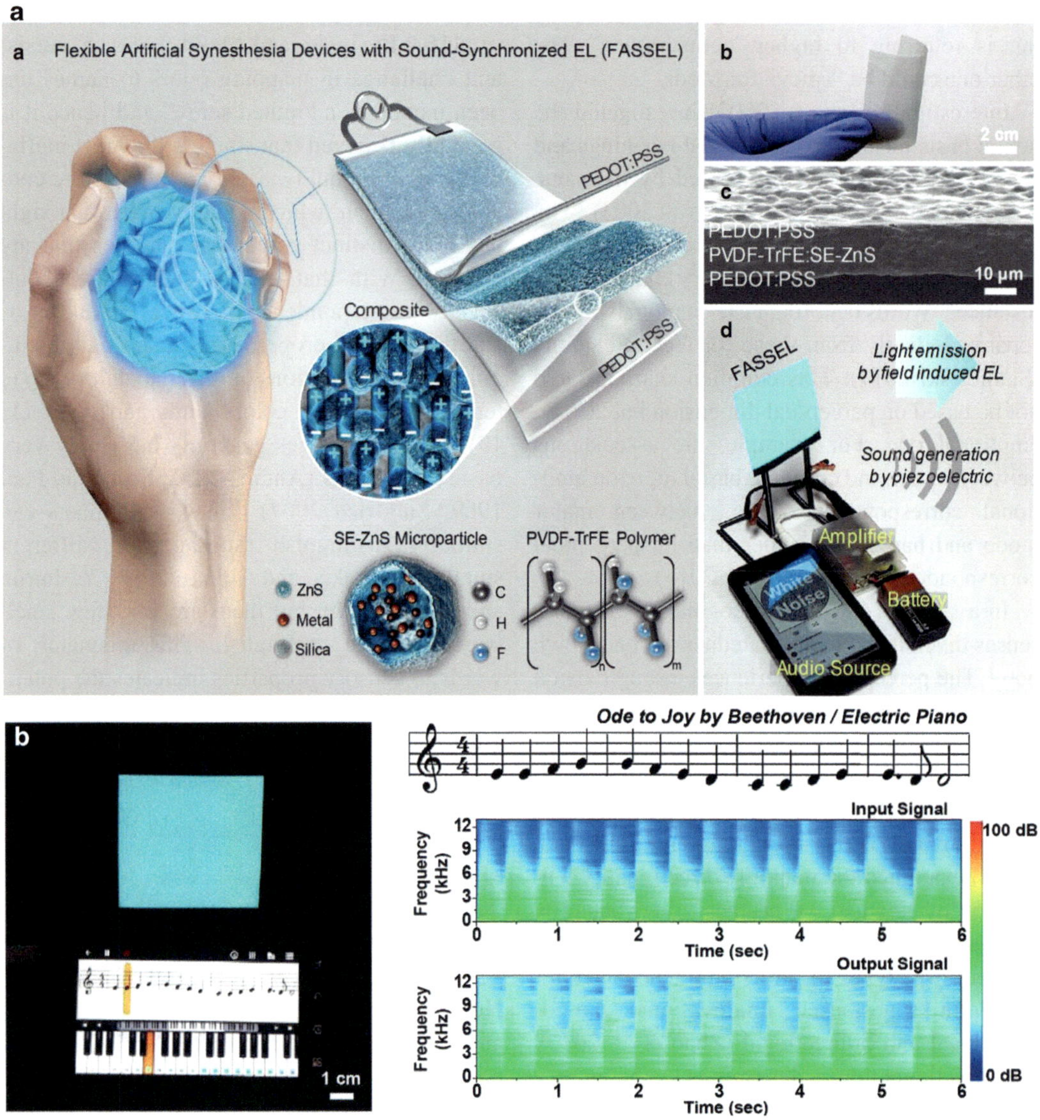

Fig. 7.5 (**a**) Thin piezoelectric film-based flexible artificial synesthesia device with sound-synchronized electroluminescence (FASSEL) (From Kim et al. (2019), Nano Energy, with permission). (**b**) Demonstration of music "Ode to Joy" by the FASSEL. (From Kim et al. (2019), Nano Energy, with permission)

working mechanism is as follows: visual luminance of FASSEL changes as a function of sound frequency, and the color-alteration is frequency-dependent. As a demonstration, while playing the music of "Ode to Joy," the FASSEL exhibited various colors with different brightness (Fig. 7.5b).

7.9 Summary: New Directions for Synesthetic Correspondences

7.9.1 Synesthetic Correspondence in Embodied Cognition and Human-Machine (AI) Interaction

The synesthetic correspondence so far has been mainly materialized in perceptual domain in which multiple sensory events/stimuli interact. The sensory stimuli in interest are less relevant with humans' active actions. However, our brain could distinguish between self- and other—generated sensory signals which lead to effects such as sensory attenuation (such as one cannot tickle himself/herself) or intentional binding (temporal interval between one's own voluntary action and its sensory consequences is subjectively compressed) (Zeng and Chen 2019). There remains the possibility that one's own action could correspond to the sensory consequences from his/her own action, or from others (Legaspi et al. 2019). Indeed, observers' uncertainty about their reported perceptual estimate reflects their perceived causal uncertainty. The framework of embodied or grounded cognition suggests that creative thoughts (such as improvising a piano solo) are partially served by simulations of motor activity associated with tools and their use (Matheson and Kenett 2020). Some types of neurodegenerative disorders, including Parkinson disorder can be interpreted within an embodied cognition framework (Bocanegra et al. 2015). At its core, higher order cognitive functions are grounded in the sensory-motor system (Gallese and Cuccio 2018). Therefore, a potential new form of correspondence between action semantics and sensory outputs (representations) within the sensorimotor loop, though uncharted so far, will surely help to address more fundamental question in consciousness and even neuroscience as a whole.

7.9.2 Knowledge (Concepts) Representation

The correspondence between sensory properties across different levels suggests that our concepts or knowledge (representation) of senses need to be enriched and updated. The representation of each sensory event (category) may not be independent and the knowledge of sensory properties could be well understood in terms of the "interaction" of semantic vocabularies. With the corpus exploration method, Lynott and Connell (2009) provided a set of norms for 423 adjectives, each describing an object property. They mapped the subjective sensory ratings (experiences) of the "word," according to the five perceptual modalities (visual, haptic, auditory, olfactory, and gustatory). Likewise, the data set also included the "exclusivity" of sensory modalities—that one adjective could be only measured/experienced through unisensory perception rather than cross-modally (Lynott and Connell 2009).

This approach of corpus lexicon exploration gives a hint to instruct how well properties from two sensory modalities are corresponded.

In the following study, Lynott and Connell (2013) further revealed that noun concepts are more multimodal than adjective concepts, as nouns tend to subsume multiple adjective property concepts. For example, as a noun, "baby" involves auditory, haptic, olfactory, and visual properties or elicits imagination of those properties—leading to multimodal perceptual strength

(Lynott and Connell 2013). By discovering modality-specific norms of perceptual strength, it is useful for exploring not just the nature of grounded concepts, but also the nature of *form–meaning* relationships, which should further determine the correspondence between two identities (nouns).

Iatropoulos et al. (2018) developed a novel computational method to characterize the olfaction-related semantic content of words. They used a large text corpus of internet sites in English. Two new metrics: olfactory association index (OAI) and olfactory specificity index (OSI) were employed. OAI is targeting on the "strength" that a word is associated with olfaction, while OSI is measuring the specificity of describing given odors. Iatropoulos et al. (2018) further validated the utilities of using OAI and OSI by showing the higher correlations between seemingly scores of given terms and subjective reports of odor association/familiarity. Therefore, this piece of study suggested that OAI and OSI could be good tools to investigate olfactory perception and cognition, including setting up guideline for correspondences between (semantic) odors and even across different sensory stimuli (vocabularies) (Iatropoulos et al. 2018).

Most people find it profoundly difficult to name familiar smells or tactile perception. This difficulty persists even when they received the test in nearly undisturbed conditions, with normal performance of naming visual objects. The essential scientific question is that through what stages and neural (language) pathways that the correspondence is achieved? A recent neuroimaging study suggested they may contain three processes: the object perception, lexical-semantic integration, and verbalization (Olofsson and Gottfried 2015). This line of framework also helps to elucidate the naming difficulty in other sensory modalities (including tactile modality) and find potential solutions. Broadly speaking, human sensory experience also shapes what and how knowledge is stored/represented in our brain. Wang et al. (2020) have proposed sensory-derived and language- and cognition-derived knowledge representations in the human brain: with different underlying brain systems (Wang et al. 2020). Specifically, they recruited congenitally (or early) blind individuals and normal sighted individuals and compared the brain basis of color knowledge in these two groups. For the congenitally blind individuals, their color knowledge can only be obtained through language descriptions and/or cognitive inference, while for the normal peer, their knowledge benefits from both sensory experience and language. Wang et al. (2020) revealed evidence of a sensory-independent knowledge coding system in both groups. Therefore, a comprehensive approach to integrate knowledge graph and artificial intelligence, with neuroscience methodology (neuroimaging), will help to establish the principle of knowledge representations in the human brain, and decipher the intricate relationship between knowledge representation and (synesthetic) sensory correspondence in brain systems.

7.9.3 Synesthetic Correspondence, Metacognition, and Multisensory Integration

As implicated above, our knowledge about properties (relationship) among different sensory stimuli may be abstract. The underlying cognitive ability of knowledge (inference) could be included as one facet of metacognition.

Metacognition—knowing about the ability to monitor one's own decisions and representations, their accuracy and uncertainty—is considered a hallmark of intelligent behavior. The crossmodal correspondences between different sensory attributes, usually relate to one's inferring of the world's causal structure. On the one hand, the underlying perceptual inference is detailed and concrete though with uncertainty and sometime ambiguous in "one-to-one" mapping between sensory properties from different domains (Deroy et al. 2016a, b). On the other hand, the multiple sorts of intersensory correspondences (Spence and Deroy 2013), such as spatiotemporal coincidence, semantic and other higher order correspondences, can inform/teach the brain about whether signals are likely to stem from a common source or independent ones (with the possibility of "averaging" different sensory cues). Metacognition hence provides a constructive tool to integrate information from various sensory channels, supported by the framework of causal inference and cue combination. Specifically, the causal metacognition emphasizes the fundamental links between elementary perception and higher level cognitive processing including social and abstract reasoning. In the framework of Bayesian inference, metacognition has been deciphered with three components that could gain access to monitoring human estimates as well as associated uncertainties: forced-fusion and full-segregation spatial estimates; the inferred causal structure; and the final cross-modal (spatial/temporal) estimates with Bayesian causal inference (Deroy et al. 2016a, b).

Causal metacognition hence is very promising for our understanding of a higher, common level of multisensory processing and could accommodate well the atypical developmental related sensory processing such as in schizophrenia, who had malfunctions in multisensory binding, causal inference, and metacognitive control. That said, to the best of our knowledge, no report of computational modeling for synesthetic correspondence with metacognition component has been given.

7.9.4 Cross-modal Correspondence in Virtual Reality

For the present, most cross-modal (synesthetic) correspondence has been implemented in 2-D scenarios even with higher (semantic) constructs. To extend the exploration in 3-D environment, Huang et al. (2020) implemented a virtual reality (VR) study to examine the color-flavor associations. In this study, participants were not given any visual cues of the green or red tea during drinking but were encouraged to report the color that immediately came to their mind. There remained relatively stable correspondence: green tea was more greenish than red tea, suggesting this correspondence could be contributed by the relevant properties of the bitterness and astringency along with the color dimension of green/red (Huang et al. 2020). Sensory correspondence can be effectively investigated in virtual environment and largely resembling the actual life conditions, indicating a feasible direction of sensory branding (with necessary sensory properties modulation in VR) research and practice.

7.10 Concluding Remarks

Synesthetic correspondence in cross-modal processing represents a vibrant topic in multisensory studies. The "correspondence" concept was originally proposed and henceforth extensively exploited in visual domain (typically in visual apparent motion). It has now been extended and points to the mapping of all discovered sensory properties, explicitly or implicitly across different sensory domains. The inherent "correspondence" contains at least three levels, from concrete levels to abstract ones: structural, statistical, and semantic. With the burgeon of artificial intelligence technology (including virtual reality) in human-machine interaction, as well as the our deepened endogenous understanding—such as metacognition (Deroy et al. 2016a, b) and hierarchical Bayesian inference (Rohe et al. 2019) in multisensory perception, the studies of synesthetic correspondence and potential applications will surely burgeon, and help us to discover the hidden knowledge of sensory representations, as well as to enrich the vocabulary for deciphering the intricate relationship between sensory events during multisensory life.

Acknowledgments This work was supported by the STI2030-Major Project 2021ZD0202600.

References

Agapakis CM, Tolaas S (2012) Smelling in multiple dimensions. Curr Opin Chem Biol 16(5–6):569–575. https://doi.org/10.1016/j.cbpa.2012.10.035

Andrea G (2007) Color names and dynamic imagery. In speaking of colors and odors. In: Plümacher M, Holz P (eds) Converging evidence in language and communication research, vol 8. John Benjamins, Amsterdam, pp 129–140

Barton DN, Halberstadt J (2018) A social Bouba/Kiki effect: a bias for people whose names match their

faces. Psychon Bull Rev 25(3):1013–1020. https://doi.org/10.3758/s13423-017-1304-x
Bertelson P, Aschersleben G (2003) Temporal ventriloquism: crossmodal interaction on the time dimension. 1. Evidence from auditory-visual temporal order judgment. Int J Psychophysiol 50(1–2):147–155. https://doi.org/10.1016/s0167-8760(03)00130-2
Bertelson P, Pavani F, Ladavas E, Vroomen J, de Gelder B (2000) Ventriloquism in patients with unilateral visual neglect. Neuropsychologia 38(12):1634–1642. https://doi.org/10.1016/s0028-3932(00)00067-1
Bien N, ten Oever S, Goebel R, Sack AT (2012) The sound of size: crossmodal binding in pitch-size synesthesia: a combined TMS. EEG and psychophysics study Neuroimage 59(1):663–672. https://doi.org/10.1016/j.neuroimage.2011.06.095
Blomert L(2011) The neural signature of orthographic-phonological binding in successful and failing reading development. Neuroimage 57:695–703. https://doi.org/10.1016/j.neuroimage.2010.11.003
Bocanegra, Y., Garcia, A. M., Pineda, D., Buritica, O., Villegas, A., Lopera, F., . . . Ibanez, A. (2015). Syntax, action verbs, action semantics, and object semantics in Parkinson's disease: Dissociability, progression, and executive influences. Cortex, 69, 237–254. https://doi.org/10.1016/j.cortex.2015.05.022
Bottini R, Barilari M, Collignon O (2019) Sound symbolism in sighted and blind. The role of vision and orthography in sound-shape correspondences. Cognition 185:62–70. https://doi.org/10.1016/j.cognition.2019.01.006
Brang D, Ghiam M, Ramachandran VS (2013) Impaired acquisition of novel grapheme-color correspondences in synesthesia. Front Hum Neurosci 7:717. https://doi.org/10.3389/fnhum.2013.00717
Bremner AJ, Caparos S, Davidoff J, de Fockert J, Linnell KJ, Spence C (2013) "Bouba" and "Kiki" in Namibia? A remote culture make similar shape-sound matches, but different shape-taste matches to westerners. Cognition 126(2):165–172. https://doi.org/10.1016/j.cognition.2012.09.007
Brent B, Paul K (1969) Basic color terms: their universality and evolution. University of California Press, Berkeley
Bueti D, Walsh V (2009) The parietal cortex and the representation of time, space, number and other magnitudes. Philosophical Transactions of the Royal Society B-Biological Sciences 364(1525):1831–1840. https://doi.org/10.1098/rstb.2009.0028
Cassirer E (1953) Language and myth. Dover Publications. Dover ed edition (June 1, 1953)
Chen L, Vroomen J (2013) Intersensory binding across space and time: a tutorial review. Atten Percept Psychophys 75(5):790–811. https://doi.org/10.3758/s13414-013-0475-4
Chen LH, Shi ZH, Muller HJ (2010) Influences of intra- and crossmodal grouping on visual and tactile Ternus apparent motion. Brain Res 1354:152–162. https://doi.org/10.1016/j.brainres.2010.07.064
Chen LH, Zhang ML, Ai F, Xie WY, Meng XZ (2016) Crossmodal synesthetic congruency improves visual timing in dyslexic children. Res Dev Disabil 55:14–26. https://doi.org/10.1016/j.ridd.2016.03.010
Chen LH, Zhou XL, Muller HJ, Shi ZH (2018) What you see depends on what you hear: temporal averaging and Crossmodal integration. Journal of Experimental Psychology-General 147(12):1851–1864. https://doi.org/10.1037/xge0000487
Cuskley C, Simner J, Kirby S (2017) Phonological and orthographic influences in the bouba-kiki effect. Psychol Res 81(1):119–130. https://doi.org/10.1007/s00426-015-0709-2
De Carolis L, Marsico E, Arnaud V, Coupe C (2018) Assessing sound symbolism: investigating phonetic forms, visual shapes and letter fonts in an implicit bouba-kiki experimental paradigm. PLoS One 13(12):e0208874. https://doi.org/10.1371/journal.pone.0208874
Deroy O, Fasiello I, Hayward V, Auvray M (2016a) Differentiated audio-tactile correspondences in sighted and blind individuals. J Exp Psychol Hum Percept Perform 42(8):1204–1214. https://doi.org/10.1037/xhp0000152
Deroy O, Spence C, Noppeney U (2016b) Metacognition in multisensory perception. Trends Cogn Sci 20(10):736–747. https://doi.org/10.1016/j.tics.2016.08.006
Ernst MO, Bülthoff HH (2004) Merging the senses into a robust percept. Trends Cogn Sci 8:162–169. https://doi.org/10.1016/j.tics.2004.02.002
Favalli S, Skov T, Spence C, Byrne DV (2013) Do you say it like you eat it? The sound symbolism of food names and its role in the multisensory product experience. Food Res Int 54(1):760–771. https://doi.org/10.1016/j.foodres.2013.08.022
Fort M, Martin A, Peperkamp S (2015) Consonants are more important than vowels in the Bouba-kiki effect. Lang Speech 58(Pt 2):247–266. https://doi.org/10.1177/0023830914534951
Fryer L, Freeman J, Pring L (2014) Touching words is not enough: how visual experience influences haptic-auditory associations in the "Bouba-Kiki" effect. Cognition 132(2):164–173. https://doi.org/10.1016/j.cognition.2014.03.015
Gallese V, Cuccio V (2018) The neural exploitation hypothesis and its implications for an embodied approach to language and cognition: insights from the study of action verbs processing and motor disorders in Parkinson's disease. Cortex 100:215–225. https://doi.org/10.1016/j.cortex.2018.01.010
Gertner L, Henik A, Reznik D, Cohen KR (2013) Implications of number-space synesthesia on the automaticity of numerical processing. Cortex 49:1352–1362. https://doi.org/10.1016/j.cortex.2012.03.019
Gold R, Segal O (2017) The bouba-kiki effect and its relation to the autism quotient (AQ) in autistic adolescents. Res Dev Disabil 71:11–17. https://doi.org/10.1016/j.ridd.2017.09.017

Goodhew SC, Freire MR, Edwards M (2015) Enhanced semantic priming in synesthetes independent of sensory binding. Conscious Cogn 33:443–456. https://doi.org/10.1016/j.concog.2015.02.019

Grossenbacher PG, Lovelace CT (2001) Mechanisms of synesthesia: cognitive and physiological constraints. Trends Cogn Sci 5(1):36–41. https://doi.org/10.1016/s1364-6613(00)01571-0

Guo L, Bao M, Guan LY, Chen LH (2017) Cognitive styles differentiate crossmodal correspondences between pitch glide and visual apparent motion. Multisens Res 30(3–5):363–385

Haryu E, Kajikawa S (2012) Are higher-frequency sounds brighter in color and smaller in size? Auditory-visual correspondences in 10-month-old infants. Infant Behavior & Development 35:727–732. https://doi.org/10.1016/j.infbeh.2012.07.015

Hein E, Cavanagh P (2012) Motion correspondence in the Ternus display shows feature bias in spatiotopic coordinates. J Vis 12(7):16. https://doi.org/10.1167/12.7.16

Hidaka S, Yaguchi A (2018) An investigation of the relationships between autistic traits and crossmodal correspondences in typically developing adults. Multisens Res 31(8):729–751. https://doi.org/10.1163/22134808-20181304

Huang FX, Qi YX, Wang CJ, Wan XA (2020) Show me the color in your mind: a study of color-flavor associations in virtual reality. Food Qual Prefer 85:103969. https://doi.org/10.1016/j.foodqual.2020.103969

Hubbard EM, Arman AC, Ramachandran VS, Boynton GM (2005) Individual differences among grapheme-color synesthetes: brain-behavior correlations. Neuron 45(6):975–985

Hubbard EM, Brang D, Ramachandran VS (2011) The cross-activation theory at 10. J Neuropsychol 5:152–177

Hung SM, Styles SJ, Hsieh PJ (2017) Can a word sound like a shape before you have seen it? Sound-shape mapping prior to conscious awareness. Psychol Sci 28(3):263–275. https://doi.org/10.1177/0956797616677313

Iatropoulos G, Herman P, Lansner A, Karlgren J, Larsson M, Olofsson JK (2018) The language of smell: connecting linguistic and psychophysical properties of odor descriptors. Cognition 178:37–49. https://doi.org/10.1016/j.cognition.2018.05.007

Judd DB, Wyszecki G (1975) Color in business, science, and industry. Wiley, New York

Karthikeyan S, Rammairone B, Ramachandra V (2016) The Bouba-Kiki phenomenon tested via schematic drawings of facial expressions: further validation of the internal simulation hypothesis. Iperception 7(1):2041669516631877. https://doi.org/10.1177/2041669516631877

Ketron S, Spears N (2019) Sounds like a heuristic! Investigating the effect of sound-symbolic correspondences between store names and sizes on consumer willingness-to-pay. J Retail Consum Serv 51:285–292. https://doi.org/10.1016/j.jretconser.2019.06.016

Kim, J. S., Cho, S. H., Kim, K. L., Kim, G., Lee, S. W., Kim, E. H., . . . Park, C. (2019). Flexible artificial synesthesia electronics with sound-synchronized electroluminescence. Nano Energy, 59, 773–783. doi:https://doi.org/10.1016/j.nanoen.2019.03.006

Köhler W (1929) Gestalt psychology. Liveright, New York

Kohler (1947) Gestalt psychology, 2nd edn. Liveright Publishing, New York, NY

Legaspi R, He ZQ, Toyoizumi T (2019) Synthetic agency: sense of agency in artificial intelligence. Curr Opin Behav Sci 29:84–90. https://doi.org/10.1016/j.cobeha.2019.04.004

Levinson SC, Majid A (2014) Differential ineffability and the senses. Mind Lang 29(4):407–427. https://doi.org/10.1111/mila.12057

Lynott D, Connell L (2009) Modality exclusivity norms for 423 object properties. Behav Res Methods 41(2):558–564. https://doi.org/10.3758/Brm.41.2.558

Lynott D, Connell L (2013) Modality exclusivity norms for 400 nouns: the relationship between perceptual experience and surface word form. Behav Res Methods 45(2):516–526. https://doi.org/10.3758/s13428-012-0267-0

Macdermott MM (1940) Vowel sounds in poetry: their music and tone-colour. Kegan Paul, London

Marks LE (2014) The Unity of the senses: interrelations among the modalities. Academic Press

Marks L, Mulvenna C (2013) Synesthesia, at and near its borders. Front Psychol 4:651. https://doi.org/10.3389/fpsyg.2013.00651

Martino G, Marks LE (1999) Perceptual and linguistic interactions in speeded classification: tests of the semantic coding hypothesis. Perception 28(7):903–923. https://doi.org/10.1068/p2866

Matheson HE, Kenett YN (2020) The role of the motor system in generating creative thoughts. NeuroImage 213:116697. https://doi.org/10.1016/j.neuroimage.2020.116697

Maurer D, Pathman T, Mondloch CJ (2006) The shape of boubas: sound-shape correspondences in toddlers and adults. Dev Sci 9(3):316–322. https://doi.org/10.1111/j.1467-7687.2006.00495.x

McCarthy JD, Caplovitz GP (2014) Color synesthesia improves color but impairs motion perception. Trends Cogn Sci 18(5):224–226. https://doi.org/10.1016/j.tics.2014.02.002

Morein-Zamir S, Soto-Faraco S, Kingstone A (2003) Auditory capture of vision: examining temporal ventriloquism. Brain Res Cogn Brain Res 17(1):154–163. https://doi.org/10.1016/s0926-6410(03)00089-2

Ngo MK, Velasco C, Salgado A, Boehm E, O'Neill D, Spence C (2013) Assessing crossmodal correspondences in exotic fruit juices: the case of shape and sound symbolism. Food Qual Prefer 28(1):361–369. https://doi.org/10.1016/j.foodqual.2012.10.004

Occelli V, Esposito G, Venuti P, Walker P, Zampini M (2012). Audiovisual crossmodal correspondences in Autism Spectrum Disorders (ASDs). Seeing and Perceiving, 25:44–44 https://doi.org/10.1163/187847612X646668

Ogulmus C, Karacaoglu M, Kafaligonul H (2018) Temporal ventriloquism along the path of appar-

ent motion: speed perception under different spatial grouping principles. Exp Brain Res 236(3):629–643. https://doi.org/10.1007/s00221-017-5159-1
Olofsson JK, Gottfried JA (2015) The muted sense: neurocognitive limitations of olfactory language. Trends Cogn Sci 19(6):314–321. https://doi.org/10.1016/j.tics.2015.04.007
Orchard-Mills E, Van der Burg E, Alais D (2016) Crossmodal correspondence between auditory pitch and visual elevation affects temporal ventriloquism. Perception 45(4):409–424. https://doi.org/10.1177/0301006615622320
Palmer SE, Schloss KB, Xu Z, Prado-Leon LR (2013) Music-color associations are mediated by emotion. Proc Natl Acad Sci U S A 110(22):8836–8841. https://doi.org/10.1073/pnas.1212562110
Parise CV (2016) Crossmodal correspondences: standing issues and experimental guidelines. Multisens Res 29(1–3):7–28. https://doi.org/10.1163/22134808-00002502
Parise CV, Ernst MO (2016) Correlation detection as a general mechanism for multisensory integration. Nat Commun 7:11543. https://doi.org/10.1038/ncomms11543
Parise C, Spence C (2008) Synesthetic congruency modulates the temporal ventriloquism effect. Neurosci Lett 442(3):257–261. https://doi.org/10.1016/j.neulet.2008.07.010
Parise CV, Spence C (2009) 'When birds of a feather flock together': synesthetic correspondences modulate audiovisual integration in non-synesthetes. PLoS One 4(5):e5664. https://doi.org/10.1371/journal.pone.0005664
Parise CV, Spence C (2012) Audiovisual crossmodal correspondences and sound symbolism: a study using the implicit association test. Exp Brain Res 220(3–4):319–333. https://doi.org/10.1007/s00221-012-3140-6
Parise CV, Spence C, Ernst MO (2012) When correlation implies causation in multisensory integration. Curr Biol 22(1):46–49. https://doi.org/10.1016/j.cub.2011.11.039
Peiffer-Smadja N, Cohen L (2019) The cerebral bases of the bouba-kiki effect. NeuroImage 186:679–689. https://doi.org/10.1016/j.neuroimage.2018.11.033
Piqueras-Fiszman B, Spence C (2011) Crossmodal correspondences in product packaging. Assessing color-flavor correspondences for potato chips (crisps). Appetite 57(3):753–757. https://doi.org/10.1016/j.appet.2011.07.012
Rakova M (2003) The extent of the literal: metaphor, polysemy and theories of concepts. Palgrave Macmillan, New York
Ramachandran V, Hubbard EM (2001a) Synaesthesia - a window into perception, thought and language. J Conscious Stud 8(12):3–34
Ramachandran VS, Hubbard EM (2001b) Psychophysical investigations into the neural basis of synaesthesia. Proceedings of the Royal Society B-Biological Sciences 268(1470):979–983
Ramachandran VS, Hubbard EM (2003) Hearing colors, tasting shapes. Sci Am 288(5):52–59. https://doi.org/10.1038/scientificamerican0503-52
Ramachandra V (2016) The linguistic and cognitive factors associated with lexical-gustatory synesthesia: A case study. Brain and cognition 106:23–32. https://doi.org/10.1016/j.bandc.2016.04.005
Roblee L, Washburn MF (1912) Minor studies from the psychological laboratory of Vassar college XX the affective values of articulate sounds. Am J Psychol 23:579–583. https://doi.org/10.2307/1413063
Rohe T, Ehlis AC, Noppeney U (2019) The neural dynamics of hierarchical Bayesian causal inference in multisensory perception. Nat Commun 10(1):1907. https://doi.org/10.1038/s41467-019-09664-2
Ross C (2018) Unlocking consciousness- lessons from the convergence of computing and cognitive psychology. World Scientific
Rothen N, Schwartzman DJ, Bor D, Seth AK (2018) Coordinated neural, behavioral, and phenomenological changes in perceptual plasticity through overtraining of synesthetic associations. Neuropsychologia 111:151–162. https://doi.org/10.1016/j.neuropsychologia.2018.01.030
Rouw R, van Driel J, Knip K, Richard RiK (2013) Executive functions in synesthesia. Consciousness and cognition 22:184–202. https://doi.org/10.1016/j.concog.2012.11.008
Rouw R, Scholte HS (2016) Personality and cognitive profiles of a general synesthetic trait. Neuropsychologia 88:35–48. https://doi.org/10.1016/j.neuropsychologia.2016.01.006
Sadaghiani S, Maier JX, Noppeney U (2009) Natural, metaphoric, and linguistic auditory direction signals have distinct influences on visual motion processing. J Neurosci 29(20):6490–6499. https://doi.org/10.1523/Jneurosci.5437-08.2009
Saenz M, Koch C (2008) The sound of change: visually-induced auditory synesthesia. Curr Biol 18(15):R650–R651. https://doi.org/10.1016/j.cub.2008.06.014
Sakamoto M, Watanabe J (2018) Bouba/Kiki in touch: associations between tactile perceptual qualities and Japanese phonemes. Front Psychol 9:295. https://doi.org/10.3389/fpsyg.2018.00295
Shen Y (2008) Metaphor and poetic figures. The Cambridge handbook of metaphor and thought. Cambridge University Press. 295–307
Siegfried W (2007) Color terms between elegance and beauty: the verbalization of color with textiles and cosmetics. In: Plümacher M, Holz P (eds) Speaking of colors and odors. [converging evidence in language and communication research 8]. John Benjamins, Amsterdam, pp 113–128
Slutsky DA, Recanzone GH (2001) Temporal and spatial dependency of the ventriloquism effect. Neuroreport 12(1):7–10. https://doi.org/10.1097/00001756-200101220-00009
Spence C (2011) Crossmodal correspondences: a tutorial review. Atten Percept Psychophys 73(4):971–995. https://doi.org/10.3758/s13414-010-0073-7

Spence C (2012) Managing sensory expectations concerning products and brands: capitalizing on the potential of sound and shape symbolism. J Consum Psychol 22(1):37–54. https://doi.org/10.1016/j.jcps.2011.09.004

Spence C, Deroy O (2013) How automatic are crossmodal correspondences? Conscious Cogn 22(1):245–260. https://doi.org/10.1016/j.concog.2012.12.006

Staniewski P (2017) Geschmack und smak– sprachliche Aspekte der gustatorischen Wahrnehmung im Deutschen und Polnischen. In: Szczek J, Kalasznik M (eds) Beitrage zur Fremdsprachenvermittlung. Verlag Empirische Pädagogik, Landau, pp 223–248

Stein BE, Meredith MA (1993) The merging of the senses. The MIT Press, Cambridge, MA

Tian Y, Liu X, Chen L (2020) Mindfulness meditation biases visual temporal order discrimination but not under conditions of temporal ventriloquism. Front Psychol 11:1937. https://doi.org/10.3389/fpsyg.2020.01937

Tzeng CY, Nygaard LC, Namy LL (2017) The Specificity of Sound Symbolic Correspondences in Spoken Language. Cognitive science 41:2191–2220. https://doi.org/10.1111/cogs.12474

Ullman S (1979) The interpretation of visual motion. MIT Press, Cambridge, MA

van Petersen E, Altgassen M, van Lier R, van Leeuwen TM (2020) Enhanced spatial navigation skills in sequence-space synesthetes. Cortex 130:49–63. https://doi.org/10.1016/j.cortex.2020.04.034

Vroomen J, Keetels M (2006) The spatial constraint in intersensory pairing: no role in temporal ventriloquism. J Exp Psychol Hum Percept Perform 32(4):1063–1071. https://doi.org/10.1037/0096-1523.32.4.1063

Vroomen J, de Gelder B, Vroomen J (2004) Temporal ventriloquism: sound modulates the flash-lag effect. J Exp Psychol Hum Percept Perform 30(3):513–518. https://doi.org/10.1037/0096-1523.30.3.513

Walsh V (2003) A theory of magnitude: common cortical metrics of time, space and quantity. Trends Cogn Sci 7(11):483–488. https://doi.org/10.1016/j.tics.2003.09.002

Wan X, Woods AT, van den Bosch JJ, McKenzie KJ, Velasco C, Spence C (2014) Cross-cultural differences in crossmodal correspondences between basic tastes and visual features. Front Psychol 5:1365. https://doi.org/10.3389/fpsyg.2014.01365

Wang QC, Bao M, Chen LH (2014) The role of spatio-temporal and spectral cues in segregating short sound events: evidence from auditory Ternus display. Exp Brain Res 232(1):273–282. https://doi.org/10.1007/s00221-013-3738-3

Wang XY, Men WW, Gao JH, Caramazza A, Bi YC (2020) Two forms of knowledge representations in the human brain. Neuron 107(2):383. https://doi.org/10.1016/j.neuron.2020.04.010

Westbury C (2005) Implicit sound symbolism in lexical access: evidence from an interference task. Brain Lang 93(1):10–19. https://doi.org/10.1016/j.bandl.2004.07.006

Westbury C, Hollis G, Sidhu DM, Pexman PM (2018) Weighing up the evidence for sound symbolism: distributional properties predict cue strength. J Mem Lang 99:122–150

Winter B, Marghetis T, Matlock T (2015) Of magnitudes and metaphors: explaining cognitive interactions between space, time, and number. Cortex 64:209–224. https://doi.org/10.1016/j.cortex.2014.10.015

Yeshurun Y, Sobel N (2010) An odor is not worth a thousand words: from multidimensional odors to unidimensional odor objects. Annu Rev Psychol 61(219–241):C211–C215. https://doi.org/10.1146/annurev.psych.60.110707.163639

Zeljko M, Kritikos A, Grove PM (2019) Lightness/pitch and elevation/pitch crossmodal correspondences are low-level sensory effects. Atten Percept Psychophys 81(5):1609–1623. https://doi.org/10.3758/s13414-019-01668-w

Zeng HK, Chen LH (2019) Robust temporal averaging of time intervals between action and sensation. Front Psychol 10:511. https://doi.org/10.3389/fpsyg.2019.00511

Zhang Y, Chen L (2016) Crossmodal statistical binding of temporal information and stimuli properties recalibrates perception of visual apparent motion. Front Psychol 7:434. https://doi.org/10.3389/fpsyg.2016.00434

Zhao QQ, Huang CR, Ahrens K (2019) Directionality of linguistic synesthesia in mandarin: a corpus-based study. Lingua 232:102744. https://doi.org/10.1016/j.lingua.2019.102744

Neural Oscillations and Multisensory Processing

8

Yanfang Zuo and Zuoren Wang

Abstract

Neural oscillations play a role in sensory processing by coordinating synchronized neuronal activity. Synchronization of gamma oscillations is engaged in local computation of feedforward signals and synchronization of alpha-beta oscillations is engaged in feedback processing over long-range areas. These spatially and spectrally segregated bi-directional signals may be integrated by a mechanism of cross-frequency coupling. Synchronization of neural oscillations has also been proposed as a mechanism for information integration across multiple sensory modalities. A transient stimulus or rhythmic stimulus from one modality may lead to phase alignment of ongoing neural oscillations in multiple sensory cortices, through a mechanism of cross-modal phase reset or cross-modal neural entrainment. Synchronized activities in multiple sensory cortices are more likely to boost stronger activities in downstream areas. Compared to synchronized oscillations, asynchronized oscillations may impede signal processing, and may contribute to sensory selection by setting the oscillations in the target-related cortex and the oscillations in the distractor-related cortex to opposite phases.

Keywords

Neural oscillation · Multisensory processing · Phase reset · Neural entrainment · Cross-frequency coupling · Sensory selection · Phase locking · Spike phase

Y. Zuo (✉)
Department of Neurology, Guangzhou First People's Hospital, School of Medicine, South China University of Technology, Guangzhou, China

Center for Medical Research on Innovation and Translation, Institute of Clinical Medicine, Guangzhou First People's Hospital, School of Medicine, South China University of Technology, Guangzhou, China

Z. Wang
Institute of Neuroscience, State Key Laboratory of Neuroscience, CAS Center for Excellence in Brain Science & Intelligence Technology, Chinese Academy of Sciences, Shanghai, China

University of Chinese Academy of Sciences, Beijing, China
e-mail: zuorenwang@ion.ac.cn

8.1 Neural Oscillations and Sensory Processing

Neural oscillations are rhythmic changes in neuronal activity, reflecting rhythmic fluctuations in neuronal excitability (Fries 2005). Sensory information processing requires related neurons to coordinate their activities with great temporal and spatial precision. Neural oscillations provide

Y. Gu, A. Zaidel (eds.), *Advances of Multisensory Integration in the Brain*, Advances in Experimental Medicine and Biology 1437, https://doi.org/10.1007/978-981-99-7611-9_8

such a fundamental mechanism of information processing by enabling synchronous activity. The timing of neuronal activity induced by stimuli relative to the phase of ongoing oscillations affects how fast and efficiently the stimuli are processed. If neurons have high excitability at the trough of an oscillation, then stimuli time-locked to the trough might be processed faster and more efficiently, compared to stimuli time-locked to the peak of the oscillatory waveform (Fries 2005).

Neural oscillations have been commonly reported and classified as delta- (<4 Hz), theta- (4–8 Hz), alpha- (8–12 Hz), beta- (12–30 Hz), and gamma- (>30 Hz) bands. The oscillations in distinct frequency bands reflect different mechanisms of multisensory processing. Due to longer period of slow oscillations than fast oscillations, slow oscillations are well suited for communication between distal areas in long temporal windows, and fast oscillations are more suited for local processing over small spatial regions in short temporal windows (Demiralp et al. 2007; Euston et al. 2007). Specifically, in sensory processing, gamma oscillations underlie bottom-up processing of sensory stimuli in a feedforward hierarchy, and alpha-beta oscillations mediate top-down processing in a feedback hierarchy (Bastos et al. 2015a; Michalareas et al. 2016).

During sensory perception, gamma oscillations occur transiently upon sensory input. Gamma oscillations are generated through the recurrent interactions between excitatory pyramidal cells and inhibitory GABAergic interneurons, such as parvalbumin-positive (PV) basket cells (Kann et al. 2014; Tamas et al. 2000). The basket cells generate spikes at high frequency and synchronize the firing of pyramidal cells by rhythmic inhibition (Kann et al. 2014). According to the characteristic laminar patterns of bottom-up hierarchy, bottom-up projections originate primarily in supragranular layers, and terminate predominantly in layer 4 (Felleman and Van Essen 1991). After neurons in layer 4 receive feedforward input, they send intracolumnar projections to supragranular layers (layers 3, 2, and 1) (Douglas and Martin 2004), which send output projections feedforward to layer 4 neurons of the next higher cortical area. Thus, within same cortical area, supragranular gamma oscillations display a systematic delay from layer 4 toward layer 1 (1 ms per 100 micron) (Livingstone 1996; van Kerkoerle et al. 2014). Supragranular neurons can also send output projections back forward to nearby areas, for example, from supragranular layers of V2 to supragranular layers of V1 (Markov et al. 2014) (Fig. 8.1). In contrast to bottom-up projections, top-down projections originate predominantly in infragranular layers, where neurons show synchronization primarily at frequencies of alpha and beta bands (Buffalo et al. 2011; Maier et al. 2010; Markov et al. 2014; Smith et al. 2013). Thus, feedback top-down

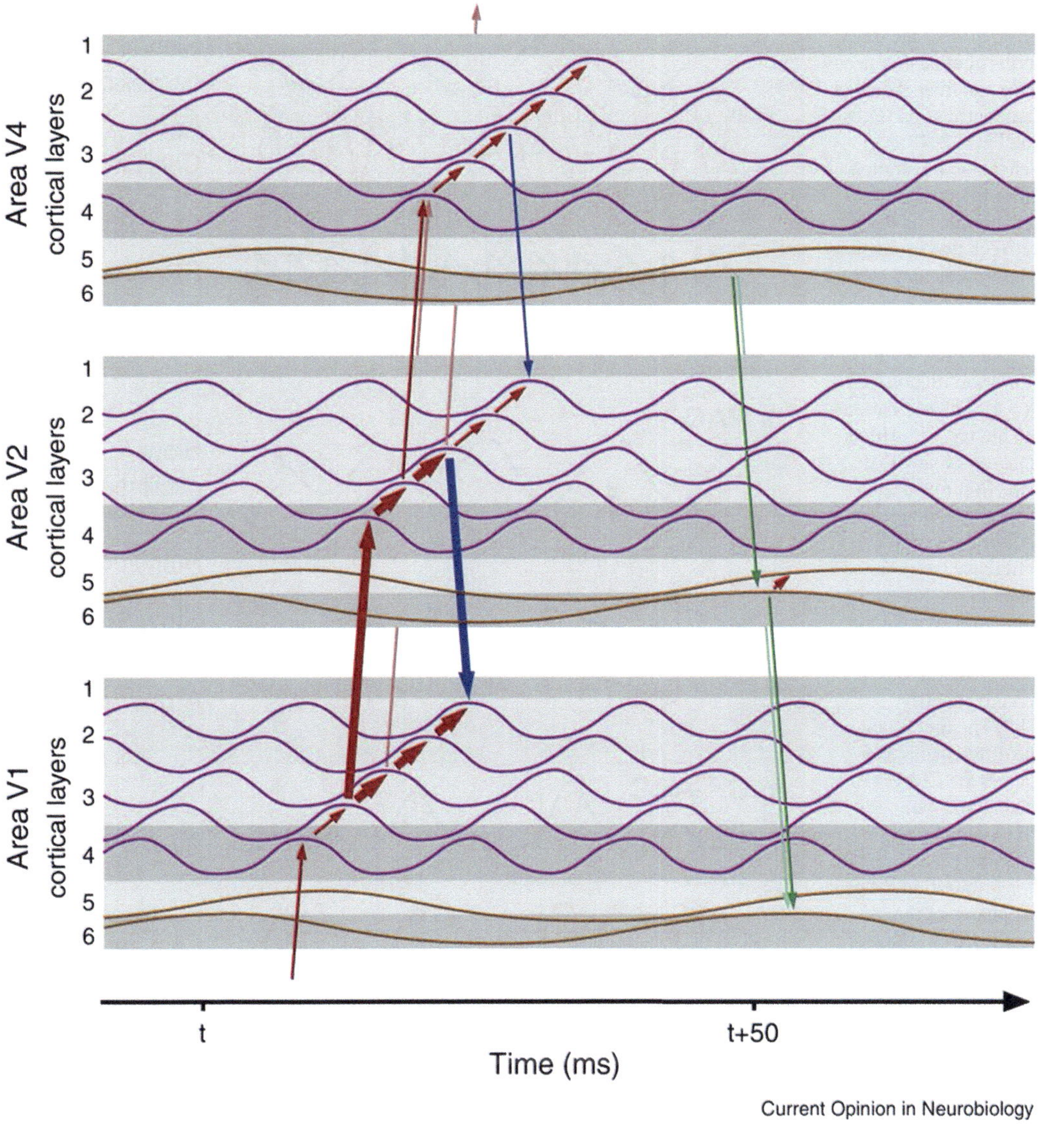

Fig. 8.1 Oscillations in feedforward and feedback hierarchies. Adopted from permission (Bastos et al. 2015b). Schematic diagram to illustrate how gamma and beta oscillations propagate between layers in a given area and between hierarchical areas. Gamma oscillations in supragranular layers are illustrated with purple lines. A systematic delay is found from layer 4 toward layer 1. Feedforward signals are illustrated with red arrows. They enter into layer 4 in the lower left and proceed onwards to layer 4 of the next higher area through supragranular layers. Supragranular feedback between neighboring areas are indicated with blue arrows. One complete cycle of feedforward and reentrant feedback signaling are highlighted with thick red and blue arrows. Green arrows indicate infragranular feedback. Brown lines indicate infragranular alpha-beta oscillations

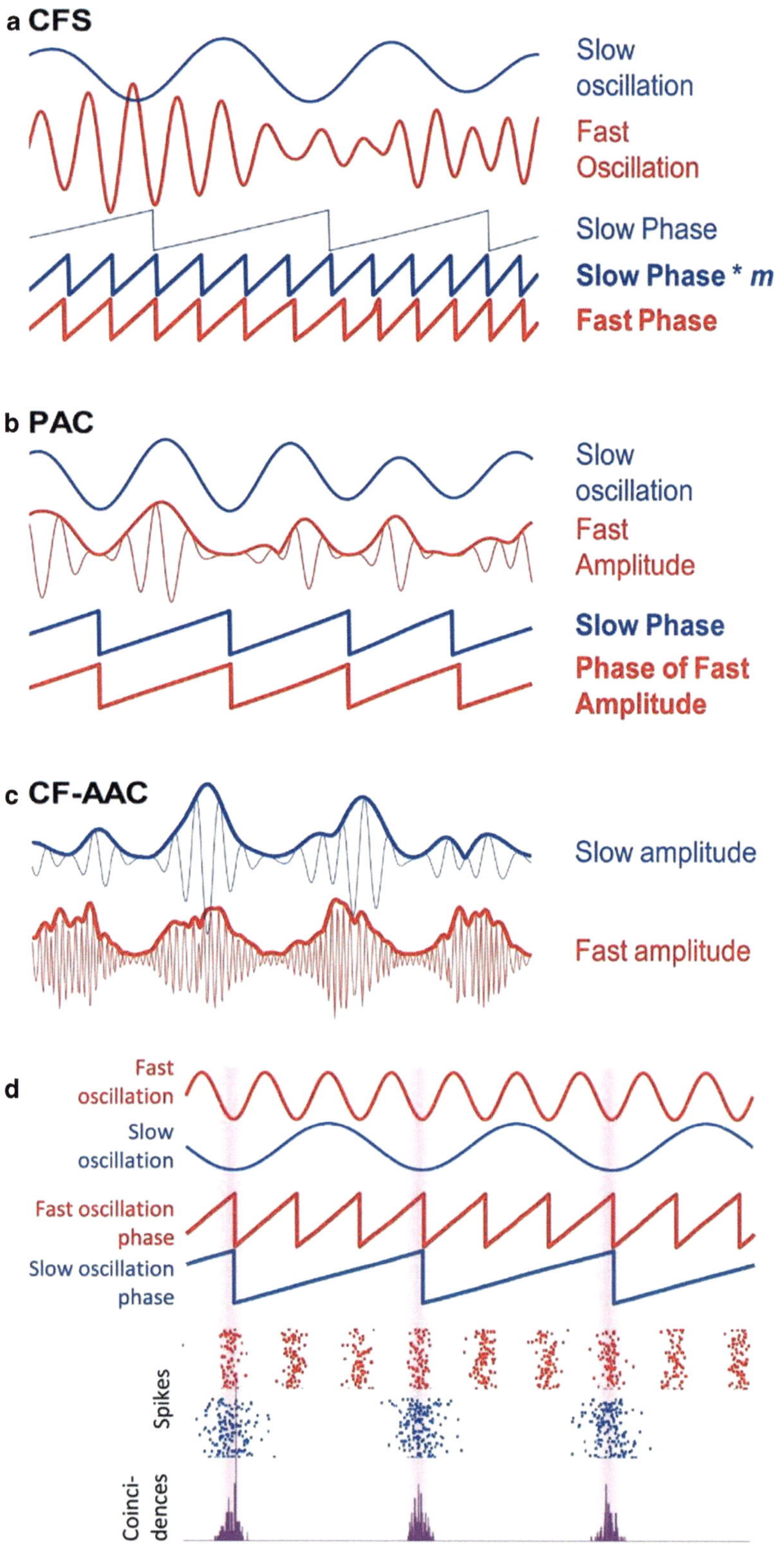

Fig. 8.2 Schematic illustration of the three forms of cross-frequency (CF) coupling. Adopted from permission (Palva and Palva 2018). (**a**) Schematic illustration of cross-frequency phase synchronization (CFS) between two oscillations whose frequency ratio is n:m. Here the faster oscillation is four times as fast as that of the slower oscillation ($n = 1$, $m = 4$). The oscillations are filtered with Hilbert transform to get real parts (top rows) and their phase time series (bottom rows). The difference between the phase of the faster oscillation and the m:n multiplied phase of the slower oscillation can be used to quantify CFS. The more stable the difference is, the greater the CFS is (Palva et al. 2005). (**b**) Phase-amplitude coupling (PAC) is phase synchronization between the phases of the amplitude envelope of the faster oscillation and the slower oscillation. (**c**) Cross-frequency amplitude-amplitude coupling (CF-AAC) signals the correlation of the amplitude envelopes between the slower and faster oscillations. (**d**) Schematic illustration of spike coincidences between the two neuronal assemblies with stable CFS. These two neuronal assemblies are oscillating with a frequency ratio of 1:3, accompanied with consistent spike-time relationships

communication derived from infragranular layers mainly uses alpha-beta frequencies (Fig. 8.2, right side) (Bastos et al. 2015a; Bressler and Richter 2015; van Kerkoerle et al. 2014) (see Bastos et al. 2015a, for review).

8.2 Cross-Frequency Coupling and Sensory Processing

As shown that signal streams in the feedforward bottom-up and feedback top-down directions are anatomically and spectrally segregated (Bastos et al. 2015a), it raises the question of how the bi-directional streams are integrated in a given area they meet. Cross-frequency (CF) coupling may subserve processing and integration of these signal streams (Lee et al. 2013; Roopun et al. 2008). Cross-frequency coupling refers to interaction and association between oscillations at different frequencies, and it has three forms: Cross-frequency phase synchrony (CFS), phase-amplitude coupling (PAC), and cross-frequency amplitude-amplitude correlation (CFAAC) (see Palva and Palva (2018) and Voytek et al. (2010) for reviews).

CFS is a possible mechanism to coordinate communication across fast and slow oscillations. CFS is referred to as a form of phase synchrony when two neuronal assemblies oscillate with different frequencies and a stable phase difference exists between them (Rosenblum et al. 1998; Palva et al. 2005). Suppose the frequencies of the slower and faster oscillations are n and m, respectively, we can quantify CFS by testing whether there is a stable difference between the m:n multiplied phases of the slower and the faster oscillations (Fig. 8.2a) (Palva and Palva 2018). For two oscillations with cross-frequency synchrony, given that their neuronal assemblies have high excitability and fire intensively at a specific phase of the corresponding oscillations, consistent spike-time relationships will be observed between the neuronal assemblies, in a manner determined by their frequency ratio. For example, as shown in Fig. 8.2d, if the faster oscillation is three times as fast as that of the slower oscillation, then the spikes of the faster and slower oscillations may coincide in every third cycle of the faster oscillation. In this way, the two spatially and spectrally segregated neuronal assemblies temporally co-localize their spikes (see Palva and Palva (2018) for review). The synchronized spikes may evoke action potentials in downstream neurons and boost greater impact of neuronal signals on post-synaptic neurons than non-synchronous input (Azouz and Gray 2003; Gutig 2014; Konig et al. 1996; Singer and Gray 1995). The synchronized spikes may promote spike-time-dependent plasticity (Gutig 2014; Jaramillo and Kempter 2017). In this way, top-down attentional control in alpha-beta band may modulate sensory processing in bottom-up gamma band (Richter et al. 2017).

In contrast to CFS, PAC signals the modulation of the amplitude (envelope) of the faster oscillations by the phase of a slower oscillation. One approach to quantify PAC is to evaluate phase synchronization between the phase of the slower oscillation and the phase of the amplitude envelope of the faster oscillation (Vanhatalo et al. 2004) (Fig. 8.2b). Since the phase of the faster oscillation is not modulated by the slower oscillation, PAC is not related to spike synchronization. Instead, it reflects a general change in population synaptic activity of the neuronal assembly in the faster oscillation, fluctuated with neuronal excitability in the slower oscillation. While high-frequency activity reflects local computation of information processing, low-frequency rhythms can be dynamically entrained by both rhythmic external sensory events (Lakatos et al. 2008; Luo and Poeppel 2007; Saleh et al. 2010) and internal cognitive processes (Rizzuto et al. 2006; Schroeder and Lakatos 2009). Therefore, PAC may provide an effective means to transfer task- and behavior-related information from large-scale brain networks to local computation and processing, thus integrating functional information across multiple spatiotemporal scales (see Canolty and Knight (2010) and Swadlow and Gusev (2001) for reviews).

The third form of cross-frequency interaction, CFAAC, refers to the coupling of amplitude envelopes between two oscillations with distinct frequencies (Fig. 8.2c). It has been observed during cognitive tasks and correlated with behavior. However, its functional role remains unclear. It has nothing to do with any phase in the fast or slow oscillations, and hence there are no stable spike-time relationships. Rather than direct integration of information across the two coupled frequencies, it may reflect that excitability of the two frequencies is co-modulated by a common factor.

These interactions between oscillations in distinct frequency bands may play a role in integration of neuronal information processing across frequencies.

8.3 Neural Oscillations and Multisensory Processing

When stimuli are simultaneously presented in different sensory modalities, the signals need to be processed and integrated. For multisensory processing, information of each modality is first processed in unisensory cortical areas, and then is fed forward to multisensory and higher order cortical areas, for example, intraparietal sulcus (IPS), superior temporal sulcus (STS), and prefrontal cortex (PFC) (Ghazanfar and Schroeder 2006; van Atteveldt et al. 2014). In these areas, the sensory information from different modalities can be merged and integrated. Integrated information on stimulus is then fed back from higher order to lower order cortical areas. This is the classical feedforward-feedback model of multisensory integration (Meredith and Stein 1996; Stein et al. 1993). In addition to this classical model, cross-modal influence can also occur at the level of the primary sensory cortices (Driver and Noesselt 2008; Ghazanfar and Schroeder 2006; Kayser et al. 2007; Keil and Senkowski 2018; Schroeder and Foxe 2005; Senkowski et al. 2008) and subcortical regions (e.g., superior colliculus and the pulvinar nucleus) (Bauer et al. 2020; Cappe et al. 2007; Hackett et al. 2007; Lakatos et al. 2007), by which sensory signals from one modality can modulate the cortical excitability to the signals in another modality (Driver and Noesselt 2008; Lakatos et al. 2007). Therefore, multisensory processing needs coordinate activities of distinct brain areas. Synchronization of neural oscillations has been reported to facilitate cross-modal influence between primary sensory cortices, such as primary auditory, visual, and somatosensory cortices (Lakatos et al. 2007; Schroeder and Lakatos 2009) and may contribute to multisensory processing and integration as a mechanism (Senkowski et al. 2008; van Atteveldt et al. 2014). In particular, neural oscillations in low frequencies (in the delta, theta, and lower alpha range) can provide optimal temporal windows for cross-modal interaction (Cecere et al. 2015; Kayser et al. 2008; Lakatos et al. 2007; Simon and Wallace 2017). There are two mechanisms involved in promoting cross-modal synchronization: phase reset and neural entrainment. These mechanisms are flexibly modulated by task demands.

Cross-modal phase reset refers to the process that stimuli in one sensory modality can reset phase of an ongoing oscillation in a cortical area of another modality (Kayser et al. 2008; Lakatos et al. 2007). In this case, a salient external or internal stimulus can reset the phase of a neural oscillation within another sensory modality to a particular state of excitability (Kayser et al. 2008) (Fig. 8.3 left). Cross-modal phase reset can modulate perceptual performance in two opposite directions (Cecere et al. 2015; Diederich et al. 2012; Fiebelkorn et al. 2013; Kayser et al. 2008; Lakatos et al. 2009; Mercier et al. 2013; Naue et al. 2011; Perrodin et al. 2015; Plass et al. 2019; Romei et al. 2012). On the one hand, optimally aligned phases promote communication between cortical areas (Mercier et al. 2015). For example, a somatosensory stimulus shortly preceding an auditory tone can change the phase of the ongoing oscillations in the primary auditory cortex (A1), rendering the auditory input arriving at a high-excitability phase, and leading to an enhanced neuronal response and faster behavioral responses (Lakatos et al. 2007). On the other hand, decreased cross-modal phase alignment will impede further processing of the paired cross-modal signals (Lakatos et al. 2007).

In addition to single transient events, phase of neural oscillations can also be reset by external

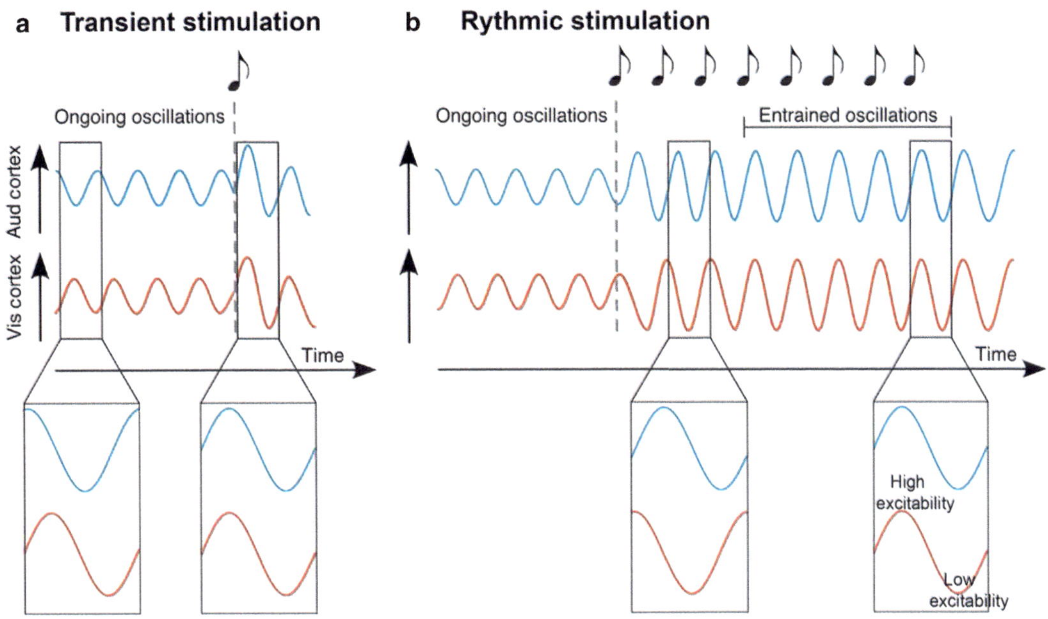

Fig. 8.3 Illustration for cross-modal phase reset and cross-modal entrainment. Adopted from permission (Bauer et al. 2020). (**a**) Schematic illustration of phase reset of ongoing neural oscillations induced by a single transient stimulus. A transient sound realigns the phases of the oscillations in the auditory cortex and visual cortex. (**b**) Schematic representation of neural entrainment induced by a rhythmic stimulus. A rhythmic auditory stimulus gradually shifts the phases of the neural oscillations in the auditory cortex and visual cortex, leading to cross-modal synchronization. For both phase reset and phase entrainment, the phase of ongoing oscillations aligns to the driving stimulus, thereby modulating the excitation-inhibition cycle of the neural oscillations

rhythmic or quasi-rhythmic stimulation. The synchronization of different oscillators is called entrainment. Neural entrainment refers to gradual synchronization of an ongoing neural oscillation to an external rhythmic or quasi-rhythmic input stream (Lakatos et al. 2008) (Fig. 8.3). Once the external input ceases, the neural oscillation will return to its intrinsic oscillatory frequency. It has been most reported in the delta and theta bands (Lakatos et al. 2008). Neural entrainment is in favor of periodic perceptual modulations. Behavioral performance is generally better when the target occurred on beat with the stimulation rhythm, relative to when the target occurred off beat (Jones et al. 2002; Mathewson et al. 2010; Miller et al. 2013). While both cross-modal phase reset and cross-modal entrainment promote neural processing and perceptual integration by temporally aligning high-excitability phases across sensory modalities to the timing of relevant events, they may transmit different types of cross-modal information. Phase reset transmits the information of the timing of an external stimulus, and neural entrainment transmits the information of an expected stimulus stream (see Bauer et al. 2020, for review).

When two stimuli are presented at the same time, stimulus congruence and temporal alignment may induce different synchronization of neural oscillations, and influence signal processing and behavioral response. For example, in a study (Krebber et al. 2015), congruent visuotactile motion was accompanied by greater gamma oscillations (50–80 Hz) in visual and somatosensory cortices, compared with incongruent motion stimuli (Fig. 8.4a). Moreover, enhanced gamma activity was correlated with shorter reaction latency in the behavioral task. In another study (Kambe et al. 2015), compared to asynchronous, synchronous audiovisual stimulation induced higher coherence in beta oscillations (Fig. 8.4b). Taken together, the enhanced power in gamma

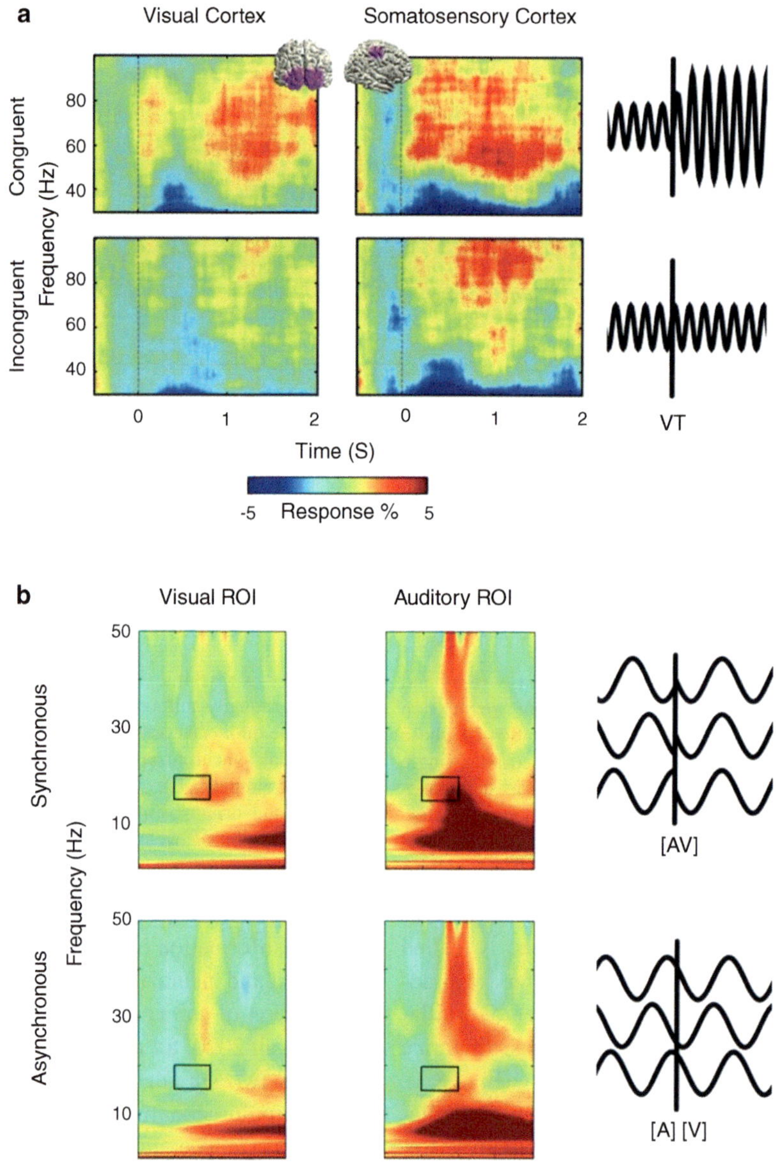

Fig. 8.4 Impact of stimulus congruence and temporal synchrony on multisensory processing. Adopted from permission (Keil and Senkowski 2018). (**a**) When visuotactile motion stimulation is congruent (top row), gamma-band power is enhanced in visual cortex (left column) and somatosensory cortex (middle column), compared to that when the stimulation is incongruent (bottom row) (adapted from Krebber et al. (2015)). The right column shows the schematic illustration of the increased gamma-band power following congruent visuotactile stimulation. (**b**) Beta-band inter-trial coherence is increased in visual (left column) and auditory (middle column) electrodes when the audiovisual stimulation is perceived as synchronous (top row) (adapted from Kambe et al. (2015)). The diagram in the right column illustrates phase alignment following a synchronous ([AV], top row) audiovisual stimulus, and no phase alignment following an asynchronous audiovisual stimulus ([A] [V], bottom row)

oscillations and increased coherence in beta oscillations reflect the temporal alignment of stimuli from different sensory modalities. Moreover, phase reset of neural oscillations in lower frequencies plays an important role in orchestrating multisensory processing and integration.

8.4 Neural Oscillations and Sensory Selection

When multiple stimuli are presented, the brain doesn't need to process all the information all the time. Instead, according to task requirements, sometimes, the brain needs to selectively process task-related signals (target) and ignores distracting information (distractor). In this process, top-down control, selective attention, is involved. In addition to integration of cross-modal information, synchronization of neural oscillations is also essential for selecting and routing cross-modal information (Fries 2005, 2015; Keil and Senkowski 2018; Senkowski et al. 2008; van Atteveldt et al. 2014; Zuo et al. 2020). Relative relationship between the time of stimulus onset (or stimulus-induced response) and local cortical excitability may affect the efficiency of sensory processing. Ongoing oscillations in the target-related and the distractor-related sensory cortices may display distinct phases relative to the time of stimulus onset. Consistent with this, in a study (Zuo et al. 2020) trained rats to selectively attend to auditory (a1 or a2) or visual stimuli (v1 or v2) when they were simultaneously presented, the authors observed significant phase reset of beta oscillations (12–30 Hz) relative to stimulus onset in A1, and the beta oscillations displayed opposite phases during auditory and visual attention (Fig. 8.5a). This makes it possible that while the signals occurring at high-excitability phases within the targeting modality are effectively transmitted, the asynchronous signals or signals occurring at low-excitability phases within the distracting modalities are likely impeded.

The posterior parietal cortex (PPC), as a higher order cortical area, is involved in top-down control. During stimulus presentation, the interaction between PPC and A1 was enhanced by increased coherence between their beta oscillations, and the phase of PPC leads A1 about 5 ms (Fig. 8.5b) (Zuo et al. 2020). Coherent oscillations between these two brain regions may facilitate the exchange of task-related information from PPC to A1. If PPC carries information about task rule, i.e., which modality to attend to, the oscillations in PPC may recruit neurons firing at modality-specific phases. The neurons in PPC and A1 tended to fire at specific phases of the beta oscillations (phase lock). The spike phase relative to PPC beta oscillations is related to the attended-to (targeting) modality, as shown that in the poor-behavior sessions, in which the attention might have in fact been directed to the distracting modality, the spike phase distribution of one modality resembled the spike phase distribution for good-behavior sessions of the distracting modality (Fig. 8.5c).

Since the neurons fired at different beta phase during auditory and visual attention, spike phase may have encoded attended-to modality during paired-modality trials. Correct information about the target must be encoded in cortices to make a correct decision. Papers have shown that neuronal responses are determined by the targeting stimulus, but not by the unattended stimulus (Moran and Desimone 1985; Desimone and Duncan 1995; Treue and Maunsell 1996). If the rats ignore the distractor and only process the target information in the paired-modality trials, the firing pattern induced by the paired-modality stimuli should be similar to that induced by the single-modality stimuli with the same target. To test whether the spike phase relative to beta oscillations could predict the target in each paired-modality trial, the authors compared the spike-LFP phase relationship in the paired-modality trials with that in the single-modality trials (Zuo et al. 2020). Figure 8.6 illustrates a representative A1 interneuron recorded during auditory attention. This neuron did not show significant differences in spike rate along time, among the four single-modality stimuli (Fig. 8.6a) or the four paired-modality stimuli (Fig. 8.6b). On the contrary, the spike phase relative to the A1 beta oscillations showed differences for the visual stimuli and auditory stimuli.

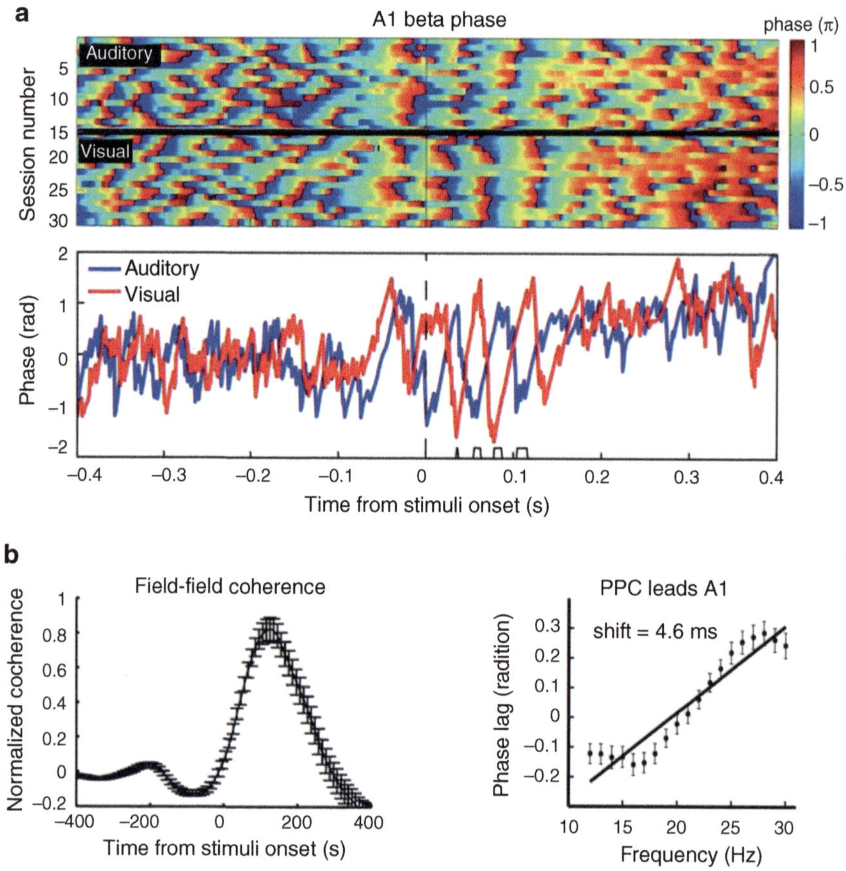

Fig. 8.5 Neural oscillations and sensory selection. (**a**) The beta oscillations in the auditory cortex displayed opposite phases during auditory and visual attention. Upper plot: phases of PPC beta oscillations aligned to stimulus onset. The duration of stimulus is ~200 ms. Each row is the averaged beta phase across all auditory (upper) or visual (lower) trials of a single session. Lower plot: instantaneous beta phase constructed as session-averaged phase. A Watson-Williams test was performed to compare LFP phases between auditory attention and visual attention in each 1-ms bin, and significances ($p < 0.05$) are shown as segments above the x-axis. (**b**) Left: coherence between PPC and A1 beta oscillations increased during stimulus presentation. Right: the phase of PPC leads A1~5 ms. To investigate the direction of interactions between PPC and A1, the phase shift between PPC and A1 (subtracting the phase of A1 from the phase of PPC) at different beta frequencies is plotted (refer to the method in Fig. 8.8). A positive correlation between the magnitude of A1 phase lag after PPC and the frequency is found, with a positive slope corresponding to 4.6 ms, suggesting top-down modulation of PPC on A1 activities (Zuo et al. 2020). (**c**) Comparison of phase-dependent firing densities relative to PPC beta oscillations during the sessions when the behavior was good or poor. Adopted from permission (Zuo et al. 2020). The neurons in PPC and A1 tended to fire at specific phases of the beta oscillations (phase-lock). The spike phase distribution in poor-behavior sessions of one modality resembled the spike phase distribution for good-behavior sessions of the distracting modality, indicating that the spike phase relative to PPC beta oscillations is related to the rat's behavior

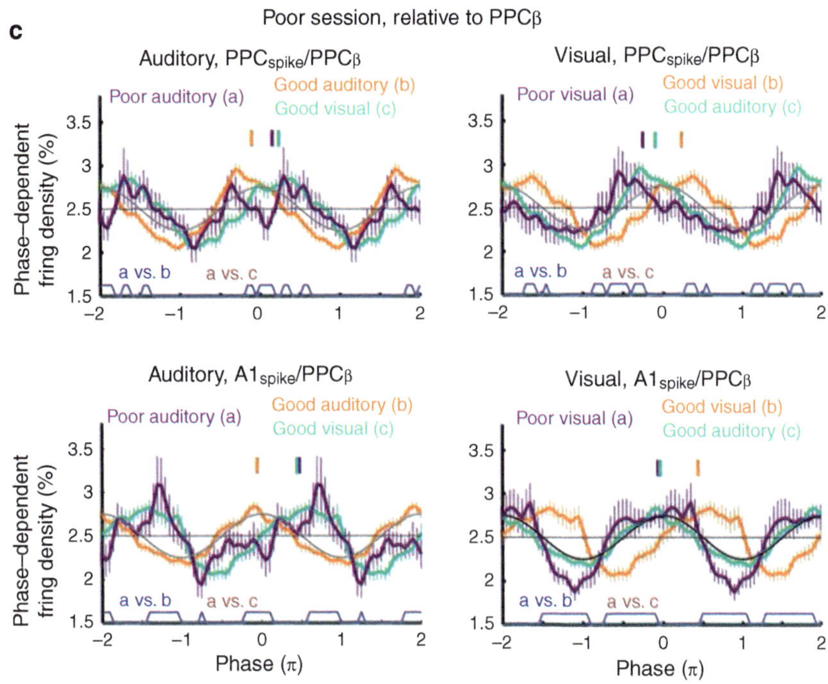

Fig. 8.5 (continued)

The neuron tended to fire at different phase ranges. When only the visual stimuli (v1 or v2) were presented, the preferred phase ranged from -0.7π to -0.2π. By contrast, when only the auditory stimuli (a1 and a2) were presented, the preferred phase ranged from -0.5π to 0.5π (Fig. 8.6c). In the paired-modality trials, similar spike-LFP phase relationship was shown as that shown in the auditory stimuli, which were the targets (Fig. 8.6d). Bayesian analysis was used to predict the target stimulus by using the timing-dependent and phase-dependent firing distribution. The "decoding rule" was built only from correct single-modality trials and then the obtained rule was used to test all paired-modality trials (correct and incorrect). The results showed that the decoding rule obtained from spike-phase distribution predicted high percentages for the target modality (Tar & SM) and significantly lower percentages for the distracting modality (SC & DD) (Fig. 8.6f), while the decoding rule obtained from spike-timing distribution could not encode the attended-to modality or stimulus (Fig. 8.6e). This result indicates that spike phase relative to beta oscillations can carry information about sensory modality, regardless of whether spike timing or rate does so. The average decoding results performed on the spike phase of the A1 interneurons relative to A1 beta oscillations showed similar results to those exhibited by the representative neuron shown in Fig. 8.6 for good-behavior sessions (Fig. 8.7a, purple bars). By contrast, poor-behavior sessions did not show good prediction for the target modality, the spike phase encoded the distracting modality instead (Fig. 8.7a, green bars). Thus, during modality selection the spike-LFP phase relationship is related to behavioral choice. Moreover, the prediction accuracy in incorrect trials was significantly lower than that in correct trials, indicating that spike phase in incorrect trials did not represent the target stimulus correctly (Fig. 8.7b).

Spike phase not only reflects attended-to modality, but also reflects firing sequence in the cycle of an oscillation. Different neuronal subpopulations in both areas were entrained by either PPC or A1 beta oscillations, or to both. The recorded neurons were assigned to 8 subpopula-

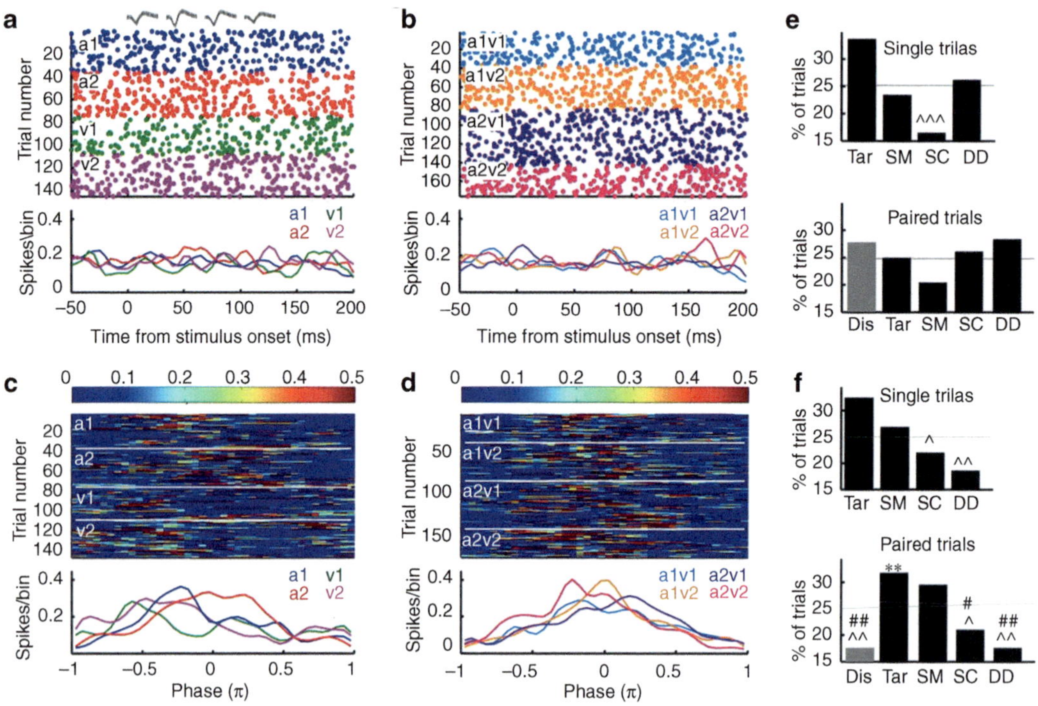

Fig. 8.6 Spike phase of a sample A1 interneuron relative to A1 beta oscillations encode the attended-to stimulus. Modified and Reprinted with permission from Zuo et al. (2020). This neuron was recorded with a tetrode and the waveforms on the four electrodes are shown at top. (**a** and **b**) Raster plot (upper) and peristimulus time histogram (PSTH, bottom) of the neuronal firing are aligned to stimulus onset. The firing over a time window from −50 to 200 ms in correct single-modality (**a**) and paired-modality (**b**) trials are shown, respectively. (**c** and **d**). Neuronal firing rate in each phase bin (phase-dependent firing rate, bin size = π/20, 40 bins in total from −π to π) in correct trials for single-modality (**c**) and paired-modality (**d**) stimuli. Phase-dependent firing rate in each trial (upper) and trial-averaged phase-dependent firing rate (bottom) are shown for single- (**c**) and paired-modality (**d**) trials. (**e**) Decoding results predicted by spike timing. Percentage of trials in which the presented stimulus was predicted as the candidate stimuli by spike timing are shown. The candidate stimuli include Tar (the presented target in a single trial), SM (the other stimulus of the same modality as Tar), SC (the stimulus of different modality but same choice side as Tar), and DD (the stimulus of different modality and different choice side of Tar). Dis is the presented distractor in that trial, which could be SC or DD. Three sets of chi-square tests were, respectively, used to compare the percentages of the candidate stimuli with chance level (25%, the gray line, statistical significance indicted by*), Tar (statistical significance indicted by^), and SM Tar (statistical significance indicted by #). ^^^ $p < 0.001$. (**f**) Decoding results predicted by spike phase. Percentage of trials in which the presented stimulus was predicted by spike phase as the candidate stimuli are shown. Same analyses were used as in (**e**). ** $p < 0.01$, ^ $p < 0.05$, ^^ $p < 0.01$, # $p < 0.05$, ## $p < 0.01$

tions: PPC pyramidal neurons or interneurons entrained by PPC or A1 beta oscillations (PPC_{Pys}/PPC_{β}, PPC_{INs}/PPC_{β}, $PPC_{Pys}/A1_{\beta}$, $PPC_{INs}/A1_{\beta}$), A1 pyramidal neurons or interneurons entrained by PPC or A1 beta oscillations ($A1_{Pys}/PPC_{\beta}$, $A1_{INs}/PPC_{\beta}$, $A1_{Pys}/A1_{\beta}$, $A1_{INs}/A1_{\beta}$). By plotting the phase shift between pyramidal neurons and interneurons in the PPC and A1 against frequency, firing order of neuronal subpopulations can be computed. The logic is that a fixed time interval (the transmission time from signal A to B, τ) translates into different multiples of π at different frequencies (Fig. 8.8a). For example, a theoretical transmission time of $\tau = 0.01$ s between two brain regions corresponds to a phase shift of 0.94 rad at 15 Hz, and a phase shift of 1.88 rad at 30 Hz. In general, the phase lag ($\varphi(f)$) between two signals (φ_B-φ_A) at a frequency f is linearly increasing with increasing frequency and can be calculated by formula: $\varphi(f) = 2\pi f\tau$ (Fig. 8.8a). As

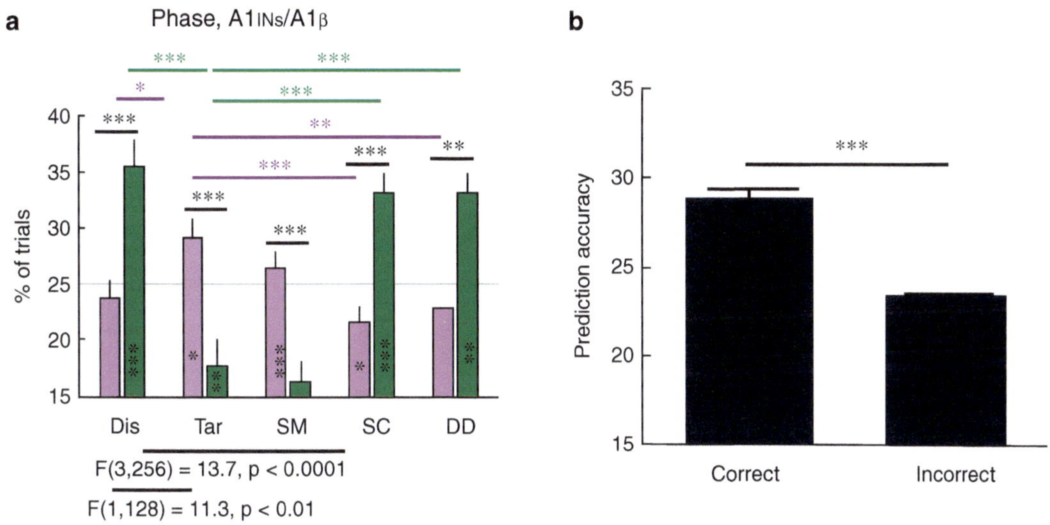

Fig. 8.7 Spike phase relative to beta oscillations decodes the attended-to modality. Adopted from permission (Zuo et al. 2020). (**a**) Percentage of trials in which the presented stimulus was decoded as Tar, SM, SC, DD, or Dis by the spike-LFP phase relationship between A1 interneurons and A1 beta oscillations ($A1_{INs}/A1_{\beta}$). One-sample t tests were used to compare the percentages with chance level (25%, the gray line, statistical significance indicated by stars over bars). A 2 × 4 repeated-measures ANOVA was performed to compare the differences of the trial percentages on behavioral performance (good, poor) and the candidate stimuli (Tar, SM, SC, DD). A significant interaction effect was found and is shown below the plot, which indicates that during good-behavior and poor-behavior sessions, the spike-LFP phase had different decoding capacities. Pairwise post-tests were used to compare the decoding capacities between good- (purple) and poor-behavior (green) sessions and the statistical significances are indicated by black stars. Pairwise post-test between Tar and SM, SC, or DD of good-behavior sessions or poor-behavior sessions were performed and the statistical significances are respectively shown by purple stars (good) and green stars (poor). Another 2 × 2 repeated-measures ANOVA with Bonferroni post-test was performed to compare the trial percentages on behavior performance (good, poor) and the candidate stimuli (Tar, Dis). A significant interaction effect was found and is shown below the plot. Significantly better prediction for Tar than Dis was revealed during good-behavior sessions. By contrast, significant lower prediction for Tar than Dis was revealed during poor-behavior sessions. Error bars are SEM across neurons. For all the significances: * $p < 0.05$, ** $p < 0.01$, *** $p < 0.0001$. (**b**) Averaged prediction accuracy across all the neurons for the correct trials and incorrect trials, respectively. A paired t test was used to compare the difference between them. *** $p < 0.001$

the deduction in Fig. 8.8b, the time lag can be estimated as $\tau = slope/(2\pi)$, where the *slope* is the slope of the regression of phase lag changing (in radians) on frequency (Fig. 8.8b) (Schoffelen et al. 2005; Baldauf and Desimone 2014; Zuo et al. 2020). To do so, the LFP signals were band-pass-filtered to 19 frequency bands from 12 to 30 Hz. The instantaneous phases of these bands were calculated by using Hilbert transformation and then neurons' spikes were assigned a phase for each of the 19 frequency bands. For any two subpopulations (e.g., A and B), phase lags were calculated by subtracting the phase of A from the phase of B between all possible pairs of individual neurons for each beta band, generating 19 phase shift distributions for the 19 beta bands. The circular mean of each phase shift distribution was calculated to estimate the mean phase shift at each band. A linear regression of mean phase shifts on the 19 beta bands was performed and the slope was calculated if the regression was significant ($p < 0.05$) (An example in Fig. 8.8c). The direction of interaction was determined by the direction of slope: a positive slope means A leads B, and a negative slope means B leads A. The slope of the regression line is indicative of the magnitude of time lag.

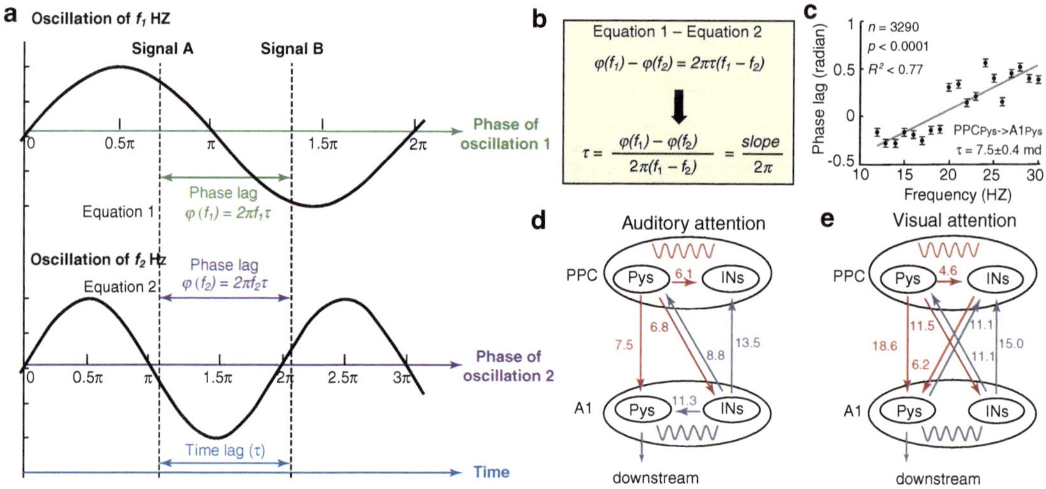

Fig. 8.8 Directionality of neuronal subpopulations in good-behavior sessions. (**a**) Schematic illustration of phase lags at two frequency bands. Suppose signal B occurs after signal A with a time lag of τ. The time lag corresponds to different phase lags relative to different frequencies. The phase lag ($\varphi(f)$) relative to an oscillation of f Hz is $\varphi(f) = 2\pi f\tau$, which is linearly increasing with increasing frequency. (**b**) Estimation of time lag τ from phase lag and frequency. The time lag can be estimated as $\tau = \text{slope}/(2\pi)$, where the *slope* is the slope of the regression of phase lag changing (in radians) on frequency. (**c**) Directionality of the PPC_{Pys} and $A1_{Pys}$ entrained by PPC beta oscillations (PPC_{Pys}/PPC_β and $A1_{Pys}/PPC_\beta$) during auditory attention. The phase shift between the two neuronal subpopulations is plotted against frequency. The dots represent the averaged phase lags by subtracting the phase of PPC_{Pys}/PPC_β from the phase of $A1_{Pys}/PPC_\beta$ at the 19 beta frequencies. Error bars are SEM across neuron pairs. Statistical values are shown in the corner: n is the number of neuron pairs, p is the p value of the linear regression, and R^2 is the coefficient of determination of the linear regression. Mean ± SEM of the time lags between neuron pairs is shown in number of ms. (**d**) Phase lags among PR and IN subpopulations entrained by either PPC beta oscillations (red) or A1 beta oscillations (blue) during good auditory sessions. Lags are only shown for pairs of subpopulations that showed significant regression with beta frequencies ($p < 0.05$). Arrows show the direction of firing sequence. (**e**) Same as (**d**) during visual attention. (**d**) and (**e**) are modified and reprinted with permission from Zuo et al. (2020)

Figure 8.8c plots the phase shift between the subpopulations of PPC_{Pys}/PPC_β and $A1_{Pys}/PPC_\beta$ against frequency. A significant positive regression was observed for the phase shift ($\varphi_{A1Pys} - \varphi_{PPCPys}$) at different beta frequencies, indicating that PPC_{Pys} leads $A1_{Pys.}$ The slope of the regression line corresponded to a transmission time of 7.5 ms. Through the same method, directionality between any two of the four subpopulations entrained by PPC beta oscillations (PPC_{Pys}/PPC_β, PPC_{INs}/PPC_β, $A1_{Pys}/PPC_\beta$, $A1_{INs}/PPC_\beta$) or any two of the four subpopulations entrained by A1 beta oscillations ($PPC_{Pys}/A1_\beta$, $PPC_{INs}/A1_\beta$, $A1_{Pys}/A1_\beta$, $A1_{INs}/A1_\beta$) was computed (Fig. 8.8d and e).

For the neurons entrained by PPC beta oscillations either during auditory (Fig. 8.8d) or visual attention (Fig. 8.8e), PPC_{Pys} consistently spiked before $A1_{Pys}$ and $A1_{INs}$, consistent with top-down modulation. The transmission time from PPC to A1 during auditory attention (Fig. 8.8d, 7.5, and 6.8 ms) was shorter than that during visual attention (Fig. 8.8e, 18.6, and 11.5 ms), suggesting that top-down signaling from the PPC to A1 is mediated via an additional relay during visual attention. The additional relay may involve cross-modal inhibitory interaction from visual areas to A1 (Ibrahim et al. 2016; Kayser et al. 2008). For the neurons entrained by A1 beta oscillations, the $A1_{INs}$ spiked before the PPC_{Pys} and PPC_{INs} either during auditory (Fig. 8.8d) or visual attention (Fig. 8.8e), consistent with bottom-up signaling. During auditory attention, the $A1_{INs}$ fired with a time lead of 11.3 ms over the $A1_{Pys}$ (Fig. 8.8d), supporting disinhibition by di-synaptic interneuron connections in the auditory cortex (Zhang et al. 2014). While during visual attention, no phase lead of $A1_{INs}$ over $A1_{Pys}$, suggesting that disinhibitory input is not recruited during visual attention.

Overall, all the above results show that neural oscillations play a role in selecting sensory signals by either synchronize LFP oscillations or neuronal spikes of different modalities in different phases.

8.5 Summary

To summarize, neural oscillations reflect rhythmic fluctuations of neuronal excitability. Signals from the stimuli time-locked to the phases of high-excitability get better processed and evoke stronger post-synaptic activities. Internal and external stimuli may reset phases of neural oscillations, not only in one sensory cortex, but also in other sensory cortices involved in the ongoing task. According to task requirement, phases in different cortices may be aligned or mis-aligned to modulate signal processing of multisensory modalities, to promote cross-modal integration, or to select task-related signals and ignore unrelated signals.

Acknowledgments This work was supported by National Science and Technology Innovation 2030 Major Program of China (grant No. STI2030-Major Projects 2021ZD0203700/2021ZD0203703) and National Nature Science Foundation of China (Grant No. 32271077) to Yanfang Zuo.

References

Azouz R, Gray CM (2003) Adaptive coincidence detection and dynamic gain control in visual cortical neurons in vivo. Neuron 37:513–523

Baldauf D, Desimone R (2014) Neural mechanisms of object-based attention. Science 344:424–427

Bastos AM, Vezoli J, Bosman CA, Schoffelen JM, Oostenveld R, Dowdall JR, De Weerd P, Kennedy H, Fries P (2015a) Visual areas exert feedforward and feedback influences through distinct frequency channels. Neuron 85:390–401

Bastos AM, Vezoli J, Fries P (2015b) Communication through coherence with inter-areal delays. Curr Opin Neurobiol 31:173–180

Bauer AR, Debener S, Nobre AC (2020) Synchronisation of neural oscillations and cross-modal influences. Trends Cogn Sci 24:481–495

Bressler SL, Richter CG (2015) Interareal oscillatory synchronization in top-down neocortical processing. Curr Opin Neurobiol 31:62–66

Buffalo EA, Fries P, Landman R, Buschman TJ, Desimone R (2011) Laminar differences in gamma and alpha coherence in the ventral stream. Proc Natl Acad Sci U S A 108:11262–11267

Canolty RT, Knight RT (2010) The functional role of cross-frequency coupling. Trends Cogn Sci 14:506–515

Cappe C, Morel A, Rouiller EM (2007) Thalamocortical and the dual pattern of corticothalamic projections of the posterior parietal cortex in macaque monkeys. Neuroscience 146:1371–1387

Cecere R, Rees G, Romei V (2015) Individual differences in alpha frequency drive crossmodal illusory perception. Curr Biol 25:231–235

Demiralp T, Bayraktaroglu Z, Lenz D, Junge S, Busch NA, Maess B, Ergen M, Herrmann CS (2007) Gamma amplitudes are coupled to theta phase in human EEG during visual perception. Int J Psychophysiol 64:24–30

Desimone R, Duncan J (1995) Neural Mechanisms of Selective Visual Attention. Annual Review of Neuroscience 18(1):193–222

Diederich A, Schomburg A, Colonius H (2012) Saccadic reaction times to audiovisual stimuli show effects of oscillatory phase reset. PLoS One 7:e44910

Douglas RJ, Martin KA (2004) Neuronal circuits of the neocortex. Annu Rev Neurosci 27:419–451

Driver J, Noesselt T (2008) Multisensory interplay reveals crossmodal influences on 'sensory-specific' brain regions, neural responses, and judgments. Neuron 57:11–23

Euston DR, Tatsuno M, McNaughton BL (2007) Fast-forward playback of recent memory sequences in prefrontal cortex during sleep. Science 318:1147–1150

Felleman DJ, Van Essen DC (1991) Distributed hierarchical processing in the primate cerebral cortex. Cereb Cortex 1:1–47

Fiebelkorn IC, Snyder AC, Mercier MR, Butler JS, Molholm S, Foxe JJ (2013) Cortical cross-frequency coupling predicts perceptual outcomes. NeuroImage 69:126–137

Fries P (2005) A mechanism for cognitive dynamics: neuronal communication through neuronal coherence. Trends Cogn Sci 9:474–480

Fries P (2015) Rhythms for cognition: communication through coherence. Neuron 88:220–235

Ghazanfar AA, Schroeder CE (2006) Is neocortex essentially multisensory? Trends Cogn Sci 10:278–285

Gutig R (2014) To spike, or when to spike? Curr Opin Neurobiol 25:134–139

Hackett TA, De La Mothe LA, Ulbert I, Karmos G, Smiley J, Schroeder CE (2007) Multisensory convergence in auditory cortex, II. Thalamocortical connections of the caudal superior temporal plane. J Comp Neurol 502:924–952

Ibrahim LA, Mesik L, Ji X-Y, Fang Q, Li H-F, Li Y-T, Zingg B, Zhang LI, Tao HW (2016) Cross-modality sharpening of visual cortical processing through layer-1-mediated inhibition and disinhibition. Neuron 89:1031–1045

Jaramillo J, Kempter R (2017) Phase precession: a neural code underlying episodic memory? Curr Opin Neurobiol 43:130–138

Jones MR, Moynihan H, MacKenzie N, Puente J (2002) Temporal aspects of stimulus-driven attending in dynamic arrays. Psychol Sci 13:313–319

Kambe J, Kakimoto Y, Araki O (2015) Phase reset affects auditory-visual simultaneity judgment. Cogn Neurodyn 9:487–493

Kann O, Papageorgiou IE, Draguhn A (2014) Highly energized inhibitory interneurons are a central element for information processing in cortical networks. J Cereb Blood Flow Metab 34:1270–1282

Kayser C, Petkov CI, Augath M, Logothetis NK (2007) Functional imaging reveals visual modulation of specific fields in auditory cortex. J Neurosci 27:1824–1835

Kayser C, Petkov CI, Logothetis NK (2008) Visual modulation of neurons in auditory cortex. Cereb Cortex 18:1560–1574

Keil J, Senkowski D (2018) Neural oscillations orchestrate multisensory processing. Neuroscientist 24:609–626

Konig P, Engel AK, Singer W (1996) Integrator or coincidence detector? The role of the cortical neuron revisited. Trends Neurosci 19:130–137

Krebber M, Harwood J, Spitzer B, Keil J, Senkowski D (2015) Visuotactile motion congruence enhances gamma-band activity in visual and somatosensory cortices. NeuroImage 117:160–169

Lakatos P, Chen CM, O'Connell MN, Mills A, Schroeder CE (2007) Neuronal oscillations and multisensory interaction in primary auditory cortex. Neuron 53:279–292

Lakatos P, Karmos G, Mehta AD, Ulbert I, Schroeder CE (2008) Entrainment of neuronal oscillations as a mechanism of attentional selection. Science 320:110–113

Lakatos P, O'Connell MN, Barczak A, Mills A, Javitt DC, Schroeder CE (2009) The leading sense: supramodal control of neurophysiological context by attention. Neuron 64:419–430

Lee JH, Whittington MA, Kopell NJ (2013) Top-down beta rhythms support selective attention via interlaminar interaction: a model. PLoS Comput Biol 9:e1003164

Livingstone MS (1996) Oscillatory firing and interneuronal correlations in squirrel monkey striate cortex. J Neurophysiol 75:2467–2485

Luo H, Poeppel D (2007) Phase patterns of neuronal responses reliably discriminate speech in human auditory cortex. Neuron 54:1001–1010

Maier A, Adams GK, Aura C, Leopold DA (2010) Distinct superficial and deep laminar domains of activity in the visual cortex during rest and stimulation. Front Syst Neurosci 4

Markov NT, Vezoli J, Chameau P, Falchier A, Quilodran R, Huissoud C, Lamy C, Misery P, Giroud P, Ullman S et al (2014) Anatomy of hierarchy: feedforward and feedback pathways in macaque visual cortex. J Comp Neurol 522:225–259

Mathewson KE, Fabiani M, Gratton G, Beck DM, Lleras A (2010) Rescuing stimuli from invisibility: inducing a momentary release from visual masking with pre-target entrainment. Cognition 115:186–191

Mercier MR, Foxe JJ, Fiebelkorn IC, Butler JS, Schwartz TH, Molholm S (2013) Auditory-driven phase reset in visual cortex: human electrocorticography reveals mechanisms of early multisensory integration. NeuroImage 79:19–29

Mercier MR, Molholm S, Fiebelkorn IC, Butler JS, Schwartz TH, Foxe JJ (2015) Neuro-oscillatory phase alignment drives speeded multisensory response times: an electro-corticographic investigation. J Neurosci 35:8546–8557

Meredith MA, Stein BE (1996) Spatial determinants of multisensory integration in cat superior colliculus neurons. J Neurophysiol 75:1843–1857

Michalareas G, Vezoli J, van Pelt S, Schoffelen JM, Kennedy H, Fries P (2016) Alpha-Beta and Gamma rhythms subserve feedback and feedforward influences among human visual cortical areas. Neuron 89:384–397

Miller JE, Carlson LA, McAuley JD (2013) When what you hear influences when you see: listening to an auditory rhythm influences the temporal allocation of visual attention. Psychol Sci 24:11–18

Moran J, Desimone R (1985) Selective Attention Gates Visual Processing in the Extrastriate Cortex. Science 229(4715):782–784

Naue N, Rach S, Struber D, Huster RJ, Zaehle T, Korner U, Herrmann CS (2011) Auditory event-related response in visual cortex modulates subsequent visual responses in humans. J Neurosci 31:7729–7736

Palva JM, Palva S (2018) Functional integration across oscillation frequencies by cross-frequency phase synchronization. Eur J Neurosci 48:2399–2406

Palva JM, Palva S, Kaila K (2005) Phase synchrony among neuronal oscillations in the human cortex. J Neurosci 25:3962–3972

Perrodin C, Kayser C, Logothetis NK, Petkov CI (2015) Natural asynchronies in audiovisual communication signals regulate neuronal multisensory interactions in voice-sensitive cortex. Proc Natl Acad Sci U S A 112:273–278

Plass J, Ahn E, Towle VL, Stacey WC, Wasade VS, Tao J, Wu S, Issa NP, Brang D (2019) Joint encoding of auditory timing and location in visual cortex. J Cogn Neurosci 31:1002–1017

Richter CG, Thompson WH, Bosman CA, Fries P (2017) Top-down beta enhances bottom-up gamma. J Neurosci 37:6698–6711

Rizzuto DS, Madsen JR, Bromfield EB, Schulze-Bonhage A, Kahana MJ (2006) Human neocortical oscillations exhibit theta phase differences between encoding and retrieval. NeuroImage 31:1352–1358

Romei V, Gross J, Thut G (2012) Sounds reset rhythms of visual cortex and corresponding human visual perception. Curr Biol 22:807–813

Roopun AK, Kramer MA, Carracedo LM, Kaiser M, Davies CH, Traub RD, Kopell NJ, Whittington MA (2008) Period concatenation underlies interactions between gamma and beta rhythms in neocortex. Front Cell Neurosci 2:1

Rosenblum MG, Weule J, Kurths J, Pickovsky A, Volkmann J, Schnitzler A, Freund HJ (1998) Detection of n:m phase locking from noisy data: application to magnetoencephalography. Phys Rev Lett 81:3291–3294

Saleh M, Reimer J, Penn R, Ojakangas CL, Hatsopoulos NG (2010) Fast and slow oscillations in human primary motor cortex predict oncoming behaviorally relevant cues. Neuron 65:461–471

Schoffelen J, Oostenveld R, Fries P (2005) Neuronal coherence as a mechanism of effective corticospinal interaction. Science 308:111–114

Schroeder CE, Foxe J (2005) Multisensory contributions to low-level, 'unisensory' processing. Curr Opin Neurobiol 15:454–458

Schroeder CE, Lakatos P (2009) Low-frequency neuronal oscillations as instruments of sensory selection. Trends Neurosci 32:9–18

Senkowski D, Schneider TR, Foxe JJ, Engel AK (2008) Crossmodal binding through neural coherence: implications for multisensory processing. Trends Neurosci 31:401–409

Simon DM, Wallace MT (2017) Rhythmic modulation of entrained auditory oscillations by visual inputs. Brain Topogr 30:565–578

Singer W, Gray CM (1995) Visual feature integration and the temporal correlation hypothesis. Annu Rev Neurosci 18:555–586

Smith MA, Jia X, Zandvakili A, Kohn A (2013) Laminar dependence of neuronal correlations in visual cortex. J Neurophysiol 109:940–947

Stein BE, Meredith MA, Wallace MT (1993) The visually responsive neuron and beyond: multisensory integration in cat and monkey. Prog Brain Res 95:79–90

Swadlow HA, Gusev AG (2001) The impact of 'bursting' thalamic impulses at a neocortical synapse. Nat Neurosci 4:402–408

Tamas G, Buhl EH, Lorincz A, Somogyi P (2000) Proximally targeted GABAergic synapses and gap junctions synchronize cortical interneurons. Nat Neurosci 3:366–371

Treue S, Maunsell J H (1996) Attentional modulation of visual motion processing in cortical areas MT and MST. Nature 382(6591):539–541

van Atteveldt N, Murray MM, Thut G, Schroeder CE (2014) Multisensory integration: flexible use of general operations. Neuron 81:1240–1253

van Kerkoerle T, Self MW, Dagnino B, Gariel-Mathis MA, Poort J, van der Togt C, Roelfsema PR (2014) Alpha and gamma oscillations characterize feedback and feedforward processing in monkey visual cortex. Proc Natl Acad Sci U S A 111:14332–14341

Vanhatalo S, Palva JM, Holmes MD, Miller JW, Voipio J, Kaila K (2004) Infraslow oscillations modulate excitability and interictal epileptic activity in the human cortex during sleep. Proc Natl Acad Sci U S A 101:5053–5057

Voytek B, Canolty RT, Shestyuk A, Crone NE, Parvizi J, Knight RT (2010) Shifts in gamma phase-amplitude coupling frequency from theta to alpha over posterior cortex during visual tasks. Front Hum Neurosci 4:191

Zhang S, Xu M, Kamigaki T, Hoang Do JP, Chang WC, Jenvay S, Miyamichi K, Luo L, Dan Y (2014) Long-range and local circuits for top-down modulation of visual cortex processing You only see what you want to see. Science 345(6197):660–665

Zuo Y, Huang Y, Wu D, Wang Q, Wang Z (2020) Spike phase shift relative to Beta oscillations mediates modality selection. Cereb Cortex 30:5431–5448

9 Multisensory Calibration: A Variety of Slow and Fast Brain Processes Throughout the Lifespan

Adam Zaidel

Abstract

From before we are born, throughout development, adulthood, and aging, we are immersed in a multisensory world. At each of these stages, our sensory cues are constantly changing, due to body, brain, and environmental changes. While integration of information from our different sensory cues improves precision, this only improves accuracy if the underlying cues are unbiased. Thus, multisensory calibration is a vital and ongoing process. To meet this grand challenge, our brains have evolved a variety of mechanisms. First, in response to a systematic discrepancy between sensory cues (without external feedback) the cues calibrate one another (unsupervised calibration). Second, multisensory function is calibrated to external feedback (supervised calibration). These two mechanisms superimpose. While the former likely reflects a lower level mechanism, the latter likely reflects a higher level cognitive mechanism. Indeed, neural correlates of supervised multisensory calibration in monkeys were found in higher level multisensory cortical area VIP, but not in the relatively lower level multisensory area MSTd. In addition, even without a cue discrepancy (e.g., when experiencing stimuli from different sensory cues in series) the brain monitors supra-modal statistics of events in the environment and adapts perception cross-modally. This too comprises a variety of mechanisms, including confirmation bias to prior choices, and lower level cross-sensory adaptation. Further research into the neuronal underpinnings of the broad and diverse functions of multisensory calibration, with improved synthesis of theories is needed to attain a more comprehensive understanding of multisensory brain function.

Keywords

Unsupervised · Supervised · Recalibration · Adaptation · Serial dependence · Visual · Vestibular · Self-motion · Accurate · Perception

9.1 Introduction

When reaching for your cup of coffee, a slight deviation from the intended movement can have the undesirable outcome of knocking it over. To avoid coffee-soaked laptops, stained couches, and visits to the emergency room (if you like boiling-hot coffee) our brains have acquired the ability to perform accurate movements (Schwartz 2016). Note that accuracy is not the same as pre-

A. Zaidel (✉)
Gonda Multidisciplinary Brain Research Center, Bar-Ilan University, Ramat Gan, Israel
e-mail: adam.zaidel@biu.ac.il

Y. Gu, A. Zaidel (eds.), *Advances of Multisensory Integration in the Brain*, Advances in Experimental Medicine and Biology 1437, https://doi.org/10.1007/978-981-99-7611-9_9

cision (although these terms are often confused). The difference can be explained by the example of throwing darts. *Precise* performance means that the darts are tightly clustered (Fig. 9.1, top row). Thus, precision reflects consistency of execution across multiple attempts (the term reliability is also used, interchangeably with precision). But, precision is not enough for accurate function. The result also needs to be *unbiased*—such that on average, the darts are centered on target (Fig. 9.1, left column). However, precision and bias are two independent properties, and performance can reflect any combination of the two (Fig. 9.1). *Accurate* actions need to be both precise and unbiased (Fig. 9.1, top left).

Note that this definition of accuracy differs slightly from our previous disambiguation of the terms accuracy and precision, where we defined accuracy as being unbiased (Zaidel et al. 2011, 2013, 2021). The rationale for this refinement is that unbiased but imprecise performance (Fig. 9.1, bottom left) will still lead to many misses (i.e., low success rate); whereas the term accurate implies a high success rate. In other words, we think that accuracy should reflect the average level of success, rather than the level of success of the average outcome. According to this definition, both properties, being unbiased and precise, are needed for accurate performance (high success hitting the target).

In certain sports (like shooting), precision is often considered a more important property than being unbiased. That is because the tool (e.g., a rifle) can be calibrated by adjusting the aiming apparatus to correct for an individual's bias. By contrast, precision reflects an innate limitation of the individual—one's ability to consistently and reliably reproduce the same result. Thus, precision is often the limiting factor of performance, and therefore broadly considered to reflect skill (Yon and Frith 2021). However, in the absence of a mechanism for calibration (whether external or internal), precise but biased performance (Fig. 9.1, top right) is worse than the reverse. An

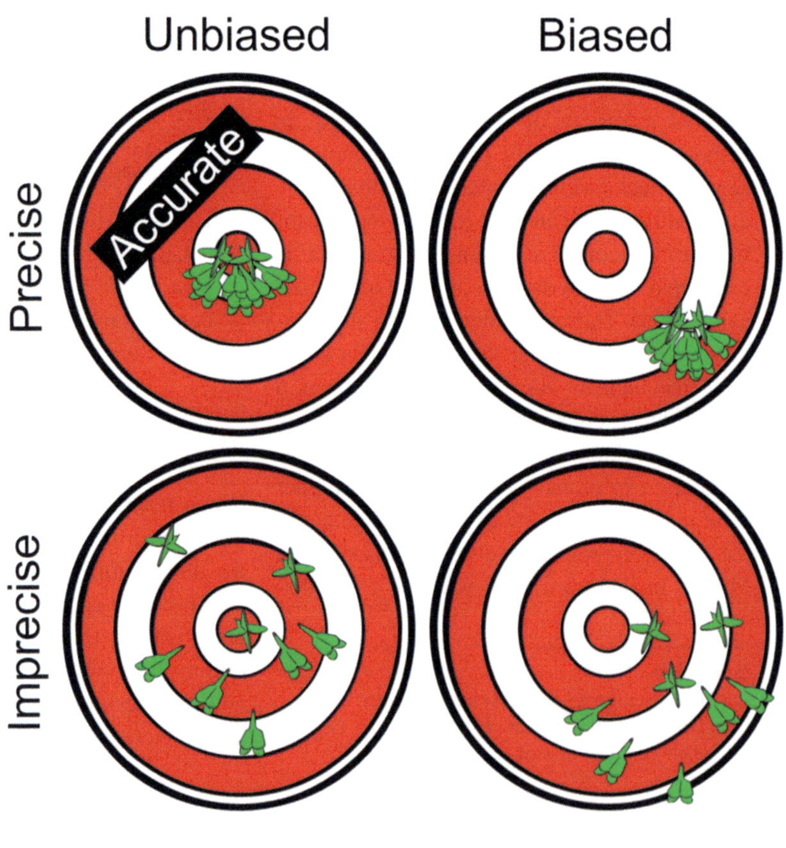

Fig. 9.1 Disambiguation of terms. When performance is "precise," the darts are tightly clustered (top row). "Unbiased" performance means that *on average* the darts are centered on target (left column). "Accurate" performance is both precise and unbiased (top left)

outcome that is consistently biased (always miss) is worse than one that is imprecise but (on average) unbiased (Fig. 9.1, bottom left). Therefore, without mechanisms of calibration, precision might be worthless.

In the same way that we consider precision and bias in terms of action, these concepts relate also to perception. High perceptual precision means that the same stimulus and environmental conditions will give rise to a consistent and reliable percept (the perceptual counterpart of tightly clustered darts). Also, being perceptually unbiased means that the average percept is in line with the real world. Once again, these are two different properties that can be dissociated, and both contribute to accurate perception. Yet, whereas multiple studies of multisensory function have explored how combining cues from different sensory modalities ("multisensory integration") leads to improved precision (Ernst and Banks 2002; van Beers et al. 2002; Alais and Burr 2004; Knill and Pouget 2004; Fetsch et al. 2009; Butler et al. 2010), attaining unbiased multisensory perception, which is perhaps more important than precision for overall accuracy, has received relatively little attention.

The question of multisensory calibration is much broader than appreciated at face value. While multisensory integration defines the rules by which information is combined for present perception, multisensory calibration occurs at multiple time-scales. Long-term (or slow) calibration begins with development and continues through adulthood (with different characteristics of and levels of plasticity depending on life-stage; discussed in Sect. 9.2). On the opposite end of the time-scale, short-term (fast) calibration happens in seconds. This includes adaptation (e.g., motion aftereffects) and serial dependence (discussed in Sect. 9.3). Between these two extremes, the brain needs to continuously recalibrate to multisensory changes in the environment. These changes can be temporary (e.g., sea or space travel) or permanent (e.g., loss of sensory function), sudden (e.g., injury), or gradual (e.g., aging). Therefore, they can take place over seconds, minutes, hours, days, months, or years. Multisensory calibration resulting from internal sensory discrepancies (discussed in Sect. 9.4) and/or feedback from the environment (discussed in Sect. 9.5) lead to very different phenomena. Thus, multisensory calibration is a multifaceted and multidimensional problem, requiring a diversity of tools.

A general overarching theory of multisensory calibration may be unfeasible to attain, as it would need to account for (at minimum): calibration at different ages (with different levels of plasticity), different time-scales (short, medium, and long), and the type of error signal available (e.g., internal sensory discrepancy vs. external feedback). Furthermore, it may function differently for different combinations of sensory cues, and models would need to be designed and understood within broader cognitive frameworks (such as reinforcement learning, attention, and Bayesian inference). Thus, an understanding of multisensory calibration requires further research both theoretically and experimentally. Finally, research into the neural bases of multisensory calibration (discussed in Sect. 9.6) offers an important and open frontier into large areas of unchartered territory, necessary for understanding (multisensory) brain function.

In this chapter, we discuss diverse developments in this field, and highlight gaps of knowledge. By presenting many different aspects of multisensory calibration together in one chapter, with some current theories, we hope to trigger further investigation into developing new theories, and hypotheses for experiments, in order to improve our understanding of these important functions in the brain.

9.2 Multisensory Calibration Across the Lifespan

From the moment we are born (and even before, in uterus), we experience the world in a multisensory manner (Bremner et al. 2012; De Klerk et al. 2021). Therefore, to fully understand the development of sensory (and motor) systems one cannot ignore the fact that other senses are present and provide complementary learning signals along the way (Noppeney 2021). Yet, studies of sensory development (and even more so, didactic curricula and textbooks) often focus on individual sensory systems in isolation. Making this more complicated, the learning signals themselves change as the different sensory systems develop within the fast-changing body and brain of a child. Even different sensory functions within a specific modality develop at different rates (Olsho 1984; Kovács et al. 1999; Atkinson 2008). Moreover, external feedback that we gain from the environment is received and interpreted via our senses. Therefore, sensory calibration during development is inherently a multisensory question.

Multisensory integration is immature in childhood and emerges only late in development, after the maturation of individual senses (Misceo et al. 1999; Wallace 2004; Ernst 2008; Gori et al. 2008; Nardini et al. 2008; Barutchu et al. 2009; although certain multisensory processes are present already at an early age, e.g., Streri and Gentaz 2003; Lewkowicz and Ghazanfar 2009). It has been argued that this delay may be to allow the systems to first calibrate during development (Burr and Gori 2012). The rationale is that multisensory integration would not be useful if the individual senses were themselves not calibrated. Thus, during development, measurements from different sensory modalities can act as complementary error signals to calibrate one another. Gori et al. (2008) found that, unlike adults, children do not integrate visual and haptic cues optimally; rather, touch dominates discriminations of size, and vision dominates discriminations of orientation (even when the other cue is more precise). They propose that this dominance during development may allow cross-modal calibration, in which the less-biased sense in children (touch for size and vision for orientation) calibrates the other. In support of this notion, they found that children with visual impairment have impaired haptic orientation discrimination (Gori et al. 2010), and children with movement disorders have impaired visual size discrimination (Gori et al. 2012b).

But, cross-sensory calibration is not limited to childhood. Neural reorganization in the brain lasts well into adolescence (Paus et al. 1999; Paus 2005; Brandwein et al. 2011), and some multisensory brain functions only stabilize in adulthood (Hillock-Dunn and Wallace 2012). In fact, certain multisensory calibration processes can occur only after the development multisensory integration (Rohlf et al. 2020), for example, when the integrated (combined) percept is used for calibration (Zaidel et al. 2013). Thus, beyond development, behavioral adaptation and neural plasticity are normal capacities of the mature brain, and present throughout our lifespan (Pascual-Leone et al. 2005; Shams and Seitz 2008).

As adults, we encounter changes in the environment that put our sensory cues in conflict (requiring multisensory calibration) on a regular basis. Travel or use of virtual-reality induces visual-vestibular conflict (Akiduki et al. 2003; Garzorz and MacNeilage 2019) associated with motion-sickness (Alcañiz Escandell and Román Ivorra 2019). Remarkably, people adapt to continued visual-vestibular conflict, such as at sea or in space (Black et al. 1995; Shupak and Gordon 2006), and even need to readapt when returning to land, experiencing symptoms of mal de debarquement ("sea legs," Nachum et al. 2004; Van Ombergen et al. 2016). Asynchronous audio-visual, audio-tactile and visuo-tactile cues all lead to recalibration of timing perception (Hanson et al. 2008; Van der Burg et al. 2013; Noel et al. 2016). And spatial perception is also shifted in response to audio-visual spatial conflict (known as the ventriloquism aftereffect; Recanzone 1998). Thus, multisensory calibration is an ongoing process in the adult brain (in the following sections we expand on different aspects of adult multisensory calibration).

Naturally, also as we age, the brain and body change. Here too, the different sensory systems change according to their own time-course, in a way that is unique for different people (Cole et al. 2018). Aging affects peripheral sensory systems, for example, hair-cell loss in the auditory (Frisina 2009) and vestibular (Rauch et al. 2001) organs, changes in the eyes (Klein and Klein 2013), and broad changes throughout the body, including proprioceptive and skeletomuscular (Timiras 2002). Aging also leads to central brain changes in sensory processing (Humes et al. 2013) and changes in body representation (Sorrentino et al. 2021). Thus, in an aging body, the brain needs to adapt to, and compensate for, the loss of sensory, motor, and (paradoxically, also central) brain functions. Hence, adaptation to aging also requires and elicits multisensory calibration.

To compensate for the deterioration of sensory and cognitive systems (and making use of substantial life experience) aging adults integrate multisensory information to a greater extent (Laurienti et al. 2006; Diederich et al. 2008; Hernández et al. 2019). Enhanced multisensory integration in aging is associated with better function, for example, greater visual–somatosensory integration is related to better balance and reduced risk of falls (Mahoney et al. 2019). Thus, in healthy aging, individuals accumulate sensory evidence (over longer periods) while preserving multisensory functions of cue integration and causal inference (Jones et al. 2019).

By contrast, poor adjustment to altered sensory states may damage function in disorders of aging. We recently found that (unlike young adults and healthy older adults, Fetsch et al. 2009; Ramkhalawansingh et al. 2018) individuals with Parkinson's disease (PD) do not integrate visual and vestibular cues in a near-optimal (Bayesian) manner (Yakubovich et al. 2019). Rather, despite impairments of visual (but not vestibular) perception of self-motion in the PD cohort, they consistently overweighted their visual (vs. vestibular) cues. This suggests that the PD patients did not update their multisensory function to their new (altered) state. Namely, they did not down-weight their visual cue weights according to their current level of function. This resulted in inappropriate visual-vestibular integration, which can contribute to balance issues in PD (Halperin et al. 2020). We found support for this notion in another experiment that tested meta-cognitive confidence of visual performance (using a task of coherent motion perception)—indeed PD patients were significantly overconfident in their visual performance vs. controls (Halperin et al. 2021). Thus, an important and challenging aspect of aging with a neurodegenerative disease might be inappropriate updating (or calibration) of multisensory function.

Multisensory adaptation and calibration are thus vital functions throughout the lifespan—from development until (and including) old age. Also in mid-life adulthood, multisensory function is not static. We are constantly confronted with changes that require multisensory calibration. Multisensory changes can be due to (permanent or temporary) loss of sensory function, for example, following vestibular loss, the brain learns to compensate using the other senses, with remarkable functional recovery (Smith and Curthoys 1989; Sadeghi et al. 2012). Also everyday living, such as going on a boat ride, using sensory-related devices (like hearing aids or glasses), or even just wearing gloves, require dynamic multisensory adaptation. We tested adult multisensory (visual-vestibular) calibration in a series of experiments and summarize the results in the sections below.

9.3 Unsupervised Multisensory Calibration

Before delving into the different types of calibration, we would first like to differentiate calibration from integration. According to Bayesian theories of perception (and based on signal detection theory) when two different cues measure the same event in the world, a discrepancy between the cues' measurements is always expected, even when both cues are perfectly calibrated (Angelaki et al. 2009; Ma 2019). That is because each cue's measurement is noisy, on a trial-to-trial basis. For example, when a single multisensory (visual-vestibular) motion stimulus is presented at heading 10°, if both cues are calibrated then only *on average* will both cues measure 10° heading. Assuming conditionally independent Gaussian noise distributions, the cues' measurements will differ on each trial (e.g., the visual cue measurement may be 12°, while the vestibular measurement might be 7°). Thus, on a single trial, reliability-based (Bayesian) integration improves precision, and, if both cues are unbiased, it therefore also improves trial-by-trial accuracy (by reducing variance).

In the face of discrepant sensory cues (which, as explained above is the standard situation for a single trial/event), the brain needs to perform causal inference (Kording et al. 2007)—namely, to infer whether the sensory cues arose from the same source in the environment (in which case, they should be integrated), or from two different sources (in which case they should not be integrated). Causal inference is indeed performed on a single-trial basis: if the cues are very different, the brain infers that they most likely don't originate from the same source, and integration is reduced; whereas, if the cues are similar (e.g., relatively small discrepancy) they are integrated (De Winkel et al. 2017; Acerbi et al. 2018; Dokka et al. 2019). When multisensory cues have a *systematic* discrepancy (i.e., over many repetitions), despite coming from the same source—this indicates that at least one of the cues is biased, and multisensory calibration is needed.

We differentiate between two main types of calibration: unsupervised and supervised. By "unsupervised" we mean that there is no external feedback regarding the correctness of our perception. By contrast, "supervised" calibration means that feedback from the environment provides information regarding correctness of the cues. Take for example, the ventriloquist effect, in which visual and auditory cues are discrepant in space. On a single trial, the visual and auditory cues are integrated in a near-optimal Bayesian manner (Alais and Burr 2004). However, the audio-visual spatial discrepancy also leads to a shift in perception, known as the ventriloquist aftereffect (Recanzone 1998; Lewald 2002; Wozny and Shams 2011; Bosen et al. 2018). This is a case of unsupervised calibration because the brain cannot know which cue is biased, and which is not.

The rationale and need for unsupervised multisensory calibration is that if two cues are systematically discrepant, at least one must be biased and needs calibration. Accordingly, the brain aims to maintain a homeostatic state of "internal consistency," in which the cues' estimates are in accordance with one another, on average. But this is not enough. Beyond "internal consistency," the cue estimates also need to be in alignment with the world ("externally unbiased") for perception to be functionally useful. Thus, the brain is left with the difficult task of estimating the extent to which each cue is biased, and to correct for that. The best way to do this is via feedback from the environment (discussed in the following section on supervised calibration). But, in the absence of external feedback, i.e., during unsupervised calibration, the brain still needs to attribute the discrepancy to one (or both cues) to decide which cue(s) to calibrate.

Classic studies on multisensory calibration proposed that vision dominates (e.g., audio localization adapts to vision, Knudsen 2002) due to an inherent advantage of vision over other cues (Recanzone 1998). While this may be the case for certain brain circuits and functions, this does not seem to be a general rule because substantial visual recalibration was exposed in many conditions (Atkins et al. 2003; Di Luca et al. 2009; Adams et al. 2010; Burge et al. 2010; Zaidel et al. 2011). On the heels of reliability-based cue-combination (multisensory integration), it was suggested that also calibration should be reliability-based (Ghahramani et al. 1997; Witten and Knudsen 2005; Burge et al. 2010). According to this notion, the less reliable cue should be calibrated more than the more reliable cue. However, as noted by Ernst and Di Luca (2011), the more reliable cue might not necessarily be the less biased, and in the case of a reliable but biased cue vs. an unreliable but unbiased cue the former should be calibrated to attain better accuracy.

We found (in humans and monkeys) that unsupervised visual-vestibular calibration of heading perception (in response to a systematic heading discrepancy) was not reliability-based (Zaidel et al. 2011). We found that both vestibular and visual cues were calibrated—however, the relative magnitude of calibration was roughly fixed. Specifically, the vestibular shifts were approximately twice as large as the visual shifts, irrespective of which cue was more reliable. These results suggest that we have a default estimate regarding how likely it is that each cue is biased, and (in the absence of any other information) we calibrate our cues according to these estimates.

Furthermore, the estimated likelihood of being biased is not set by cue reliability. This notion is in line with the findings that haptic cues calibrate visual cues during development, presumably because naïve haptic size estimates are less biased than visual during development, even though the latter are more reliable (Gori et al. 2012a). Also in adults, visual cues are calibrated by haptic estimates when the former are deemed erroneous (Ernst et al. 2000; Atkins et al. 2003).

Thus, when confronted with a systematic bias between two sensory cues, in the absence of additional information, the brain should (theoretically) calibrate the cues in accordance with their estimated likelihoods of being biased. For visual-vestibular heading cues, we found this to follow an empirical ~2:1 ratio—suggesting that, by default, primates estimate vestibular headings to be approximately twice as likely to be biased vs. visual. Confirmation of this theory would require manipulation and measurement of the likelihoods of cue bias (or the relative likelihood that each cue is biased), and comparison thereof to the extent of each cue's subsequent calibration.

9.4 Supervised Multisensory Calibration

Feedback from the environment is highly valuable for multisensory calibration. Theoretically, it should solve the ambiguity problem regarding which cue is biased (which unsupervised calibration cannot). Accordingly, the biased cue alone should be calibrated, while the unbiased cue should not be shifted to remain unbiased. We tested supervised visual-vestibular calibration by inducing a systematic conflict between the cues' heading, and providing feedback (for correct heading discriminations) in accordance with one of the cues—either visual or vestibular (Zaidel et al. 2013).

Our initial prediction that only the biased cue (i.e., the one not aligned to external feedback) would be calibrated, while the other (already unbiased) cue would not be shifted (to remain unbiased) was disproven by the data. Rather, the results showed an interesting phenomenon of "cue yoking." As the biased cue was calibrated (to become less biased) the initially unbiased cue was "yoked" (calibrated in the same direction as the first cue)—with the surprising result that it was shifted *away* from the feedback, becoming biased. While this result was puzzling at first, a simple assumption, tested by modeling the results, could explain the finding. This assumption was that supervised calibration works on the combined cue estimate, and shifts the individual (unisensory) cues together in a manner to reduce the bias of the combined-cue estimate (Zaidel et al. 2013).

This exposes a high-level cognitive mechanism that changes our behavior to attain better performance vis-à-vis the world (i.e., "external accuracy"). It further suggests that this mechanism only has access to (or at least only calibrates) the combined (multisensory) percept, and therefore it does not calibrate the underlying (unisensory) cues each according to their individual bias. But, the separate lower level (unsupervised) calibration mechanism that calibrates cues towards one another, towards "internal consistency" (described in the previous section; Zaidel et al. 2011), works in parallel and superimposed on the supervised calibration mechanism (Zaidel et al. 2013). Thus, these results expose two separate mechanisms that together eventually attain both internal consistency and external accuracy.

Support for the notion that high-level cognitive mechanisms are applied supra-modally is found in experiments that manipulated visual and tactile priors separately (Spence et al. 2000; Mengotti et al. 2018). Those results showed that observers depended on the joint distributions of both sensory modalities and did not learn independent spatial priors for the separate cues, even when they were informed of the identity of the upcoming stimulus.

9.5 Rapid Calibration: Adaptation and Serial Dependence

Auditory shifts in response to a visual-auditory offset (the ventriloquism aftereffect) occur even after sub-second presentation of a discrepant stimulus (Wozny and Shams 2011; Bosen et al. 2018). Watson et al. (2019) recently found (also using the ventriloquist aftereffect) that when long-term adaptation was followed by brief period with a discrepancy to the opposite side, the initial effects were cancelled out, but subsequently reappeared. This suggests that audio-visual (unsupervised) calibration comprises at least two mechanisms with different time-scales (slow and fast; reminiscent of similar results in sensorimotor adaptation, Smith et al. 2006).

But, just as we divided multisensory calibration on the order of minutes/hours into separate supervised and unsupervised components (in the previous sections), rapid calibration (on the order of seconds) likely also comprises at least two different types—low-level (e.g., unsupervised calibration in response to a concurrent discrepancy between cues) and higher level (e.g., supervised calibration in response to feedback, influences of one's own choices and supra-modal stimulus statistics).

Many recent studies have found that when performing a series of perceptual discriminations across trials, the decisions are not based solely on the current stimulus features (as is often assumed). Rather, they are biased towards the recently experienced stimuli (Fischer and Whitney 2014; Liberman et al. 2014; Taubert et al. 2016; Alexi et al. 2018; Liberman and Manassi 2018; Feigin et al. 2021a, b). This effect is often termed "serial dependence," and its direction is mostly "positive" or "attractive"—meaning that perceptual decisions are more likely to be the same as the preceding one. For example, after a discriminating a rightward (leftward) oriented visual stimulus, a subsequent ambiguous stimulus is more likely to also be discriminated rightward (leftward). This effect (serial dependence) seems to be different from stimulus adaptation aftereffects (Gordon et al. 1995; Anstis et al. 1998; Dalton 2000; Di Lorenzo and Lemon 2000; Kohn 2007; Schweinberger et al. 2008; Thompson and Burr 2009; Crane 2012), which are "negative" or repulsive, meaning that they shift perception away from the previous stimuli.

We recently performed several experiments to study serial dependence and adaptation to preceding stimuli by inducing short-term biases (Feigin et al. 2021a, b; Shalom-Sperber et al. 2021). Specifically, trials were grouped into batches comprising several "prior" trials that were slightly biased (either to the right or left), followed by an unbiased "test" trial. Multiple repetitions of these batches were then presented, interleaved, across the block, and the effects of the prior trials on the test trial discriminations were studied. First, we studied this in a unisensory task of visual location discrimination with short (0.25 s) stimulus durations (Feigin et al. 2021a). We used a logistic-regression model to disambiguate the influences of prior stimuli and prior choices, and several control conditions to account for other factors, such as motor actions and task relevance. We found a positive (attractive) serial dependence, which was driven by prior choices, not prior stimuli or motor actions (interestingly we also found that the effect of prior choices was increased for individuals with autism spectrum disorder; Feigin et al. 2021b). Thus, we suggest that the positive effects of serial dependence reflect a "confirmation bias" of prior choices.

We next set out to investigate the effects of short-term biases, in a multisensory setting (heading discrimination of visual or vestibular self-motion stimuli, Shalom-Sperber et al. 2021). We tested four conditions—in which the prior (biased) stimuli and test stimuli were from the same cue (visual or vestibular) or cross-sensory (the biased prior stimuli were visual, while the test stimuli were vestibular, and vice-versa). Here, we found significant adaptation (a negative effect of prior trials) in all four conditions (both cross-sensory and unisensory). Furthermore, the logistic-regression model revealed that the effect in this case was elicited by the prior stimuli (not prior choices). Thus, we suggest that while prior choices lead to a positive (attractive) effect; prior stimuli lead to a negative (repulsive) effect. We speculate that positive (choice) effects were best exposed in the visual location discrimination task because of short stimulus durations and short inter-trial intervals (i.e., rapid succession of choices); and that negative (repulsive) adaptation effects were best exposed in the heading discrimination experiment, with longer (1 s) stimulus durations (while the long inter-trial intervals associated with self-motion stimuli might diminish choice effects).

Two other (previous) studies of visual-vestibular aftereffects did not find cross-modal adaptation: Crane (2013) found that short-duration (1.5 s) self-motion stimuli (either visual or vestibular) do not cause cross-sensory adaptation. And, Cuturi and MacNeilage (2014) demonstrated that longer duration visual self-motion stimuli (15 s) are needed to cause vestibular adaptation. Several differences between our study and these previous studies might explain the different results: (1) the adaptation in our study could reflect a compound effect of several biased stimuli (only one adapting stimulus was presented in the other studies). (2) In our study, both the adapting (prior) and the test stimuli were discriminated, whereas in both of the other studies the participants were instructed to discriminate only the "test" stimulus (and the act of discrimination could be important). (3) We used a task of fine (heading) discrimination to test cross-modal adaption, whereas the other two studies used a task of coarse (forward vs. backward) motion detection (Crane 2013; Cuturi and MacNeilage 2014). Fine heading discrimination might be more susceptible to cross-modal adaptation. Further research is required to dissociate these factors.

Findings of rapid cross-sensory calibration effects, when the (unisensory) stimuli are presented in succession, suggest that the brain monitors supra-modal statistics of events in the environment, leading to functional (supra-modal) adaptation of perception. This notion is in line with the findings that changes in visual and tactical priors affect perception cross-modally, i.e., also of the other cue (Spence et al. 2000; Mengotti et al. 2018).

9.6 Neural Correlates of Multisensory Calibration

Seminal studies by Stein et al. on the superior colliculus (see Stein and Stanford 2008, for review) revealed fundamental principles of multisensory integration in single neurons. Coincidence of multisensory stimuli enhances event saliency, improves discriminability, and speeds up responses. Interestingly, multisensory responses in superior colliculus neurons are not innate, but rather emerge with development, in response to multisensory experience (Stein et al. 1973; Wallace and Stein 2001). In complementary work, classic studies on owls by Knudsen et al. (see Knudsen 2002 for review) revealed great plasticity in the optic tectum, both during development, and in adulthood. Thus, multisensory responses, and plastic changes thereof, are seen already in the brainstem (and in multiple species).

Beyond the brainstem, multisensory experience leads to changes in cortical function (Powers et al. 2012; Zierul et al. 2017) and cortico-cerebellar connectivity (Lee and Noppeney 2011). In fact, sensory cortices can even functionally change to process other modalities (Pascual-Leone and Hamilton 2001; Merabet et al. 2005; Amedi et al. 2010), with multisensory regions demonstrating the greatest capacity for plasticity (Fine 2008). The vestibular sense is inherently a "multisensory" sense in that it does not have its own primary vestibular cortex (unlike other senses). Rather, vestibular inputs project to widespread multisensory cortical areas (Lopez et al. 2012; Rancz et al. 2015). Thus, visual-vestibular paradigms of self-motion offer a valuable substrate to study multisensory calibration.

We recently searched for neuronal correlates of supervised calibration in the dorsal medial superior temporal area (MSTd) and the ventral intraparietal area (VIP) of adult monkeys (Zaidel et al. 2021). We found little plasticity in neural responses in MSTd. By contrast, neural correlates of plasticity were found in VIP. Although the two areas have similar stimulus-driven responses (Colby et al. 1993; Bremmer et al. 1997, 2002; Duffy 1998; Page and Duffy 2003; Zhang et al. 2004; Gu et al. 2006, 2007; Takahashi et al. 2007; Zhang and Britten 2010; Maciokas and Britten 2010; Chen et al. 2011a), VIP has been shown to have stronger correlations with behavioral choices than MSTd (Chen et al. 2013;

Zaidel et al. 2017). Further research is required in order to identify whether VIP plays an active role in supervised calibration, or simply correlates with the resulting shifts. Furthermore, unsupervised calibration needs to be investigated in these areas.

The visual posterior sylvian (VPS) area also exhibits visual and vestibular self-motion signals (Chen et al. 2011b; Frank et al. 2016). Interestingly, visual and vestibular self-motion responses in VPS are primarily tuned in the opposite direction (180° apart, Chen et al. 2011b). Although neurons with "opposite" visual and vestibular tuning have also been described in MSTd (Gu et al. 2008) and VIP (Chen et al. 2013), VPS stands out as an area with *predominantly* opposite tuning (the others have comparable proportions). Opposite tuning has been proposed to support segregation of multisensory signals (Zhang et al. 2019), which would be invaluable for cross-sensory calibration. The brain has been shown to demonstrate flexible integrative behavior in terms of fusing and segregating sensory information with different spatiotemporal dynamics (Cao et al. 2019). Thus, further research should investigate the involvement of opposite tuning (e.g., in area VPS), and segregation, in multisensory calibration.

Beyond developmental changes, and multisensory calibration over hours, rapid cross-sensory calibration over seconds, even when the stimuli are presented serially and not in conjunction (Shalom-Sperber et al. 2021) exposes exquisitely flexible skills of the brain in adapting to dynamic contexts and monitoring supra-modal statistics. Some of these effects likely involve top-down influences such as attention (Zuanazzi and Noppeney 2019) and other cognitive effects, such as prior choices. How the brain monitors statistics of perceptual events for specific functions (e.g., self-motion) supra-modally, and how the brain mediates the subsequent biasing of perception of specific functions, irrespective of its presented modality (e.g., visual or vestibular) requires future investigation.

9.7 Concluding Remarks

Multisensory calibration comprises a family of brain functions that work on time-scales ranging from sub-second to years. Multisensory calibration is important in development of the young brain, as well as in healthy aging. But, mechanisms of multisensory integration are active throughout our lifespan, including adulthood. Besides the time domain—multisensory calibration spans a range of functions from low-level (e.g., unsupervised calibration) to high-level (e.g., supervised calibration, choice-effects, and rapid monitoring of supra-modal statistics in the environment). Research into its neuronal underpinnings have already exposed fundamental principles. However, as the diversity of multisensory calibration phenomena is broad, its neuronal bases require a lot of further research. Finally, overarching theories that can explain multiple aspects of multisensory calibration, generically (i.e., for different combinations of modalities) would be a boon to the field and our understanding of multisensory brain function.

Acknowledgments This research was supported by the Israel Science Foundation (ISF grant No. 1291/20 and ISF-NSFC grant no. 3318/20) to A.Z.

References

Acerbi L, Dokka K, Angelaki DE, Ma WJ (2018) Bayesian comparison of explicit and implicit causal inference strategies in multisensory heading perception. PLoS Comput Biol 14:e1006110

Adams WJ, Kerrigan IS, Graf EW (2010) Efficient visual recalibration from either visual or haptic feedback: the importance of being wrong. J Neurosci 30:14745–14749

Akiduki H, Nishiike S, Watanabe H, Matsuoka K, Kubo T, Takeda N (2003) Visual-vestibular conflict induced by virtual reality in humans. Neurosci Lett 340:197–200

Alais D, Burr D (2004) The ventriloquist effect results from near-optimal bimodal integration. Curr Biol 14:257–262

Alcañiz Escandell CP, Román Ivorra JA (2019) Motion sickness: an overview. Drugs Context 8

Alexi J, Cleary D, Dommisse K, Palermo R, Kloth N, Burr D, Bell J (2018) Past visual experiences weigh in on body size estimation. Sci Rep 8:1–8

Amedi A, Raz N, Azulay H, Malach R, Zohary E (2010) Cortical activity during tactile exploration of objects in blind and sighted humans. Restor Neurol Neurosci 28:143–156
Angelaki DE, Gu Y, DeAngelis GC (2009) Multisensory integration: psychophysics, neurophysiology, and computation. Curr Opin Neurobiol 19:452–458
Anstis S, Verstraten FAJ, Mather G (1998) The motion aftereffect, vol 2. Trends Cogn, Sci, p 111
Atkins JE, Jacobs RA, Knill DC (2003) Experience-dependent visual cue recalibration based on discrepancies between visual and haptic percepts. Vis Res 43:2603–2613
Atkinson J (2008) The developing visual brain. Dev. Vis. Brain
Barutchu A, Crewther DP, Crewther SG (2009) The race that precedes coactivation: development of multisensory facilitation in children. Dev Sci 12:464–473
Black FO, Paloski WH, Doxey-Gasway DD, Reschke MF (1995) Vestibular plasticity following orbital spaceflight: recovery from postflight postural instability. Acta Otolaryngol Suppl 520(Pt 2):450–454
Bosen AK, Fleming JT, Allen PD, O'Neill WE, Paige GD (2018) Multiple time scales of the ventriloquism aftereffect 13:e0200930
Brandwein AB, Foxe JJ, Russo NN, Altschuler TS, Gomes H, Molholm S (2011) The development of audiovisual multisensory integration across childhood and early adolescence: a high-density electrical mapping study. Cereb Cortex 21:1042–1055
Bremmer F, Ilg UJ, Thiele A, Distler C, Hoffmann KP (1997) Eye position effects in monkey cortex. I. Visual and pursuit-related activity in extrastriate areas MT and MST. J Neurophysiol 77:944–961
Bremmer F, Duhamel JR, Ben HS, Graf W (2002) Heading encoding in the macaque ventral intraparietal area (VIP). Eur J Neurosci 16:1554–1568
Bremner AJ, Lewkowicz DJ, Spence C (2012) The multisensory approach to development. In: Bremner AJ, Lewkowicz DJ, Spence C (eds) Multisensory development. Oxford University Press, pp 1–26
Burge J, Girshick AR, Banks MS (2010) Visual-haptic adaptation is determined by relative reliability. J Neurosci 30:7714–7721
Burr D, Gori M (2012) Multisensory integration develops late in humans. In: Murray MM, Wallace MT (eds) The neural bases of multisensory processes, 1st edn. CRC Press, Frontiers in Neuroscience, pp 345–362
Butler JS, Smith ST, Campos JL, Bülthoff HH (2010). Bayesian integration of visual and vestibular signals for heading. Journal of vision 10(11):23–23
Cao Y, Summerfield C, Park H, Giordano BL, Kayser C (2019) Causal inference in the multisensory brain. Neuron 102:1076–1087.e8
Chen A, DeAngelis GC, Angelaki DE (2011a) A comparison of vestibular spatiotemporal tuning in macaque parietoinsular vestibular cortex, ventral intraparietal area, and medial superior temporal area. J Neurosci 31:3082–3094
Chen A, DeAngelis GC, Angelaki DE (2011b) Convergence of vestibular and visual self-motion signals in an area of the posterior sylvian fissure. J Neurosci 31:11617–11627
Chen A, DeAngelis GC, Angelaki DE (2013) Functional specializations of the ventral intraparietal area for multisensory heading discrimination. J Neurosci 33:3567–3581
Colby CL, Duhamel JR, Goldberg ME (1993) Ventral intraparietal area of the macaque: anatomic location and visual response properties. J Neurophysiol 69:902–914
Cole JH, Marioni RE, Harris SE, Deary IJ (2018) (2018) brain age and other bodily 'ages': implications for neuropsychiatry. Mol Psychiatry 242(24):266–281
Crane BT (2012) Direction specific biases in human visual and vestibular heading perception. PLoS One 7:e51383
Crane BT (2013) Limited interaction between translation and visual motion aftereffects in humans. Exp Brain Res 224:165–178
Cuturi LF, MacNeilage PR (2014) Optic flow induces nonvisual self-motion aftereffects. Curr Biol 24:2817–2821
Dalton P (2000) Psychophysical and behavioral characteristics of olfactory adaptation. Chem Senses 25:487–492
De Klerk CCJM, Filippetti ML, Rigato S (2021) The development of body representations: an associative learning account. Proc R Soc B 288:20210070
De Winkel KN, Katliar M, Bülthoff HH (2017) Causal inference in multisensory heading estimation. PLoS One 12:e0169676
Di Lorenzo PM, Lemon CH (2000) The neural code for taste in the nucleus of the solitary tract of the rat: effects of adaptation. Brain Res 852:383–397
Di Luca M, Machulla TK, Ernst MO (2009) Recalibration of multisensory simultaneity: cross-modal transfer coincides with a change in perceptual latency. J Vis 9:7–7
Diederich A, Colonius H, Schomburg A (2008) Assessing age-related multisensory enhancement with the time-window-of-integration model. Neuropsychologia 46:2556–2562
Dokka K, Park H, Jansen M, DeAngelis GC, Angelaki DE (2019) Causal inference accounts for heading perception in the presence of object motion. Proc Natl Acad Sci U S A 116:9060–9065
Duffy CJ (1998) MST neurons respond to optic flow and translational movement. J Neurophysiol 80:1816–1827
Ernst MO, Banks MS (2002). Humans integrate visual and haptic information in a statistically optimal fashion. Nature 415(6870):429–433.
Ernst MO (2008) Multisensory integration: a late bloomer. Curr Biol 18:R519–R521
Ernst MO, Di Luca M (2011) Multisensory perception: from integration to remapping. In: Trommershauser J, Kording K, Landy MS (eds) Sensory Cue integration, 1st edn. Oxford University Press, Computational Neuroscience Series, pp 224–250

Ernst MO, Banks MS, Bülthoff HH (2000) (2000) touch can change visual slant perception. Nat Neurosci 31(3):69–73
Feigin H, Baror S, Bar M, Zaidel A (2021a) Perceptual decisions are biased toward relevant prior choices. Sci Rep 11:1–16
Feigin H, Shalom-Sperber S, Zachor DA, Zaidel A (2021b) Increased influence of prior choices on perceptual decisions in autism. elife 10:e61595
Fetsch CR, Turner AH, DeAngelis GC, Angelaki DE (2009) Dynamic reweighting of visual and vestibular cues during self-motion perception. J Neurosci 29:15601–15612
Fine I (2008) The behavioral and neurophysiological effects of sensory deprivation. In: Rieser JJ, Ashmead DH, Ebner FF, Corn AL (eds) Blindness and brain plasticity in navigation and object perception. Taylor & Francis, New York, pp 127–155
Fischer J, Whitney D (2014) Serial dependence in visual perception. Nat Neurosci 17:738–743
Frank SM, Wirth AM, Greenlee MW (2016) Visual-vestibular processing in the human Sylvian fissure. J Neurophysiol 116:263–271
Frisina RD (2009) Age-related hearing loss: ear and brain mechanisms. Ann N Y Acad Sci 1170:708–717
Garzorz IT, MacNeilage PR (2019) Towards dynamic modeling of visual-vestibular conflict detection. Prog Brain Res 248:277–284
Ghahramani Z, Wolpert DM, Jordan MI (1997) Computational models of sensorimotor integration. In: Morasso PG, Sanguineti V (eds) Self-organization, computational maps and motor control, advances in psychology. Elsevier, Amsterdam, pp 117–147
Gordon CR, Fletcher WA, Melvill Jones G, Block EW (1995) Adaptive plasticity in the control of locomotor trajectory. Exp Brain Res 102:540–545
Gori M, Del VM, Sandini G, Burr DC (2008) Young children do not integrate visual and haptic form information. Curr Biol 18:694–698
Gori M, Sandini G, Martinoli C, Burr D (2010) Poor haptic orientation discrimination in nonsighted children may reflect disruption of cross-sensory calibration. Curr Biol 20:223–225
Gori M, Giuliana L, Sandini G, Burr D (2012a) Visual size perception and haptic calibration during development. Dev Sci 15:854–862
Gori M, Tinelli F, Sandini G, Cioni G, Burr D (2012b) Impaired visual size-discrimination in children with movement disorders. Neuropsychologia 50:1838–1843
Gu Y, Watkins PV, Angelaki DE, DeAngelis GC (2006) Visual and nonvisual contributions to three-dimensional heading selectivity in the medial superior temporal area. J Neurosci 26:73–85
Gu Y, DeAngelis GC, Angelaki DE (2007) A functional link between area MSTd and heading perception based on vestibular signals. Nat Neurosci 10:1038–1047
Gu Y, Angelaki DE, DeAngelis GC (2008) Neural correlates of multisensory cue integration in macaque MSTd. Nat Neurosci 11:1201–1210
Halperin O, Israeli-Korn S, Yakubovich S, Hassin-Baer S, Zaidel A (2020) Self-motion perception in Parkinson's disease. Eur. J. Neurosci 53:2376
Halperin O, Karni R, Israeli-Korn S, Hassin-Baer S, Zaidel A (2021) Overconfidence in visual perception in parkinson's disease. Eur J Neurosci 53:2027–2039
Hanson JVM, Heron J, Whitaker D (2008) Recalibration of perceived time across sensory modalities. Exp Brain Res 185:347–352
Hernández B, Setti A, Kenny RA, Newell FN (2019) Individual differences in ageing, cognitive status, and sex on susceptibility to the sound-induced flash illusion: a large-scale study. Psychol Aging 34:978–990
Hillock-Dunn A, Wallace MT (2012) Developmental changes in the multisensory temporal binding window persist into adolescence. Dev Sci 15:688–696
Humes LE, Busey TA, Craig J, Kewley-Port D (2013) Are age-related changes in cognitive function driven by age-related changes in sensory processing? Attention, perception. Psychophysiology 75:508–524
Jones SA, Beierholm U, Meijer D, Noppeney U (2019) Older adults sacrifice response speed to preserve multisensory integration performance. Neurobiol Aging 84:148–157
Klein R, Klein BEK (2013) The prevalence of age-related eye diseases and visual impairment in aging: current estimates. Invest Ophthalmol Vis Sci 54:ORSF5–ORSF13
Knill DC, Pouget A (2004). The Bayesian brain: the role of uncertainty in neural coding and computation. TRENDS in Neurosciences 27(12):712–719
Knudsen EI (2002) Instructed learning in the auditory localization pathway of the barn owl. Nature 417:322–328
Kohn A (2007) Visual adaptation: physiology, mechanisms, and functional benefits. J. Neurophysiol. 97:3155
Kording KP, Beierholm U, Ma WJ, Quartz S, Tenenbaum JB, Shams L (2007) Causal inference in multisensory perception. PLoS One 2:e943
Kovács I, Kozma P, Fehér Á, Benedek G (1999) Late maturation of visual spatial integration in humans. Proc Natl Acad Sci 96:12204–12209
Laurienti PJ, Burdette JH, Maldjian JA, Wallace MT (2006) Enhanced multisensory integration in older adults. Neurobiol Aging 27:1155–1163
Lee HL, Noppeney U (2011) Long-term music training tunes how the brain temporally binds signals from multiple senses. Proc Natl Acad Sci U S A 108:E1441–E1450
Lewald J (2002) Rapid adaptation to auditory-visual spatial disparity. Learn Mem 9:268–278
Lewkowicz DJ, Ghazanfar AA (2009) The emergence of multisensory systems through perceptual narrowing. Trends Cogn Sci 13:470–478
Liberman A, Manassi M (2018) Serial dependence promotes the stability of perceived emotional expression depending on face similarity. Atten Percept Psychophys 80:1461–1473

Liberman A, Fischer J, Whitney D (2014) Serial dependence in the perception of faces. Curr Biol 24:2569–2574

Lopez C, Blanke O, Mast FW (2012) The human vestibular cortex revealed by coordinate-based activation likelihood estimation meta-analysis. Neuroscience 212:159–179

Ma WJ (2019) Bayesian decision models: a primer. Neuron 104:164

Maciokas JB, Britten KH (2010) Extrastriate area MST and parietal area VIP similarly represent forward headings. J Neurophysiol 104:239–247

Mahoney JR, Cotton K, Verghese J, Newman A (2019) Multisensory integration predicts balance and falls in older adults. Journals Gerontol Ser A 74:1429–1435

Mengotti P, Boers F, Dombert PL, Fink GR, Vossel S (2018) (2018) integrating modality-specific expectancies for the deployment of spatial attention. Sci Reports 81(8):1–10

Merabet LB, Rizzo JF, Amedi A, Somers DC, Pascual-Leone A (2005) What blindness can tell us about seeing again: merging neuroplasticity and neuroprostheses. Nat Rev Neurosci 6:71–77

Misceo GF, Hershberger WA, Mancini RL (1999) Haptic estimates of discordant visual-haptic size vary developmentally. Percept Psychophys 61:608–614

Nachum Z, Shupak A, Letichevsky V, Ben-David J, Tal D, Tamir A, Talmon Y, Gordon CR, Luntz M (2004) Mal de debarquement and posture: reduced reliance on vestibular and visual cues. Laryngoscope 114:581–586

Nardini M, Jones P, Bedford R, Braddick O (2008) Development of cue integration in human navigation. Curr Biol 18:689–693

Noel JP, De Niear M, Van Der Burg E, Wallace MT (2016) Audiovisual simultaneity judgment and rapid recalibration throughout the lifespan. PLoS One 11:e0161698

Noppeney U (2021) Perceptual inference, learning, and attention in a multisensory world. Annu Rev Neurosci 44:449–473. https://doi.org/10.1146/annurev-neuro-100120-085519

Olsho LW (1984) Infant frequency discrimination. Infant Behav Dev 7:27–35

Page WK, Duffy CJ (2003) Heading representation in MST: sensory interactions and population encoding. J Neurophysiol 89:1994–2013

Pascual-Leone A, Hamilton R (2001) The metamodal organization of the brain. Prog Brain Res 134:427–445

Pascual-Leone A, Amedi A, Fregni F, Merabet LB (2005) The plastic human brain cortex. Annu Rev Neurosci 28:377–401

Paus T (2005) Mapping brain development and aggression. Can Child Adolesc Psychiatr Rev 14:10–15

Paus T, Zijdenbos A, Worsley K, Collins DL, Blumenthal J, Giedd JN, Rapoport JL, Evans AC (1999) Structural maturation of neural pathways in children and adolescents: in vivo study. Science 283:1908–1911

Powers AR, Hevey MA, Wallace MT (2012) Neural correlates of multisensory perceptual learning. J Neurosci 32:6263

Ramkhalawansingh R, Butler JS, Campos JL (2018) Visual-vestibular integration during self-motion perception in younger and older adults. Psychol Aging 33:798–813

Rancz EA, Moya J, Drawitsch F, Brichta AM, Canals S, Margrie TW (2015) Widespread vestibular activation of the rodent cortex. J Neurosci 35:5926–5934

Rauch SD, Velazquez-Villaseñor L, Dimitri PS, Merchant SN (2001) Decreasing hair cell counts in aging humans. Ann N Y Acad Sci 942:220–227

Recanzone GH (1998) Rapidly induced auditory plasticity: the ventriloquism aftereffect. Proc Natl Acad Sci 95:869–875

Rohlf S, Li L, Bruns P, Röder B (2020) Multisensory integration develops prior to Crossmodal recalibration. Curr Biol 30:1726–1732.e7

Sadeghi SG, Minor LB, Cullen KE (2012) Neural correlates of sensory substitution in vestibular pathways following complete vestibular loss. J Neurosci 32:14685–14695

Schwartz AB (2016) Movement: how the brain communicates with the world. Cell 164:1122–1135

Schweinberger SR, Casper C, Hauthal N, Kaufmann JM, Kawahara H, Kloth N, Robertson DMC, Simpson AP, Zäske R (2008) Auditory adaptation in voice perception. Curr Biol 18:684–688

Shalom-Sperber S, Chen A, Zaidel A (2021) Rapid cross-sensory adaptation of self-motion perception. Cortex 148:14–30

Shams L, Seitz AR (2008) Benefits of multisensory learning. Trends Cogn Sci 12:411–417

Shupak A, Gordon CR (2006) Motion sickness: advances in pathogenesis, prediction, prevention, and treatment. Aviat Sp Environ Med 77:1213–1223

Smith PF, Curthoys IS (1989) Mechanisms of recovery following unilateral labyrinthectomy: a review. Brain Res Brain Res Rev 14:155–180

Smith MA, Ghazizadeh A, Shadmehr R (2006) Interacting adaptive processes with different timescales underlie short-term motor learning. PLoS Biol 4:e179

Sorrentino G, Franza M, Zuber C, Blanke O, Serino A, Bassolino M (2021) How ageing shapes body and space representations: a comparison study between healthy young and older adults. Cortex 136:56–76

Spence C, Pavani F, Driver J (2000) Crossmodal links between vision and touch in covert endogenous spatial attention. J Exp Psychol Hum Percept Perform 26:1298–1319

Stein BE, Stanford TR (2008) Multisensory integration: current issues from the perspective of the single neuron 9:255–266

Stein BE, Labos E, Kruger L (1973) Sequence of changes in properties of neurons of superior colliculus of the kitten during maturation. J Neurophysiol. 36:667–679. https://doi.org/10.1152/jn.1973.36.4.667

Streri A, Gentaz E (2003) Cross-modal recognition of shape from hand to eyes in human newborns. Somatosens Mot Res 20:13–18

Takahashi K, Gu Y, May PJ, Newlands SD, DeAngelis GC, Angelaki DE (2007) Multimodal coding of three-dimensional rotation and translation in area MSTd: comparison of visual and vestibular selectivity. J Neurosci 27:9742–9756

Taubert J, Van Der Burg E, Alais D (2016) Love at second sight: sequential dependence of facial attractiveness in an on-line dating paradigm. Sci Rep 6:2–6

Thompson P, Burr D (2009) Visual aftereffects. Curr. Biol. 19:R11

Timiras PS (2002) Physiological basis of aging and geriatrics, 3rd edn. CRC Press, Boca Raton

Van der Burg E, Alais D, Cass J (2013) Rapid recalibration to audiovisual asynchrony. J Neurosci 33:14633–14637

Van Ombergen A, Van Rompaey V, Maes LK, Van de Heyning PH, Wuyts FL (2016) Mal de debarquement syndrome: a systematic review. J Neurol 263:843–854

Wallace MT (2004) The development of multisensory processes. Cogn Process 5:69–83

Wallace MT, Stein BE (2001) Sensory and multisensory responses in the newborn monkey superior colliculus. J Neurosci 21:8886–8894

Watson DM, Akeroyd MA, Roach NW, Webb BS (2019) Distinct mechanisms govern recalibration to audio-visual discrepancies in remote and recent history. Sci Rep 91(9):1–12

Witten IB, Knudsen EI (2005) Why seeing is believing: merging auditory and visual worlds. Neuron 48:489–496

Wozny DR, Shams L (2011) Computational characterization of visually induced auditory spatial adaptation. Front Integr Neurosci 5:75

Yakubovich S, Israeli-Korn S, Halperin O, Yahalom G, Hassin-Baer S, Zaidel A (2019) Visual self-motion cues are impaired in Parkinson's disease yet over-weighted during visual-vestibular integration. Brain Commun 2(1):fcaa035

Yon D, Frith CD (2021) Precision and the Bayesian brain. Curr Biol 31:R1026–R1032

van Beers RJ, Wolpert DM, Haggard P (2002). When feeling is more important than seeing in sensorimotor adaptation. Current biology 12(10):834–837.

Zaidel A, Turner AH, Angelaki DE (2011) Multisensory calibration is independent of cue reliability. J Neurosci 31:13949–13962

Zaidel A, Ma WJ, Angelaki DE (2013) Supervised calibration relies on the multisensory percept. Neuron 80:1544–1557

Zaidel A, DeAngelis GC, Angelaki DE (2017) Decoupled choice-driven and stimulus-related activity in parietal neurons may be misrepresented by choice probabilities. Nat Commun 8:715

Zaidel A, Laurens J, DeAngelis GC, Angelaki DE (2021) Supervised multisensory calibration signals are evident in VIP but not MSTd. J. Neurosci. 41:10108

Zhang T, Britten KH (2010) The responses of VIP neurons are sufficiently sensitive to support heading judgments. J Neurophysiol 103:1865–1873

Zhang T, Heuer HW, Britten KH (2004) Parietal area VIP neuronal responses to heading stimuli are encoded in head-centered coordinates. Neuron 42:993–1001

Zhang WH, Wang H, Chen A, Gu Y, Lee TS, Wong KM, Wu S (2019) Complementary congruent and opposite neurons achieve concurrent multisensory integration and segregation. elife 8:e43753

Zierul B, Röder B, Tempelmann C, Bruns P, Noesselt T (2017) The role of auditory cortex in the spatial ventriloquism aftereffect. NeuroImage 162:257–268

Zuanazzi A, Noppeney U (2019) Distinct neural mechanisms of spatial attention and expectation guide perceptual inference in a multisensory world. J Neurosci 39:2301–2312

The Development of Multisensory Integration at the Neuronal Level

10

Liping Yu and Jinghong Xu

Abstract

Multisensory integration is a fundamental function of the brain. In the typical adult, multisensory neurons' response to paired multisensory (e.g., audiovisual) cues is significantly more robust than the corresponding best unisensory response in many brain regions. Synthesizing sensory signals from multiple modalities can speed up sensory processing and improve the salience of outside events or objects. Despite its significance, multisensory integration is testified to be not a neonatal feature of the brain. Neurons' ability to effectively combine multisensory information does not occur rapidly but develops gradually during early postnatal life (for cats, 4–12 weeks required). Multisensory experience is critical for this developing process. If animals were restricted from sensing normal visual scenes or sounds (deprived of the relevant multisensory experience), the development of the corresponding integrative ability could be blocked until the appropriate multisensory experience is obtained. This section summarizes the extant literature on the development of multisensory integration (mainly using cat superior colliculus as a model), sensory-deprivation-induced cross-modal plasticity, and how sensory experience (sensory exposure and perceptual learning) leads to the plastic change and modification of neural circuits in cortical and subcortical areas.

Keywords

Multisensory integration · Cross-modal interaction · Superior colliculus (SC) · Plasticity · Perceptual learning

L. Yu (✉) · J. Xu
Key Laboratory of Brain Functional Genomics (Ministry of Education and Shanghai), School of Life Sciences, East China Normal University, Shanghai, China
e-mail: lpyu@bio.ecnu.edu.cn; jhxu@bio.ecnu.edu.cn

10.1 Introduction

We are often confronted with multiple sensory cues (primarily visual and auditory) in the real world. Different sensory information propagates via specific energy and is processed through different sensory systems. Many events and objects emit sensory information of more than one modality, each containing the source's feature signals. For instance, when we watch a speech, auditory and visual information can be emitted from a speaker. To better understand the event or object, the brain should use the proper strategy to combine all necessary sensory inputs of different modalities. The integration of sensory information from different modalities can strengthen the brain's ability to detect, locate, and identify

Y. Gu, A. Zaidel (eds.), *Advances of Multisensory Integration in the Brain*, Advances in Experimental Medicine and Biology 1437, https://doi.org/10.1007/978-981-99-7611-9_10

external events as well as improve perceptual performance (Stein and Meredith 1993; Bavelier and Neville 2002; Ernst and Banks 2002; Diederich and Colonius 2004; Rowland et al. 2007; Fetsch et al. 2011; Chandrasekaran et al. 2013). For many wild-living animals (prey and predator), multisensory integration closely correlates with their lives because better multisensory integration makes them easily find danger and food.

In the past 50 years, this process of multisensory integration or cross-modal interaction has been widely examined in many cortical and subcortical regions such as association cortices, sensory cortices, and superior colliculus (Ghazanfar and Schroeder 2006; Stein et al. 2014; Murray et al. 2016; Choi et al. 2018). For a typical multisensory neuron, its notable characteristic is that it can show significantly enhanced responses to paired sensory inputs. Physical properties of stimuli, such as spatial alignment, temporal coincidence, and intensity (or effectiveness), can jointly influence multisensory integration. Stein et al. summarized three multisensory integration principles (spatial, temporal, and inverse effectiveness) based on studies of adult cat SC neurons (Stein and Meredith 1993). Briefly, the spatial and temporal principles are that multisensory cues (e.g., visual and auditory) evoke enhanced responses if they are spatially aligned and temporally concordant and elicit depressed or non-enhanced responses if they are spatially or temporally disparate. The principle of "inverse effectiveness" describes the observation that proportionately greater gain of multisensory enhancement is obtained when individual cues are weak or less effective. Data in many other brain regions now verify these multisensory integrative principles (Gu et al. 2008; Holmes 2009; Meredith et al. 2009). Behavioral and neural evidence show that spatial, temporal, and inverse effectiveness factors do not influence multisensory integration in isolation but interact with each other (Cappe et al. 2012; Krueger Fister et al. 2016; Nidiffer et al. 2016).

10.2 Development of Multisensory Integration in the Superior Colliculus (SC)

SC, one of the midbrain areas, can combine cross-modal information to initiate motor signals, including head and eye movements (Sparks 1986). SC consists of several layers, each receiving distinct innervation sources (May 2006; Basso and May 2017). Roughly, these layers are divided into two laminae (superficial and deep). The superficial laminae (I–III layers) mainly receive visual inputs from the retina, and the deep laminae (IV-VII layers) receive multiple sensory inputs (visual, auditory, and somatosensory) from many cortical and subcortical areas and contain all possible types of multisensory neurons (see Fig. 10.1). Each sensory representation has a map-like fashion in the deep layer of SC, and distinct sensory maps are registered with each other (Stein and Meredith 1993). This general feature seems independent of species (Stein et al. 2014). The anterior SC encodes the frontal sensory space (for auditory and visual) and the face area (for somatosensory), whereas the posterior SC represents more peripheral space (for somatosensory, the rear of the body). Medial and lateral parts of SC represent superior and inferior sensory spaces, respectively. Different sensory maps are in the register with the common premotor map in SC (Wurtz and Goldberg 1971; Stein and Clamann 1981; Grantyn and Grantyn 1982; Jay and Sparks 1987a, b; Guitton and Munoz 1991; Munoz and Wurtz 1993a, b).

Due to the convergence of multiple sensory inputs, the deep layer of cats SC is rich in multisensory neurons (more than 50%) (Jay and Sparks 1987b; Perrault Jr. et al. 2003; Yu et al. 2013; Xu et al. 2017b, 2018). It has become a good model for studying the mechanism of multisensory integration and its plasticity (Meredith et al. 1987; Meredith and Stein 1996; Wallace et al. 2004; Alvarado et al. 2007; Xu et al. 2017a; Wang et al. 2020). Every SC multisensory neuron has more than two spatial receptive fields (RFs),

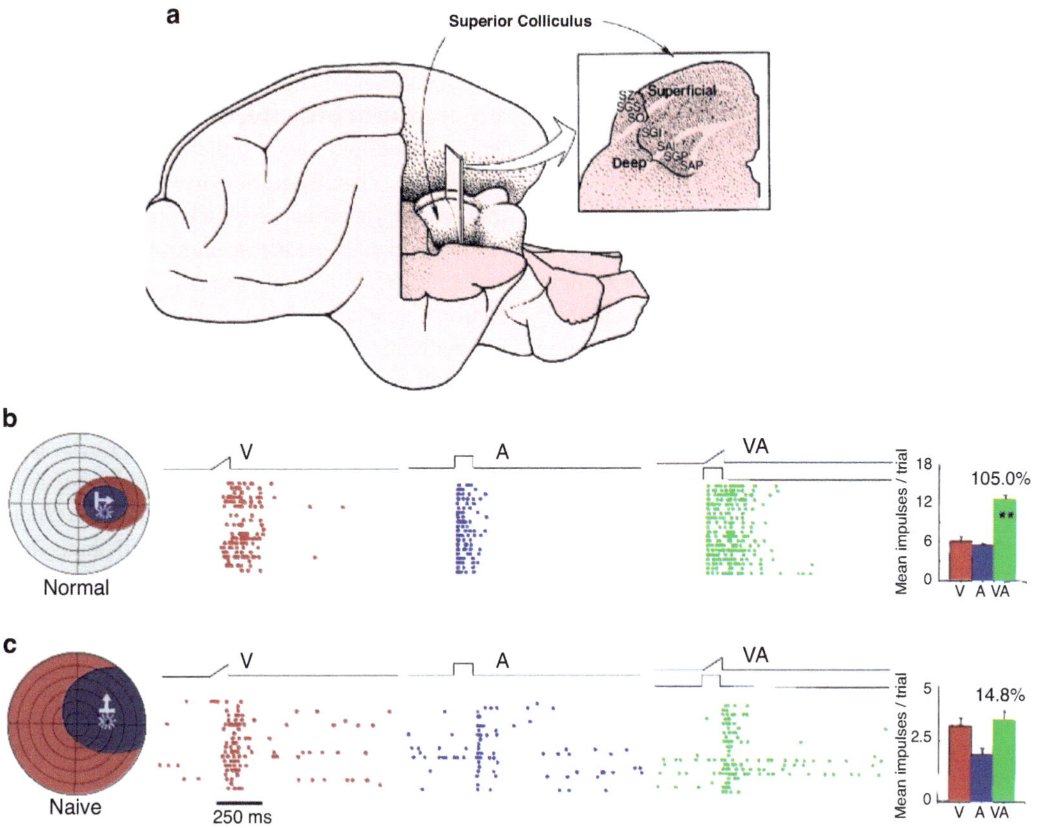

Fig. 10.1 Comparison of responses of neurons recorded in naïve and adult cats. (**a**) The coronal histological section of the cat SC. (**b**, **c**) The receptive fields (RFs) and responses of two typical neurons recorded from a normal adult cat (**b**) and a dark-reared cat (**c**). Left: the frontal sensory space. The vertical and horizontal lines denote meridians, and each circle in the sensory space represents 10°. Plots were made and modified with permission from Yu et al. (2010)

and RFs of different modalities are coincidental spatially (King et al. 1996; Meredith and Stein 1996; King 1999; Jiang and Stein 2003; Xu et al. 2018) (Fig. 10.1b). If a spatiotemporally concordant multisensory stimulus falls into a neuron's overlapping area of RFs, it usually evokes more robust responses than are elicited by the unisensory component stimulus (sometimes even greater than the sum of unisensory responses), which is called "multisensory enhancement" (Meredith and Stein 1996) (Fig. 10.1b). In contrast, multisensory inhibition usually occurs when cue components are far disparate in space (Kadunce et al. 1997). Therefore, the product of multisensory integration can well reflect the spatiotemporal arrangement of multisensory cues.

Effectively integrating multisensory inputs is not a naïve characteristic of SC (Wallace et al. 2004). A previous study showed that postnatal maturation is necessary for SC neurons to master effective multisensory integration for supporting mature behaviors (Kao et al. 1994). Sensory experience is vital for the neonatal development of multisensory enhancement. If cats were deprived of visual experience after birth, the vast majority of SC neurons failed to show multisensory enhancement (Fig. 10.1c). Multisensory neurons can be recorded soon after the ear channels and eyes are open (Wallace and Stein 2000). In kittens of postnatal 20 days, multisensory neurons responsive to visual stimuli were found (Stein et al. 1973). However, generally, they have long latencies, large RFs, weak response, and strong habituation to repeated stimuli. The integrative capabilities of these early SC multisensory neurons are immature (see Fig. 10.2). Their multisensory response was similar to the responsiveness to the modality-specific component (often close to a weighted average of those unisensory responses) (Wallace and Stein 1997). By the second postnatal month, multisensory neurons' incidence is gradually increasing, and their sensory responses are more vigorous, rapid, and selective. The electrical stimulation of the SC evoked overt orienting movements. After 4 months, the incidence of multisensory neurons reached the adult level (60–70%) (see Fig. 10.2). During postnatal development, multisensory neurons' RFs contract in size (initially decline rapidly and then gradually until the adult) and register with each other.

During the early postnatal life of cats, researchers also found that multisensory neurons are rare in the anterior ectosylvian sulcus (AES) (one of the association cortices in cats) and gradually develop with the increment of postnatal age (Wallace et al. 2006). The AES can be separated into visual, auditory, and somatosensory regions but can be recorded the high incidence of multisensory neurons (mainly in the border area between subregions) (Stein et al. 2002). AES neurons' sensory responses in different modalities appear in an orderly sequence in early life (Wallace et al. 2006). At the age of four postnatal weeks, most recorded AES neurons cannot respond to any "search" stimuli (visual, auditory, somatosensory). The somatosensory responses are recorded first at 4 weeks of age and followed by the auditory and somatosensory–auditory responses. Visually-responsive unisensory and multisensory neurons are first recorded at postnatal 12 weeks. The earliest multisensory neurons cannot exhibit multisensory enhancement. Neurons' capacity for multisensory integration gradually matures with postnatal development.

Adult cat SC receives dense afferent projection from AES. The integrative function of SC neurons appears controlled by AES. Deactivation of this association cortex compromised SC neurons' integrative abilities (Wallace and Stein 1994) and multisensory-cue-guided behaviors (Wilkinson et al. 1996). The close AES-SC connection is not an inborn characteristic of the cat brain. The sensory input from this corticotectal projection occurs on roughly postnatal day 28 (Wallace and Stein 2000). For a single SC neuron, the establishment of this projection seems abrupt. However, achieving the adult-like corticotectal projection requires 3–4 postnatal months (Wallace and Stein 2000). This long temporal period needed for the corticotectal projection's maturation corresponds well with the critical period of SC multisensory enhancement. Thus, establishing SC neurons' integrative function may depend on the experience-dependent top-down control from AES.

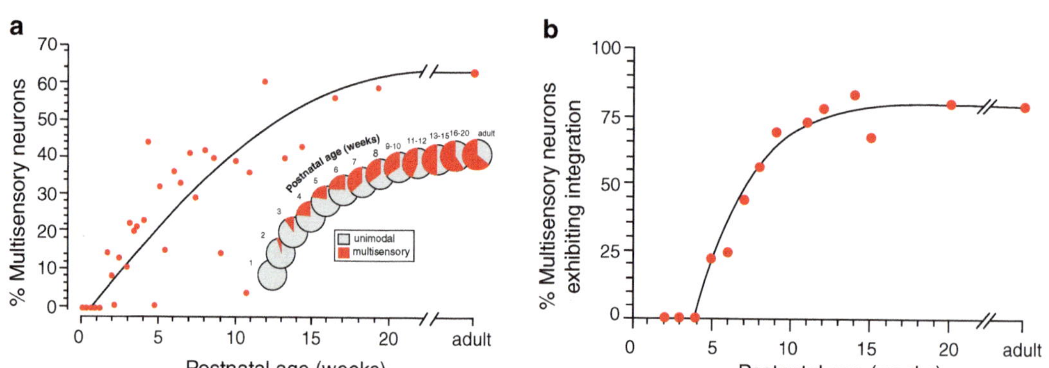

Fig. 10.2 Development of multisensory neurons and their integrative capability. (**a**) The proportion of multisensory neurons recorded at different postnatal ages in the deep layer of cat SC. (**b**) The percentage of multisensory neurons showing significant ($p < 0.05$) multisensory integration at different postnatal ages. Plots were made and modified with permission from Wallace and Stein (2000)

10.3 The Influence of Sensory Deprivation on Multisensory Integration

Like response properties in the unisensory system, multisensory neurons' integrative manner is also significantly influenced by sensory experience. Stein lab conducted long-term studies regarding the impact of sensory experience (mainly visual) on cat SC neurons' integrative capabilities. They reared newborn kittens in a dark room (deprived of visual-related experience). When examined as adults, they found that dark rearing did not affect the proportion of sensory-responsive neurons in the deep layer of SC despite a steep decline in the incidence of visual-only responsive neurons and a corresponding incidence rise in somatosensory and auditory neurons. Despite that, overall impacts on multisensory representation induced by visual deprivation were modest (Wallace et al. 2004). The absence of visual experience substantially affected the developmental contraction of RFs. The large size of RFs was not restricted to the visual modality but was also seen in the somatosensory and auditory modalities, suggesting that other sensory systems were also affected by visual deprivation. Although SC neurons in dark-reared cats had RFs of increased size and low spatial fidelity, their RFs of different modalities were still topographically organized. The critical difference in SC neurons' properties between dark-reared and normal rats was seen in responsiveness to multisensory cues. Generally, multisensory SC neurons recorded in dark-reared animals showed responsiveness to multisensory stimulation that was no greater than unisensory response (Fig. 10.1c), reflecting a lack of multisensory enhancement (Wallace et al. 2004).

Unlike visual deprivation, auditory deprivation is difficult to achieve. Xu et al. reared cats in the omnidirectional noise environment to mask effective auditory inputs for examining its effect on SC neurons (Xu et al. 2014a). They found noise rearing has little impact on the overall incidence of sensory responsive SC neurons. However, the operation resulted in abnormalities in the auditory RF of these neurons. Most audiovisual neurons showed contracted auditory RF and poor alignment between visual and auditory RFs (see Fig. 10.3). Typically, neurons bias the auditory RF center peripherally away from

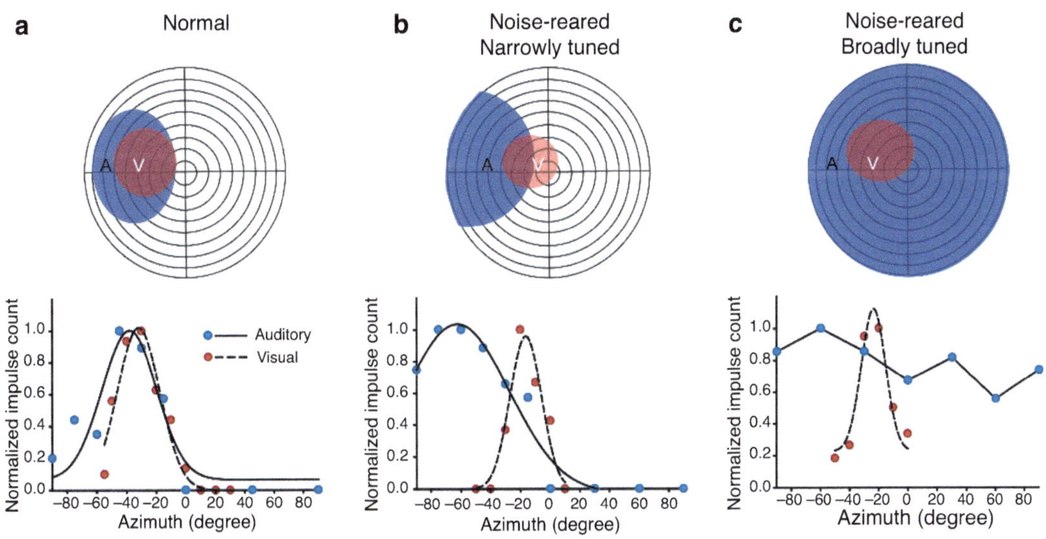

Fig. 10.3 Auditory and visual RFs of SC neurons. (**a**) Showing the auditory and visual RFs of a typical SC multisensory neurons recorded in the normal cat. Top: visual (V) and auditory (A) RFs plotted on a schematic of sensory space. Bottom: The response profile to an auditory (solid line) and a visual (dashed line) cue at different azimuths. (**b**, **c**) Auditory and visual RFs of two representative neurons recorded from a noise-reared cat. Plots were made and modified with permission from Xu et al. (2014a)

its visual counterpart. The average distance between auditory and visual RF centers is substantially expanded (see Fig. 10.3). The poor overlap between visual and auditory RFs disrupted the correlation. Thus, the spatially congruent visual and auditory cues could not fall into the overlapped RF area. Like their counterparts in kittens, some neurons recorded from noise-reared animals appeared to have no apparent spatial tuning and had a large RF. Like visual deprivation, noise rearing disrupted SC multisensory integration's normal development although a near-normal proportion of audiovisual neurons developed.

To further investigate the impact of sensory experience on multisensory neurons' integrative capacity, Xu et al. examined multisensory integration in SC trisensory neurons of three groups of cats: a control group, a "visually-deprived" group, and a "noise-reared" group reared with omnidirectional noise (Xu et al. 2015). By restricting auditory-nonauditory or visual-nonvisual experience during early life, they observed that the multisensory circuit learns to synthesize each multisensory pairing individually, and both SC trisensory and bisensory neurons became adapted to specific multisensory combinations given by each rearing environment. For instance, trisensory neurons recorded in dark-reared groups can integrate auditory and somatosensory inputs well but cannot synthesize visual-auditory or visual-somatosensory inputs (Xu et al. 2015). Thus, even in adult animals, SC trisensory neurons still failed to integrate all kinds of multisensory combinations to which they were responsive. This selective experience-dependent maturational process might determine which environmental objects or events can be the most effective targets for SC-mediated orientation. These results reveal that multisensory co-activation is necessary for multisensory maturation, in line with the role of SC in detecting and orienting critical environmental events (Fig. 10.4).

Like SC, the maturation of the effective multisensory integration in association cortices takes place in postnatal several months (Wallace et al. 2006), indicating that postnatal sensory experiences play an essential role in this developing process. The impact of visual deprivation on one of the cat association cortices (AES) was also examined (Carriere et al. 2007). Several notable changes were seen. First, dark-rearing declined the integrative capabilities of AES neurons, similar to that seen in SC. Second, the proportion of AES multisensory neurons modulated rather than driven by a second sensory modality significantly increased. Whereas in normal cats, most multisensory neurons are responsive to cues of two or more modalities (Stein et al. 2002). Third, most modulated neurons show multisensory inhibition to spatiotemporally congruent multisensory pairing (typically gives rise to multisensory enhancement in normal cats). These depressive interactions are not limited to the visual domain but to all types of multisensory neurons examined (i.e., VA, VS, AS). These results indicate that sensory experience plays a critical role in shaping the developmental process of multisensory circuits in cortices and shifting the relative excitation and inhibition balance in these circuits.

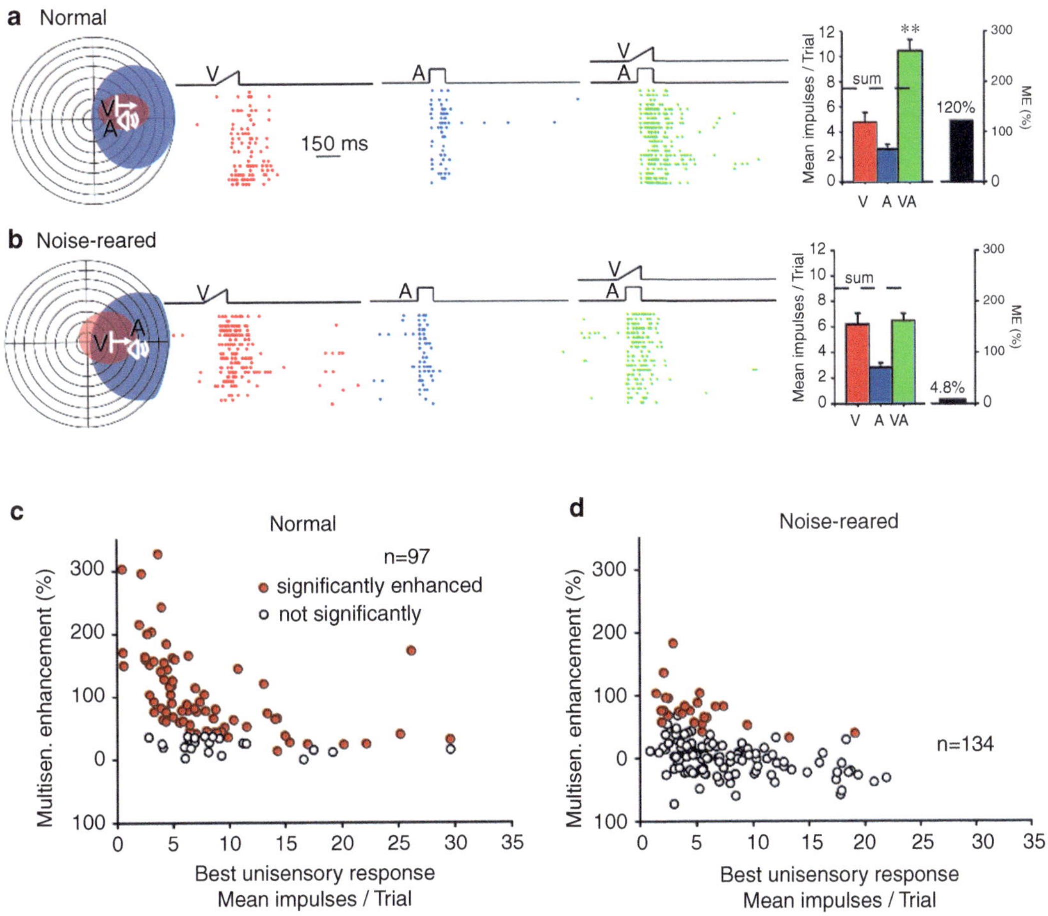

Fig. 10.4 Multisensory integration was rare and yielded less response enhancement in noise-reared cats. (**a**, **b**) Showing visual, auditory, and multisensory responses of two neurons recorded from normal (**a**) and noise-reared (**b**) cats. (**c**, **d**) The index of multisensory enhancement is plotted for each neuron as a function of the best (largest) unisensory response. Although the typically high proportion of neurons (*n*, number of neurons) in normal cats exhibited multisensory integration capabilities, few neurons in the noise-reared cats did. Plots were made and modified with permission from Xu et al. (2014a)

10.4 The Role of Multisensory Experience in the Development of Multisensory Integration

Mature SC multisensory neurons can use the multisensory enhancement model to elicit an increased response to the spatiotemporally concordant cross-modal pairing. Developing this essential feature of SC neurons relies on appropriate multisensory experience. Without such experience, a neuron's multisensory response is no more greater than the corresponding unisensory response. However, the adult brain still maintains high neural plasticity. Yu et al. exposed visual-deprivation cats to the repeated presentation of multisensory stimuli in their anesthetized state. Following multiple experimental sessions, the proportion of recorded neurons capable of multisensory enhancement increased. With the increase in multisensory exposure, the multisensory enhancement magnitude gradually rose (Yu et al. 2010) (Fig. 10.5). This evidence indicates that neurons in adult animals retain experience-

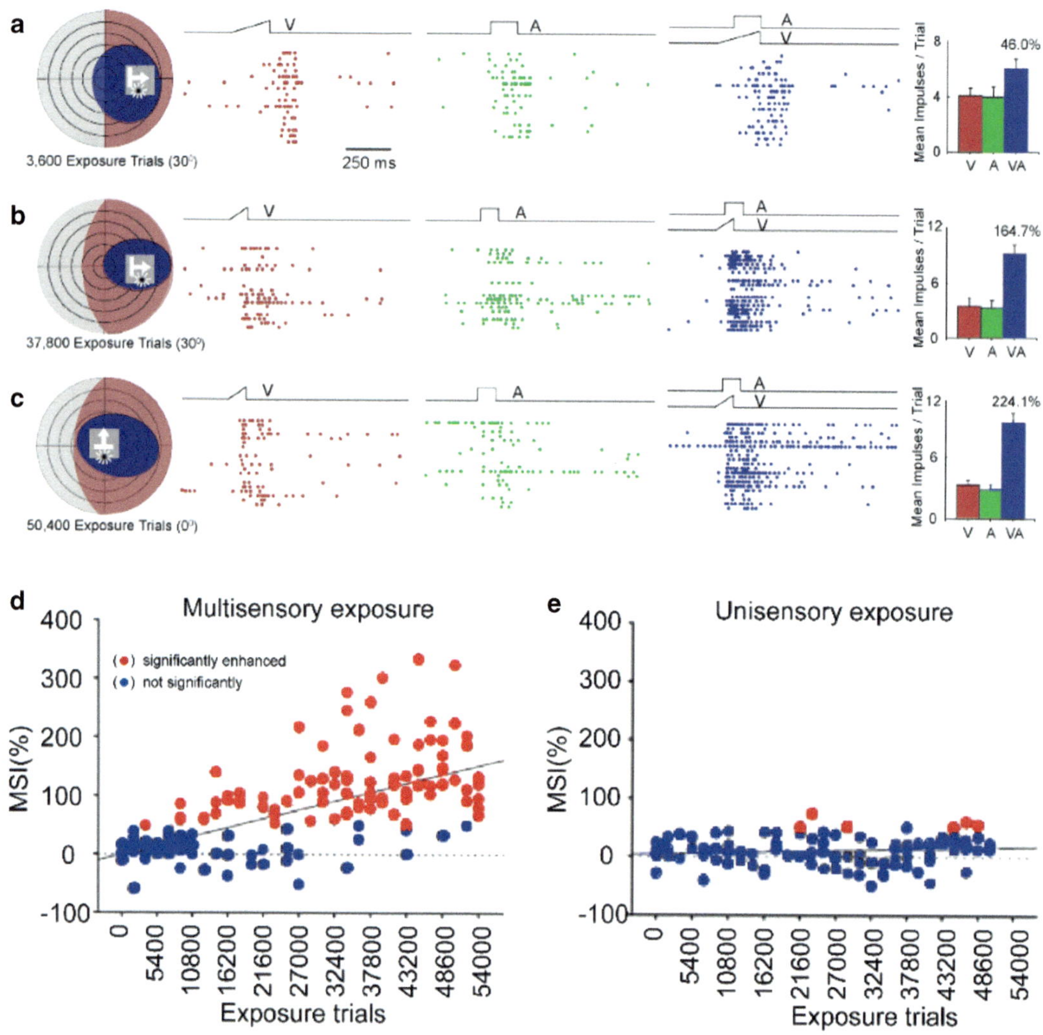

Fig. 10.5 Multisensory exposure initiates the development of SC neurons' integrative capability. (**a–c**) Three representative neurons' visual, auditory, and audiovisual responses. Left, visual (blue) and auditory (red) RFs. A transparent gray square denotes the exposure site. Middle, raster plots show visual, auditory, and audiovisual responses. Right, showing the mean magnitude of unisensory and multisensory responses. (**d**, **e**) Multisensory enhancement magnitude changed with the number of multisensory (**d**) and unisensory (**e**) exposure trials, indicating that only multisensory exposure triggered the development of multisensory enhancement. Each circle represents one neuron. Blue circles, neurons exhibited multisensory enhancement; red circles, no multisensory enhancement. Plots were made and modified with permission from Yu et al. (2010)

dependent plasticity. However, exposing an animal to multisensory cues in a specific sensory space was insufficient to drive the multisensory integration development across entire SC neurons. Only those with RFs encompassing the exposure site achieved this development. Once a neuron develops the multisensory integration ability with the increasing cross-modal exposure within its receptive fields, it can generalize this integrative capacity to other RF sites. Exposing animals to a random presentation of visual-only and auditory-only stimuli failed to initiate the multisensory development in most SC neurons (see Fig. 10.5), indicating that multisensory rather than unisensory experience plays a critical role in the development of mature multisensory circuits.

Multisensory integration can be induced in anesthetized adult cats, raising a few questions about it. First, can this integrative capability be maintained without continued exposure? Second, Is the trained multisensory integration in anesthesia equivalent to those in alert animals? Third, can neurons integrate non-exposure multisensory stimuli? To further investigate these issues, the Stein group did a similar experiment in which they periodically exposed cats to auditory and visual stimuli appearing randomly in time and space or in spatiotemporal congruence. Except for sensory exposure, animals stayed in darkness at all times. After experiencing 6 months of sensory exposure (12 h daily), animals continued to remain in the dark room without any sensory exposure. Then, they tested SC neurons' response in these animals at ~2 years of age. Their data showed that the random exposure was insufficient to spur the development of neurons' integrative ability, but concordant exposure could. The integrative capabilities of neurons in the concordance-exposed group resembled those recorded in normal animals. This result indicates that integrative capabilities, once obtained, can be retained for a long time. In addition, neurons could integrate multisensory cues of physical properties different from those used in the exposure.

Cat SC receives sensory inputs from many subcortical and cortical sources (Stein and Meredith 1993). Of them, inputs from two association cortices (the rostral lateral suprasylvian sulcus, rLS; the anterior ectosylvian sulcus, AES) are particularly vital for SC neurons' multisensory integration. Deactivation or ablation of association cortices could render SC multisensory neurons incapable of integrating multisensory cues at any stage of life (Wallace and Stein 2000; Jiang et al. 2006, 2007; Alvarado et al. 2009). Yu et al. examined the impact of AES on the development of SC multisensory integration (Yu et al. 2016). They deactivated one side of AES of dark-reared adult cats during multisensory exposure (Fig. 10.6). Their results indicated that AES deactivation on one hemisphere disrupted the development of integrative abilities of ipsilateral SC neurons. Thus, the high-order association cortices might serve as a portal through which multisensory experience could shape the functional capability of the SC multisensory circuit (Catalano and Shatz 1998; Khazipov and Luhmann 2006; Blankenship and Feller 2010).

Experimental evidence reveals that the relationship between different sensory inputs is not naturally cooperative. Instead, they are mutual competitions in a naïve state (Yu et al. 2019). All spatiotemporally congruent and disparate cross-modal configurations can cause the failure of multisensory enhancement or the response depression in this state. Cross-modal inhibition is most visible when the unisensory component stimuli induce significant imbalanced responses on the target multisensory neuron (Miller et al. 2015). Compared with SC neurons in normal animals, those recorded in sensory deprivation animals were more likely to have a significant imbalance in sensitivities to auditory and visual inputs, which amplifies their competition (Yu et al. 2019). Also, visual deprivation maximizes the inhibitory effect of auditory or somatosensory inputs on visual modality. This might be the reason why sensory-restricted circumstances corrupted the development of multisensory integration. Understanding this initial state of cross-modal sensory processes is critical for understanding the normal developmental trajectory of multisensory integration as well as the etiology of the anomalous multisensory development typically seen in autism spectrum disorder (Brandwein et al. 2013; Stevenson et al. 2014; Foxe et al. 2015), dyslexia (Hairston et al. 2005; van Laarhoven et al. 2018), schizophrenia (Williams et al. 2010), and in patients of early multisensory experience compromised (Putzar et al. 2007, 2010).

Repetitive cross-modal events appeared to drive the transition from competition to cooperation between multisensory inputs. Whether this transition is a general characteristic of multisensory neurons remains to be determined. Yu et al. raise a neuro-computational model that posits the existence of a default competitive circuit within the SC, and this model satisfactorily explains the biological observations (Yu et al. 2019). In their model, topographically organized inputs from one modality excite specific units but inhibit the neural network's other sen-

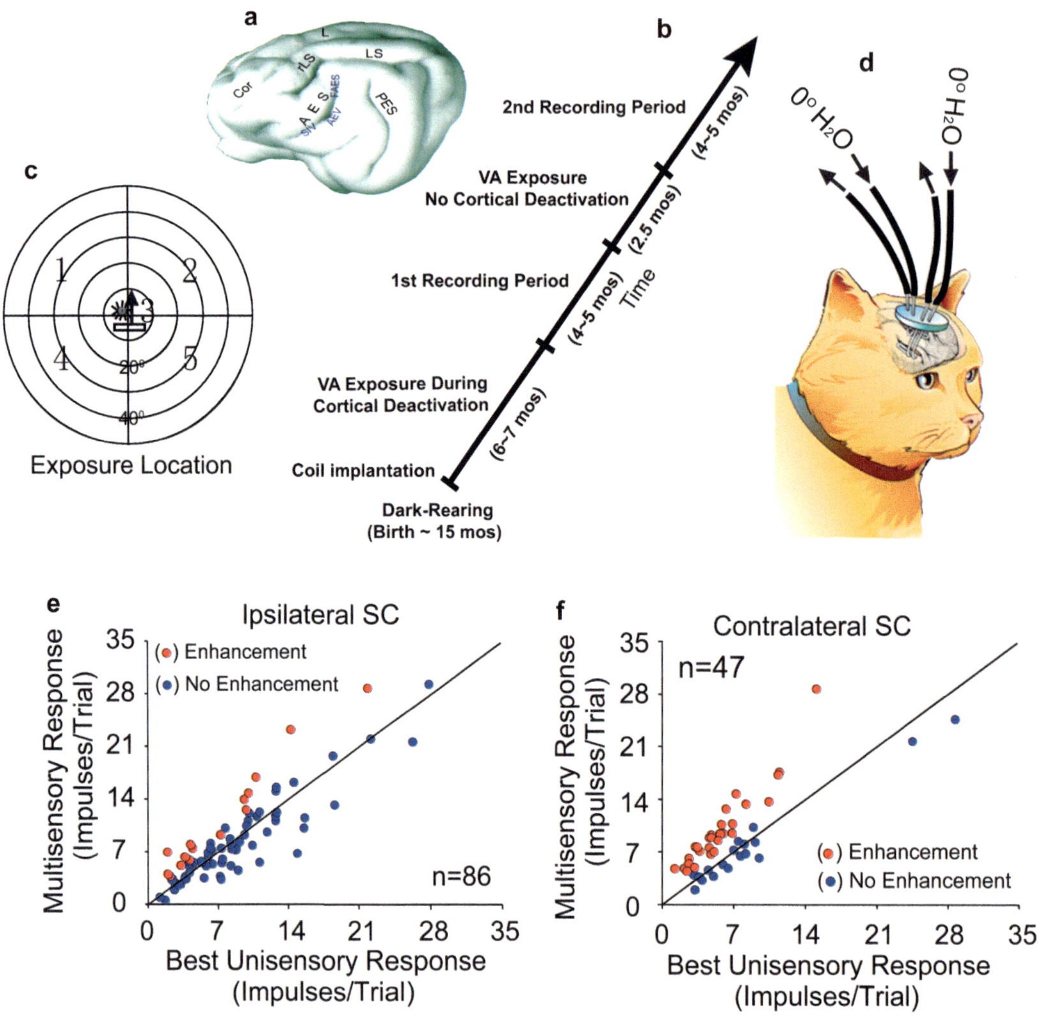

Fig. 10.6 Cortical deactivation blocked the experience-induced development of multisensory integration in ipsilateral SC neurons. (**a**) A schematic of the brain with AES (and rLS) identified and its visual (AEV), auditory (FAES), and somatosensory (SIV) subdivisions. (**b**) The diagonal timeline shows the time points of performing sequential exposure and recording sessions. (**c**) The cross-modal (visual-auditory) exposure stimulus was presented randomly at five spatial locations. (**d**) the schematic shows circulating refrigerated water in the coils to deactivate adjacent brain tissue during the period of multisensory exposure. (**e**, **f**) showing the comparison of multisensory and the best unisensory responses for neurons recorded in SC ipsilateral to deactivated AES (**e**) and for neurons recorded in SC contralateral to deactivated AES (**f**), indicating that deactivation of AES disrupted experience-dependent development of multisensory integration. Plots were made and modified with permission from Yu et al. (2016)

sory signals. Crucially, this inhibitory effect can extend across sensory modalities. In the default state, cross-modal inputs compete effectively with one another through mutual inhibition, even when convergent onto the same target neuron. When the sensory inputs are pretty balanced in efficacy, it yields minimal cross-modal inhibition in the responses of SC output unit because both modalities only partially attenuate the excitation of the other one, but the unit receives excitatory components from both modalities. When sensory inputs are imbalanced, the inhibitory effect could be maximal because the most effective modality can completely inhibit the excitatory input of the other one but is less affected by the presence of the other modality at the same time, causing the overall reduction of excitation reaching the unit.

10.5 Sensory Loss and Cross-Modal Plasticity

Decreased or abnormal sensory input can lead to cross-modal plasticity, and typically the remaining intact sensory modalities would recruit cortical regions of the deprived modality (Glick and Sharma 2017). For example, partial or complete vision loss induces significantly reorganizational changes in the brain. A common phenomenon in many studies is that visual cortical areas are occupied by other modalities, with the most evidence for somatosensory and auditory inputs. This cross-modal cortical reorganization compensates for the absence of visual information. Such cross-modal plasticity creates a different reorganizational framework for multisensory interactions. Studies in blind subjects report their perceptual abilities in the remaining senses enhanced (Neville and Lawson 1987). Compared with the normal subjects, psychophysical studies in the early blind revealed enhanced auditory functions. For example, early-blind people showed better sound localization accuracy than sighted subjects (Lessard et al. 1998). Another study showed blind persons had better sound localization abilities in the peripheral auditory space (Roder et al. 1999). A study on binocularly deprived cats shows improved auditory localization capabilities and similar somatosensory behavior compared to normal cats (Rauschecker 1995). This study also shows that somatosensory and auditory inputs completely take over the visual area of AES and the auditory RF of single units in this cortical region has better spatial tuning (Rauschecker 1995). Positron emission tomography (PET) study showed that the visual cortex in blind subjects in early life was activated during tactile discrimination tasks (Braille readers), indicating that the input from the somatosensory system enters into the visual cortex, and normal subjects showed no such deactivation in visual cortex (Sadato et al. 1996). These studies suggest that these superior sensory abilities in the non-deprivation modalities are believed to reside in the deprived cerebral cortices reorganized by the remaining sensory modalities through cross-modal plasticity. Such cross-modal plasticity is most extensive in early blindness (blinded after birth but before age 16) but also exists substantially in late blindness (blinded after the age of 16). The reason is that the brain has a critical period in early life during which cross-modal plasticity is easy to occur. In addition, late-blind subjects usually suffer from a slower progression of vision loss and often retain visual perception and memory, which might preclude cross-modal plasticity.

In some patients, their vision is partially impaired. Studies in such cases showed that the other senses could provide redundant information supporting the weak visual cues provided via the visual pathway (Targher et al. 2012, 2017). The evidence from other studies demonstrates that a paired sound can improve visual detection in the subthreshold visual condition (Frassinetti et al. 2002) or the state of induced myopia (Hairston et al. 2003). Some multisensory conceptual models, such as the model of cue reliability and maximum likelihood estimation (Ernst and Banks 2002), might explain these low vision conditions. According to cue reliability and maximum likelihood estimation mode, the product of multisensory interaction depends on the relative reliability of sensory signals. Thus, in low vision conditions the addition of the appropriate sensory information from the other modality (such as auditory) can significantly improve visual performance. Such cross-modal benefits revealed important evidence for explaining the neural mechanism of vision rehabilitation induced by multisensory training (Proulx et al. 2014). So far, we lack experimental evidence probing the links between multisensory function and low perceptual abilities. Recent evidence from two large US national data sets shed new light on this issue that vision loss is strongly associated with cognitive decline in older subjects (Chen et al. 2017). Such correlations suggest that multisensory training might be able to improve multisensory performance, benefit vision, and protect against age-related cognitive decline (especially for those with poor vision).

Much like visual deprivation, many studies have investigated cross-modal plasticity induced by hearing loss. Like vision loss, hearing loss resulted in substantial alterations in auditory cortical regions' anatomical and functional properties, characterized by sensory inputs entry from other modalities. In deaf subjects, visual stimuli and sign language can activate their auditory cortex (Nishimura et al. 1999; Finney et al. 2001, 2003; Vachon et al. 2013). No such auditory cortical activation was seen in normal persons. Studies in early-deaf humans reveal compensatory or adaptive improvements in visual detection tasks in the periphery (Bavelier et al. 2000; Bosworth and Dobkins 2002). Improved visual abilities were also seen in congenitally deaf cats, with lower thresholds of visual movement detection and superior localization in the peripheral space. Lomber et al. examined whether the behavioral compensatory effects observed in deaf cats were linked with visual processing in reorganized auditory cortices (Lomber et al. 2010). Their result showed that superior visual localization could be selectively eliminated by deactivating the posterior auditory cortex, whereas deactivating the auditory cortical dorsal area could decrease superior visual motion detection, indicating that cross-modal reorganization of auditory cortex into visual processing enhanced visual performance in the deaf. Discrete regions of the reorganized auditory cortex possibly correlate with visual functions. Cross-modal reorganization following hearing loss appears not to require a long time period. An electrophysiological study shows that restructuring the auditory cortex into visual processing can begin in the early period of hearing loss (Campbell and Sharma 2014). Fortunately, hearing loss can remarkably benefit from the cochlear implant (CI) (Kral et al. 2019).

The CI can take auditory information received by a microphone, transforms the sound signal into the corresponding electrical voltage, and then sends it to electrodes implanted within the inner ear. For early deafness, the cochlear implant is believed to be the most successful treatment. Generally, a period of hearing training is required after CI. In most cases, electrical signals generated by CI are sufficient to support speech comprehension after training. In late deaf (deaf adults), CI significantly recovers auditory speech intelligibility. In most cases, multisensory-based training could facilitate language development and speech and produce more optimal language and speech outcomes in CI subjects (Kral and Sharma 2012). Multisensory training emphasizes combining concordant auditory and visual cues as tools to build more robust multisensory representations, which is essential for the development of cognitive capability (De Niear et al. 2018; Jacobs and Xu 2019).

Also, CIs bring some deficits. A study on CI individuals shows that masking sounds impaired speech-recognition performance (Fu et al. 1998). Compared with normally hearing children, CI patients' speech outcomes often lag (Holt and Svirsky 2008; Murphy et al. 2011). Some researchers attributed these deficits to the visually occupying auditory temporal cortex in the deaf (Champoux et al. 2009; Sandmann et al. 2012). Due to this, some clinicians are not inclined to use visual speech to improve pediatric CI recipients' rehabilitation. However, a new study showed that compared with controls although CI users exhibited more robust activities to visual speech in auditory cortical regions, auditory speech processing in the same regions did not significantly differ between the groups (Mushtaq et al. 2020), indicating that this auditory cortex can normally process auditory information. In addition, this conclusion suggests that CI users might develop the skills of specific visuoauditory and visual integration to overcome the corresponding difficulties in everyday-life speech recognition. Isaiah et al. showed that sound localization improved significantly following multisensory training in the early-deafened bilateral CIs group (Isaiah et al. 2014). They attribute this improvement to increased auditory cortical activities and the enhanced sensitivity to interaural level differences. Anderson et al. examined the influence of visual speech's cross-modal activation of auditory brain regions on CI success and found that increased cross-modal activation from before to after implantation is linked with better speech understanding (Anderson et al.

2017). In the same paper, Anderson et al. also showed that elevated activation of the auditory cortex by auditory and visual speech synchronously developed after cochlear implantation. These findings indicate that cross-modal training does not exert previously assumed maladaptive influences on CI success but instead can bring substantially adaptive benefits to hearing restoration after implantation through the multisensory plasticity mechanism. We believed that visual and auditory inputs might compete with each other in the early period following CI. However, the long-term multisensory training would drive the competition into cooperation.

10.6 The Influence of Perceptual Association Learning on Cortical Cross-Modal Plasticity

Previous studies on experience-dependent learning (e.g., perceptual learning) usually focus on unisensory systems. However, our interaction with environmental events (or objects) typically involves multisensory cues. For example, in class, to better understand the teaching content, we need to integrate visual and auditory speech given by the lecturer. The brain appears to have evolved to optimally use multisensory cues to learn and operate in multisensory environments. In this respect, multisensory learning might be optimal and more effective.

Several studies have shown that multisensory learning could influence cross-modal interaction in many sensory cortices (Mezzera and Lopez-Bendito 2016). The classic view holds that sensory cortices (visual, auditory, somatosensory) only process the unisensory process, and multisensory processing is restricted to higher order brain areas. A gradually increasing number of papers demonstrated that many cortical and subcortical regions, classically believed to process unisensory information, are involved in cross-modal interaction (Fu et al. 2003; Ghazanfar and Schroeder 2006; Kayser and Logothetis 2007; Driver and Noesselt 2008). Sensory cortices, even at the primary sensory cortex level, can engage in cross-modal interaction (Vasconcelos et al. 2011; Iurilli et al. 2012; Feng et al. 2014; Kayser et al. 2017; Atilgan et al. 2018). Intracellular recording in mice reports that many neurons in the sensory cortex are unisensory in the view of spike signals but multisensory from the perspective of intracellular signals (Olcese et al. 2013; Ibrahim et al. 2016). Imaging studies in humans show that the visual cortex can be activated by task-irrelevant sounds, which improves visual discrimination (McDonald et al. 2000; Feng et al. 2014). Anatomical studies have revealed reciprocal nerve projections among the visual, auditory, and somatosensory cortices (Falchier et al. 2002, 2010; Cappe and Barone 2005; Campi et al. 2010; Stehberg et al. 2014).

Now, several studies show that multisensory learning experience could shape multisensory processing in sensory cortices. When rats learned a task associating an audiovisual cue with a water reward, the visual response could be seen in a significantly higher proportion of auditory neurons, and most of them exhibited significant multisensory enhancement (see Fig. 10.7) (Han et al. 2021). Consistently, passively exposing adult or young rats to combined auditory and visual stimuli could expand audiovisual neurons' distribution in the visual-auditory cortical transition area and the auditory cortex (Xu et al. 2014b). In the gustatory cortex, multisensory associative learning increases the number of neurons showing multisensory enhancement (Vincis and Fontanini 2016). A recent study, using genetically encoded voltage and calcium indicators, showed that coincident visual-auditory experience could modify both the supra- and subthreshold response properties of L2/3 neurons of mouse V1 (Knopfel et al. 2019). Specifically, they find that some multimodal neurons develop enhanced responses to the paired auditory stimulus after auditory-visual pairing. Taken together, these results demonstrate that associative learning can give rise to closer associations between feature-specific assemblies of multimodal neurons and leave a multisensory experience trace in the cortical network.

Multisensory training is a powerful force that can reorganize intercortical and cortical-subcortical circuits. The primary visual pathway

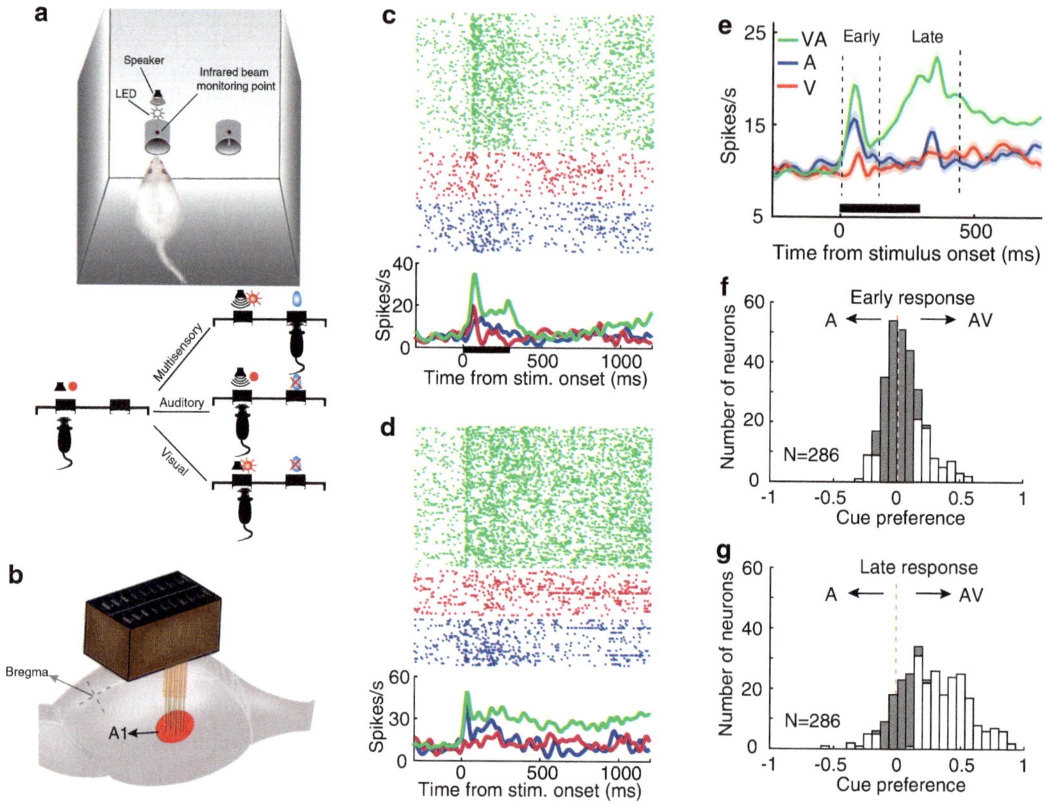

Fig. 10.7 Multisensory perceptual learning enhances multisensory representation in auditory cortex. (**a**) A schematic of a Go/No-Go task. (**b**) showing the primary auditory cortex (A1, red circle) and multichannel electrodes. (**c**, **d**) showing two typical A1 neurons' visual (red), auditory (blue), and audiovisual (green) responses in task engagement. (**e**) Mean response PSTHs across populations, indicating that visual response occurs in A1 and audiovisual response is much greater than auditory response. (**f**, **g**) showing neurons' modality preferences (multisensory vs. auditory) in the early (**f**) and late (**g**) response period. Unfilled bars indicate audiovisual response was significantly higher (>0) or lower (<0) than auditory response (permutation test, $P < 0.05$, bootstrap $n = 5000$). Plots were made and modified with permission from Han et al. (2021)

routes most retinal fibers to the lateral geniculate nucleus (LGN) of the thalamus and then, to primary visual cortex. A secondary visual pathway directly conveys a minority of retinal fibers to a midbrain structure, Superior Colliculus, which also has reciprocal projections with striate and extrastriate visual cortices (May 2006). Patients with lateralized primary visual cortex damage would develop homonymous hemianopia (the loss of conscious vision in one hemifield). Hemianopic patients fail to be able to perceive visual stimuli in the blind hemifield and cannot make effective oculomotor strategies to compensate for the visual loss (Hildebrandt et al. 1999; Tant et al. 2002). Several studies in hemianopia patients showed that auditory-visual exposure could improve visual performance, which lasts a long period of time, and increase oculomotor response to the visual stimuli in the initially blind hemifield (Bolognini et al. 2005; Passamonti et al. 2009; Tinelli et al. 2015; Grasso et al. 2016). To explore the underlying neural mechanism of hemianopia, Jiang et al. performed a similar experiment on cats (Jiang et al. 2015). They created a cat hemianopia model via ablating the unilateral visual cortex and then trained the model cats to perform a multisensory-guided localizing task (Jiang et al. 2015). They found that this simple cross-modal training could reverse the hemianopia, and visual responsiveness in the deep layers of the ipsilesional SC can be recorded again after multisensory training. Unisensory (visual or auditory) training failed to induce this recovery.

The medial prefrontal cortex (mPFC) of rats receives diverse inputs (both sensory and non-sensory) from cortical and subcortical areas and is involved in forming perceptions and arbitrating decision-making. Neurons in rat mPFC frequently showed no apparent multisensory integration when animals performed a free-choice task (Van Eden et al. 1992; Heidbreder and Groenewegen 2003). Meng et al. show that cue-discrimination two-alternative-forced choice task learning could induce the development of differential multisensory integration in mPFC neurons (Zheng et al. 2021). They trained rats to perform a perceptual task (see Fig. 10.8). During the task, rats had to make a behavioral choice according to sound frequency. They found that the presence of

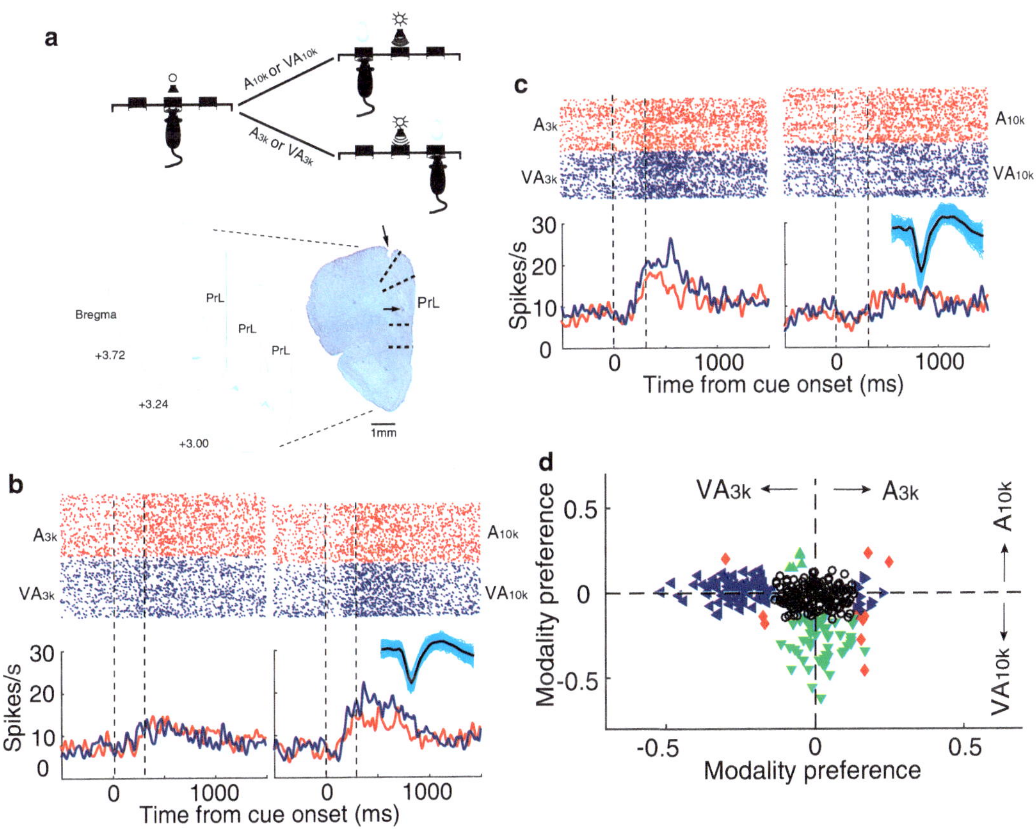

Fig. 10.8 Choice-dependent cross-modal interaction of rat mPFC neurons in a two-alternative forced-choice task. (**a**) Top: a schematic of the behavioral paradigm used. When the stimulus was either a 10 kHz pure tone (A10k) or a combination of a 10 kHz pure tone and a flash of light (VA10k), the animal would be rewarded at the left port. If the stimulus given was a 3 kHz pure tone alone (A3k) or the same tone paired with a flash of light (VA3k), the animal would receive the reward in the right port. Below: the photograph of a stained brain section showing the electrode track (top arrow) and the final location of the electrode tip (bottom arrow). (**b**, **c**) Showing two neurons' activities in A3k, VA3k, A10k, and VA10k trials, indicating that a differential multisensory integration was induced. (**d**) Modality choice preferences (ROC value) are shown. Each symbol shows the value of a single neuron. Triangles: multisensory response was significantly greater or less than the corresponding auditory response in one tone condition ($p < 0.05$; permutation test, 5000 iterations); red diamonds: multisensory response was significantly greater or less than auditory response in both auditory conditions. Open circles: no significant difference in multisensory and auditory responses. Plots were made and modified with permission from Zheng et al. (2021)

the visual cue only facilitated responsiveness to one of the auditory cues although auditory and visual cues were given in spatiotemporal congruence for each condition. If a neuron showed an auditory preference, mostly it would only exhibit enhanced multisensory responses to the preferred auditory cue. If the task was changed slightly by adding a meaningful visual cue, neurons showed a similar cue selectivity but exhibited a different integration pattern. Enhanced multisensory responsiveness could still be evoked, but usually only where both visual and auditory cues denoted the same behavioral direction. These results reveal that following perceptual task training, mPFC neurons' integrative strategy becomes task-dependent, and this model substantially differs from that we have seen in previous studies.

References

Alvarado JC, Vaughan JW, Stanford TR, Stein BE (2007) Multisensory versus unisensory integration: contrasting modes in the superior colliculus. J Neurophysiol 97:3193–3205

Alvarado JC, Stanford TR, Rowland BA, Vaughan JW, Stein BE (2009) Multisensory integration in the superior colliculus requires synergy among corticocollicular inputs. J Neurosci 29:6580–6592

Anderson CA, Wiggins IM, Kitterick PT, Hartley DEH (2017) Adaptive benefit of cross-modal plasticity following cochlear implantation in deaf adults. Proc Natl Acad Sci U S A 114:10256–10261

Atilgan H, Town SM, Wood KC, Jones GP, Maddox RK, Lee AKC, Bizley JK (2018) Integration of visual information in auditory cortex promotes auditory scene analysis through multisensory binding. Neuron 97:640–655

Basso MA, May PJ (2017) Circuits for action and cognition: a view from the superior colliculus. Annu Rev Vis Sci 3:197–226

Bavelier D, Neville HJ (2002) Cross-modal plasticity: where and how? Nat Rev Neurosci 3:443–452

Bavelier D, Tomann A, Hutton C, Mitchell T, Corina D, Liu G, Neville H (2000) Visual attention to the periphery is enhanced in congenitally deaf individuals. J Neurosci 20:RC93

Blankenship AG, Feller MB (2010) Mechanisms underlying spontaneous patterned activity in developing neural circuits. Nat Rev Neurosci 11:18–29

Bolognini N, Rasi F, Coccia M, Ladavas E (2005) Visual search improvement in hemianopic patients after audio-visual stimulation. Brain 128:2830–2842

Bosworth RG, Dobkins KR (2002) The effects of spatial attention on motion processing in deaf signers, hearing signers, and hearing nonsigners. Brain Cogn 49:152–169

Brandwein AB, Foxe JJ, Butler JS, Russo NN, Altschuler TS, Gomes H, Molholm S (2013) The development of multisensory integration in high-functioning autism: high-density electrical mapping and psychophysical measures reveal impairments in the processing of audiovisual inputs. Cereb Cortex 23:1329–1341

Campbell J, Sharma A (2014) Cross-modal re-organization in adults with early stage hearing loss. PLoS One 9:e90594

Campi KL, Bales KL, Grunewald R, Krubitzer L (2010) Connections of auditory and visual cortex in the prairie vole (Microtus ochrogaster): evidence for multisensory processing in primary sensory areas. Cereb Cortex 20:89–108

Cappe C, Barone P (2005) Heteromodal connections supporting multisensory integration at low levels of cortical processing in the monkey. Eur J Neurosci 22:2886–2902

Cappe C, Thelen A, Romei V, Thut G, Murray MM (2012) Looming signals reveal synergistic principles of multisensory integration. J Neurosci 32:1171–1182

Carriere BN, Royal DW, Perrault TJ, Morrison SP, Vaughan JW, Stein BE, Wallace MT (2007) Visual deprivation alters the development of cortical multisensory integration. J Neurophysiol 98:2858–2867

Catalano SM, Shatz CJ (1998) Activity-dependent cortical target selection by thalamic axons. Science 281:559–562

Champoux F, Lepore F, Gagne JP, Theoret H (2009) Visual stimuli can impair auditory processing in cochlear implant users. Neuropsychologia 47:17–22

Chandrasekaran C, Lemus L, Ghazanfar AA (2013) Dynamic faces speed up the onset of auditory cortical spiking responses during vocal detection. Proc Natl Acad Sci U S A 110:E4668–E4677

Chen SP, Bhattacharya J, Pershing S (2017) Association of vision loss with cognition in older adults. JAMA Ophthalmol 135:963–970

Choi I, Lee JY, Lee SH (2018) Bottom-up and top-down modulation of multisensory integration. Curr Opin Neurobiol 52:115–122

De Niear MA, Gupta PB, Baum SH, Wallace MT (2018) Perceptual training enhances temporal acuity for multisensory speech. Neurobiol Learn Mem 147:9–17

Diederich A, Colonius H (2004) Bimodal and trimodal multisensory enhancement: effects of stimulus onset and intensity on reaction time. Percept Psychophys 66:1388–1404

Driver J, Noesselt T (2008) Multisensory interplay reveals crossmodal influences on 'sensory-specific' brain regions, neural responses, and judgments. Neuron 57:11–23

Ernst MO, Banks MS (2002) Humans integrate visual and haptic information in a statistically optimal fashion. Nature 415:429–433

Falchier A, Clavagnier S, Barone P, Kennedy H (2002) Anatomical evidence of multimodal integration in primate striate cortex. J Neurosci 22:5749–5759

Falchier A, Schroeder CE, Hackett TA, Lakatos P, Nascimento-Silva S, Ulbert I, Karmos G, Smiley JF (2010) Projection from visual areas V2 and prostriata to caudal auditory cortex in the monkey. Cereb Cortex 20:1529–1538

Feng W, Stormer VS, Martinez A, McDonald JJ, Hillyard SA (2014) Sounds activate visual cortex and improve visual discrimination. J Neurosci 34:9817–9824

Fetsch CR, Pouget A, DeAngelis GC, Angelaki DE (2011) Neural correlates of reliability-based cue weighting during multisensory integration. Nat Neurosci 15:146–154

Finney EM, Fine I, Dobkins KR (2001) Visual stimuli activate auditory cortex in the deaf. Nat Neurosci 4:1171–1173

Finney EM, Clementz BA, Hickok G, Dobkins KR (2003) Visual stimuli activate auditory cortex in deaf subjects: evidence from MEG. Neuroreport 14:1425–1427

Foxe JJ, Molholm S, Del Bene VA, Frey HP, Russo NN, Blanco D, Saint-Amour D, Ross LA (2015) Severe multisensory speech integration deficits in high-functioning school-aged children with autism Spectrum disorder (ASD) and their resolution during early adolescence. Cereb Cortex 25:298–312

Frassinetti F, Pavani F, Ladavas E (2002) Acoustical vision of neglected stimuli: interaction among spatially converging audiovisual inputs in neglect patients. J Cogn Neurosci 14:62–69

Fu QJ, Shannon RV, Wang X (1998) Effects of noise and spectral resolution on vowel and consonant recognition: acoustic and electric hearing. J Acoust Soc Am 104:3586–3596

Fu KM, Johnston TA, Shah AS, Arnold L, Smiley J, Hackett TA, Garraghty PE, Schroeder CE (2003) Auditory cortical neurons respond to somatosensory stimulation. J Neurosci 23:7510–7515

Ghazanfar AA, Schroeder CE (2006) Is neocortex essentially multisensory? Trends Cogn Sci 10:278–285

Glick H, Sharma A (2017) Cross-modal plasticity in developmental and age-related hearing loss: clinical implications. Hear Res 343:191–201

Grantyn A, Grantyn R (1982) Axonal patterns and sites of termination of cat superior colliculus neurons projecting in the tecto-bulbo-spinal tract. Exp Brain Res 46:243–256

Grasso PA, Ladavas E, Bertini C (2016) Compensatory recovery after multisensory stimulation in hemianopic patients: behavioral and neurophysiological components. Front Syst Neurosci 10:45

Gu Y, Angelaki DE, Deangelis GC (2008) Neural correlates of multisensory cue integration in macaque MSTd. Nat Neurosci 11:1201–1210

Guitton D, Munoz DP (1991) Control of orienting gaze shifts by the tectoreticulospinal system in the head-free cat. I. Identification, localization, and effects of behavior on sensory responses. J Neurophysiol 66:1605–1623

Hairston WD, Laurienti PJ, Mishra G, Burdette JH, Wallace MT (2003) Multisensory enhancement of localization under conditions of induced myopia. Exp Brain Res 152:404–408

Hairston WD, Burdette JH, Flowers DL, Wood FB, Wallace MT (2005) Altered temporal profile of visual-auditory multisensory interactions in dyslexia. Exp Brain Res 166:474–480

Han X, Xu J, Chang S, Keniston L, Yu L (2021) Multisensory-guided associative learning enhances multisensory representation in primary auditory cortex. Cereb Cortex 32:1040–1054

Heidbreder CA, Groenewegen HJ (2003) The medial prefrontal cortex in the rat: evidence for a dorso-ventral distinction based upon functional and anatomical characteristics. Neurosci Biobehav Rev 27:555–579

Hildebrandt H, Giesselmann H, Sachsenheimer W (1999) Visual search and visual target detection in patients with infarctions of the left or right posterior or the right middle brain artery. J Clin Exp Neuropsychol 21:94–107

Holmes NP (2009) The principle of inverse effectiveness in multisensory integration: some statistical considerations. Brain Topogr 21:168–176

Holt RF, Svirsky MA (2008) An exploratory look at pediatric cochlear implantation: is earliest always best? Ear Hear 29:492–511

Ibrahim LA, Mesik L, Ji XY, Fang Q, Li HF, Li YT, Zingg B, Zhang LI, Tao HW (2016) Cross-modality sharpening of visual cortical processing through layer-1-mediated inhibition and disinhibition. Neuron 89:1031–1045

Isaiah A, Vongpaisal T, King AJ, Hartley DE (2014) Multisensory training improves auditory spatial processing following bilateral cochlear implantation. J Neurosci 34:11119–11130

Iurilli G, Ghezzi D, Olcese U, Lassi G, Nazzaro C, Tonini R, Tucci V, Benfenati F, Medini P (2012) Sound-driven synaptic inhibition in primary visual cortex. Neuron 73:814–828

Jacobs RA, Xu C (2019) Can multisensory training aid visual learning? A computational investigation. J Vis 19:1

Jay MF, Sparks DL (1987a) Sensorimotor integration in the primate superior colliculus. I. Motor convergence. J Neurophysiol 57:22–34

Jay MF, Sparks DL (1987b) Sensorimotor integration in the primate superior colliculus. II. Coordinates of auditory signals. J Neurophysiol 57:35–55

Jiang W, Stein BE (2003) Cortex controls multisensory depression in superior colliculus. J Neurophysiol 90:2123–2135

Jiang W, Jiang H, Stein BE (2006) Neonatal cortical ablation disrupts multisensory development in superior colliculus. J Neurophysiol 95:1380–1396

Jiang W, Jiang H, Rowland BA, Stein BE (2007) Multisensory orientation behavior is disrupted by neonatal cortical ablation. J Neurophysiol 97:557–562

Jiang H, Stein BE, McHaffie JG (2015) Multisensory training reverses midbrain lesion-induced changes and ameliorates haemianopia. Nat Commun 6:7263
Kadunce DC, Vaughan JW, Wallace MT, Benedek G, Stein BE (1997) Mechanisms of within- and cross-modality suppression in the superior colliculus. J Neurophysiol 78:2834–2847
Kao CQ, McHaffie JG, Meredith MA, Stein BE (1994) Functional development of a central visual map in cat. J Neurophysiol 72:266–272
Kayser C, Logothetis NK (2007) Do early sensory cortices integrate cross-modal information? Brain Struct Funct 212:121–132
Kayser SJ, Philiastides MG, Kayser C (2017) Sounds facilitate visual motion discrimination via the enhancement of late occipital visual representations. NeuroImage 148:31–41
Khazipov R, Luhmann HJ (2006) Early patterns of electrical activity in the developing cerebral cortex of humans and rodents. Trends Neurosci 29:414–418
King AJ (1999) Sensory experience and the formation of a computational map of auditory space in the brain. BioEssays 21:900–911
King AJ, Schnupp JW, Carlile S, Smith AL, Thompson ID (1996) The development of topographically-aligned maps of visual and auditory space in the superior colliculus. Prog Brain Res 112:335–350
Knopfel T, Sweeney Y, Radulescu CI, Zabouri N, Doostdar N, Clopath C, Barnes SJ (2019) Audio-visual experience strengthens multisensory assemblies in adult mouse visual cortex. Nat Commun 10:5684
Kral A, Sharma A (2012) Developmental neuroplasticity after cochlear implantation. Trends Neurosci 35:111–122
Kral A, Dorman MF, Wilson BS (2019) Neuronal development of hearing and language: cochlear implants and critical periods. Annu Rev Neurosci 42:47–65
Krueger Fister J, Stevenson RA, Nidiffer AR, Barnett ZP, Wallace MT (2016) Stimulus intensity modulates multisensory temporal processing. Neuropsychologia 88:92–100
Lessard N, Pare M, Lepore F, Lassonde M (1998) Early-blind human subjects localize sound sources better than sighted subjects. Nature 395:278–280
Lomber SG, Meredith MA, Kral A (2010) Cross-modal plasticity in specific auditory cortices underlies visual compensations in the deaf. Nat Neurosci 13:1421–1427
May PJ (2006) The mammalian superior colliculus: laminar structure and connections. Prog Brain Res 151:321–378
McDonald JJ, Teder-Salejarvi WA, Hillyard SA (2000) Involuntary orienting to sound improves visual perception. Nature 407:906–908
Meredith MA, Stein BE (1996) Spatial determinants of multisensory integration in cat superior colliculus neurons. J Neurophysiol 75:1843–1857
Meredith MA, Nemitz JW, Stein BE (1987) Determinants of multisensory integration in superior colliculus neurons. I. Temporal factors. J Neurosci 7:3215–3229
Meredith MA, Allman BL, Keniston LP, Clemo HR (2009) Auditory influences on non-auditory cortices. Hear Res 258:64–71
Mezzera C, Lopez-Bendito G (2016) Cross-modal plasticity in sensory deprived animal models: from the thalamocortical development point of view. J Chem Neuroanat 75:32–40
Miller RL, Pluta SR, Stein BE, Rowland BA (2015) Relative unisensory strength and timing predict their multisensory product. J Neurosci 35:5213–5220
Munoz DP, Wurtz RH (1993a) Fixation cells in monkey superior colliculus. II. Reversible activation and deactivation. J Neurophysiol 70:576–589
Munoz DP, Wurtz RH (1993b) Fixation cells in monkey superior colliculus. I. Characteristics of cell discharge. J Neurophysiol 70:559–575
Murphy J, Summerfield AQ, O'Donoghue GM, Moore DR (2011) Spatial hearing of normally hearing and cochlear implanted children. Int J Pediatr Otorhinolaryngol 75:489–494
Murray MM, Lewkowicz DJ, Amedi A, Wallace MT (2016) Multisensory processes: a balancing act across the lifespan. Trends Neurosci 39:567–579
Mushtaq F, Wiggins IM, Kitterick PT, Anderson CA, Hartley DEH (2020) The benefit of cross-modal reorganization on speech perception in pediatric Cochlear implant recipients revealed using functional near-infrared spectroscopy. Front Hum Neurosci 14:308
Neville HJ, Lawson D (1987) Attention to central and peripheral visual space in a movement detection task. III. Separate effects of auditory deprivation and acquisition of a visual language. Brain Res 405:284–294
Nidiffer AR, Stevenson RA, Krueger Fister J, Barnett ZP, Wallace MT (2016) Interactions between space and effectiveness in human multisensory performance. Neuropsychologia 88:83–91
Nishimura H, Hashikawa K, Doi K, Iwaki T, Watanabe Y, Kusuoka H, Nishimura T, Kubo T (1999) Sign language 'heard' in the auditory cortex. Nature 397:116
Olcese U, Iurilli G, Medini P (2013) Cellular and synaptic architecture of multisensory integration in the mouse neocortex. Neuron 79:579–593
Passamonti C, Bertini C, Ladavas E (2009) Audio-visual stimulation improves oculomotor patterns in patients with hemianopia. Neuropsychologia 47:546–555
Perrault TJ Jr, Vaughan JW, Stein BE, Wallace MT (2003) Neuron-specific response characteristics predict the magnitude of multisensory integration. J Neurophysiol 90:4022–4026
Proulx MJ, Brown DJ, Pasqualotto A, Meijer P (2014) Multisensory perceptual learning and sensory substitution. Neurosci Biobehav Rev 41:16–25
Putzar L, Goerendt I, Lange K, Rosler F, Roder B (2007) Early visual deprivation impairs multisensory interactions in humans. Nat Neurosci 10:1243–1245
Putzar L, Hotting K, Roder B (2010) Early visual deprivation affects the development of face recognition and of audio-visual speech perception. Restor Neurol Neurosci 28:251–257

Rauschecker JP (1995) Compensatory plasticity and sensory substitution in the cerebral cortex. Trends Neurosci 18:36–43
Roder B, Teder-Salejarvi W, Sterr A, Rosler F, Hillyard SA, Neville HJ (1999) Improved auditory spatial tuning in blind humans. Nature 400:162–166
Rowland BA, Quessy S, Stanford TR, Stein BE (2007) Multisensory integration shortens physiological response latencies. J Neurosci 27:5879–5884
Sadato N, Pascual-Leone A, Grafman J, Ibanez V, Deiber MP, Dold G, Hallett M (1996) Activation of the primary visual cortex by braille reading in blind subjects. Nature 380:526–528
Sandmann P, Dillier N, Eichele T, Meyer M, Kegel A, Pascual-Marqui RD, Marcar VL, Jancke L, Debener S (2012) Visual activation of auditory cortex reflects maladaptive plasticity in cochlear implant users. Brain 135:555–568
Sparks DL (1986) Translation of sensory signals into commands for control of saccadic eye movements: role of primate superior colliculus. Physiol Rev 66:118–171
Stehberg J, Dang PT, Frostig RD (2014) Unimodal primary sensory cortices are directly connected by long-range horizontal projections in the rat sensory cortex. Front Neuroanat 8:93
Stein BE, Clamann HP (1981) Control of pinna movements and sensorimotor register in cat superior colliculus. Brain Behav Evol 19:180–192
Stein B, Meredith A (1993) The merging of the senses. J Cogn Neurosci 5:373–374
Stein BE, Labos E, Kruger L (1973) Sequence of changes in properties of neurons of superior colliculus of the kitten during maturation. J Neurophysiol 36:667–679
Stein BE, Wallace MW, Stanford TR, Jiang W (2002) Cortex governs multisensory integration in the midbrain. Neuroscientist 8:306–314
Stein BE, Stanford TR, Rowland BA (2014) Development of multisensory integration from the perspective of the individual neuron. Nat Rev Neurosci 15:520–535
Stevenson RA, Siemann JK, Schneider BC, Eberly HE, Woynaroski TG, Camarata SM, Wallace MT (2014) Multisensory temporal integration in autism spectrum disorders. J Neurosci 34:691–697
Tant ML, Cornelissen FW, Kooijman AC, Brouwer WH (2002) Hemianopic visual field defects elicit hemianopic scanning. Vis Res 42:1339–1348
Targher S, Occelli V, Zampini M (2012) Audiovisual integration in low vision individuals. Neuropsychologia 50:576–582
Targher S, Micciolo R, Occelli V, Zampini M (2017) The role of temporal disparity on audiovisual integration in low-vision individuals. Perception 46:1356–1370
Tinelli F, Purpura G, Cioni G (2015) Audio-visual stimulation improves visual search abilities in hemianopia due to childhood acquired brain lesions. Multisens Res 28:153–171
Vachon P, Voss P, Lassonde M, Leroux JM, Mensour B, Beaudoin G, Bourgouin P, Lepore F (2013) Reorganization of the auditory, visual and multimodal areas in early deaf individuals. Neuroscience 245:50–60
Van Eden CG, Lamme VA, Uylings HB (1992) Heterotopic cortical afferents to the medial prefrontal cortex in the rat. A combined retrograde and anterograde tracer study. Eur J Neurosci 4:77–97
van Laarhoven T, Keetels M, Schakel L, Vroomen J (2018) Audio-visual speech in noise perception in dyslexia. Dev Sci 21
Vasconcelos N, Pantoja J, Belchior H, Caixeta FV, Faber J, Freire MA, Cota VR, Anibal de Macedo E, Laplagne DA, Gomes HM, Ribeiro S (2011) Cross-modal responses in the primary visual cortex encode complex objects and correlate with tactile discrimination. Proc Natl Acad Sci U S A 108:15408–15413
Vincis R, Fontanini A (2016) Associative learning changes cross-modal representations in the gustatory cortex. elife 5
Wallace MT, Stein BE (1994) Cross-modal synthesis in the midbrain depends on input from cortex. J Neurophysiol 71:429–432
Wallace MT, Stein BE (1997) Development of multisensory neurons and multisensory integration in cat superior colliculus. J Neurosci 17:2429–2444
Wallace MT, Stein BE (2000) Onset of cross-modal synthesis in the neonatal superior colliculus is gated by the development of cortical influences. J Neurophysiol 83:3578–3582
Wallace MT, Perrault TJ Jr, Hairston WD, Stein BE (2004) Visual experience is necessary for the development of multisensory integration. J Neurosci 24:9580–9584
Wallace MT, Carriere BN, Perrault TJ Jr, Vaughan JW, Stein BE (2006) The development of cortical multisensory integration. J Neurosci 26:11844–11849
Wang Z, Yu L, Xu J, Stein BE, Rowland BA (2020) Experience creates the multisensory transform in the superior colliculus. Front Integr Neurosci 14:18
Wilkinson LK, Meredith MA, Stein BE (1996) The role of anterior ectosylvian cortex in cross-modality orientation and approach behavior. Exp Brain Res 112:1–10
Williams LE, Light GA, Braff DL, Ramachandran VS (2010) Reduced multisensory integration in patients with schizophrenia on a target detection task. Neuropsychologia 48:3128–3136
Wurtz RH, Goldberg ME (1971) Superior colliculus cell responses related to eye movements in awake monkeys. Science 171:82–84
Xu J, Yu L, Rowland BA, Stanford TR, Stein BE (2014a) Noise-rearing disrupts the maturation of multisensory integration. Eur J Neurosci 39:602–613
Xu J, Sun X, Zhou X, Zhang J, Yu L (2014b) The cortical distribution of multisensory neurons was modulated by multisensory experience. Neuroscience 272:1–9
Xu J, Yu L, Stanford TR, Rowland BA, Stein BE (2015) What does a neuron learn from multisensory experience? J Neurophysiol 113:883–889
Xu J, Yu L, Rowland BA, Stein BE (2017a) The normal environment delays the development of multisensory integration. Sci Rep 7:4772

Xu J, Bi T, Keniston L, Zhang J, Zhou X, Yu L (2017b) Deactivation of association cortices disrupted the congruence of visual and auditory receptive fields in superior colliculus neurons. Cereb Cortex 27:5568–5578

Xu J, Bi T, Wu J, Meng F, Wang K, Hu J, Han X, Zhang J, Zhou X, Keniston L, Yu L (2018) Spatial receptive field shift by preceding cross-modal stimulation in the cat superior colliculus. J Physiol 596:5033–5050

Yu L, Rowland BA, Stein BE (2010) Initiating the development of multisensory integration by manipulating sensory experience. J Neurosci 30:4904–4913

Yu L, Rowland BA, Xu J, Stein BE (2013) Multisensory plasticity in adulthood: cross-modal experience enhances neuronal excitability and exposes silent inputs. J Neurophysiol 109:464–474

Yu L, Xu J, Rowland BA, Stein BE (2016) Multisensory plasticity in superior colliculus neurons is mediated by association cortex. Cereb Cortex 26:1130–1137

Yu L, Cuppini C, Xu J, Rowland BA, Stein BE (2019) Cross-modal competition: the default computation for multisensory processing. J Neurosci 39:1374–1385

Zheng M, Xu J, Keniston L, Wu J, Chang S, Yu L (2021) Choice-dependent cross-modal interaction in the medial prefrontal cortex of rats. Mol Brain 14:13

GPSR Compliance

The European Union's (EU) General Product Safety Regulation (GPSR) is a set of rules that requires consumer products to be safe and our obligations to ensure this.

If you have any concerns about our products, you can contact us on ProductSafety@springernature.com

In case Publisher is established outside the EU, the EU authorized representative is:

Springer Nature Customer Service Center GmbH
Europaplatz 3
69115 Heidelberg, Germany

Batch number: 10732801

Printed by Printforce, the Netherlands